The Handbook
of Environmental Chemistry

Volume 2   Reactions and Processes
Part L

O. Hutzinger
Editor-in-Chief

Springer-Verlag Berlin Heidelberg GmbH

Environmental Photochemistry

Volume Editor: P. Boule

With contributions by
D. Bahnemann, M. Bolte, P. Boule, R. G. Brown,
B. C. Faust, S. Flocke, G. Grabner, R. A. Larson,
S. Madronich, K. A. Marley, P. Méallier,
O. J. Nielsen, R. M. Pagni, C. Richard, J.-F. Rontani,
M. E. Sigman, T. J. Wallington

 Springer

Environmental chemistry is a rather young and interdisciplinary field of science. Its aim is a complete description of the environment and of transformations occuring on a local or global scale. Environmental chemistry also gives an account of the impact of man's activities on the natural environment by describing observed changes.

"The Handbook of Environmental Chemistry" provides the compilation of today's knowledge. Contributions are written by leading experts with practical experience in their fields. The Handbook will grow with the increase in our scientific understandig and should provide a valuable source not only for scientists, but also for environmental managers and decision makers.

ISSN 1433-6839

ISBN 978-3-662-14735-1   ISBN 978-3-540-69044-3 (eBook)
DOI 10.1007/978-3-540-69044-3

Library of Congress Cataloging-in-Publication Data
The Natural environment and the biogeochemical cycles /
with contributions by P. Craig ... [et al.].
v. <A-F > : ill. ; 25 cm. -- (The Handbook of environmental chemistry :
v. 1) Includes bibliographical refereces and indexes.
1. Biogeochemical cycles. 2. Environmental chemistry.
I. Craig. P. J., 1944- . II. Series.
QD31. H335 vol. 1   [QH344]   628.5 s

Production Editor: ProduServ GmbH Verlagsservice, Berlin
Cover design: E. Kirchner, Springer-Verlag
Typesetting: Fotosatz-Service Köhler GmbH, Würzburg
SPIN:10569983   52/3020 - 5 4 3 2 1 0 - Printed on acid-free paper

Editor-in-Chief

Prof. Dr. em. Otto Hutzinger
Universität Bayreuth
Postfach 10 12 51
D-95440 Bayreuth, Germany
E-mail: otto.hutzinger@uni-bayreuth.de

Volume Editor

Dr. Pierre Boule
Université Blaise Pascal
Laboratoire de Photochimie Moléculaire
et Macromoléculaire
UMR CNRS 6505
F-63177 Aubière Cedex, France
E-mail: boule@cicsun.univ-bpclermont.fr

Preface

Environmental Chemistry is a relatively young science. Interest in this subject, however, is growing very rapidly and, although no agreement has been reached as yet about the exact content and limits of this interdisciplinary discipline, there appears to be increasing interest in seeing environmental topics which are based on chemistry embodied in this subject. One of the first objectives of Environmental Chemistry must be the study of the environment and of natural chemical processes which occur in the environment. A major purpose of this series on Environmental Chemistry, therefore, is to present a reasonably uniform view of various aspects of the chemistry of the environment and chemical reactions occurring in the environment.

The industrial activities of man have given a new dimension to Environmental Chemistry. We have now synthesized and described over five million chemical compounds and chemical industry produces about hundred and fifty million tons of synthetic chemicals annually. We ship billions of tons of oil per year and through mining operations and other geophysical modifications, large quantities of inorganic and organic materials are released from their natural deposits. Cities and metropolitan areas of up to 15 million inhabitants produce large quantities of waste in relatively small and confined areas. Much of the chemical products and waste products of modern society are released into the environment either during production, storage, transport, use or ultimate disposal. These released materials participate in natural cycles and reactions and frequently lead to interference and disturbance of natural systems.

Environmental Chemistry is concerned with reactions in the environment. It is about distribution and equilibria between environmental compartments. It is about reactions, pathways, thermodynamics and kinetics. An important purpose of this Handbook, is to aid understanding of the basic distribution and chemical reaction processes which occur in the environment.

Laws regulating toxic substances in various countries are designed to assess and control risk of chemicals to man and his environment. Science can contribute in two areas to this assessment; firstly in the area of toxicology and secondly in the area of chemical exposure. The available concentration ("environmental exposure concentration") depends on the fate of chemical compounds in the environment and thus their distribution and reaction behaviour in the environment. One very important contribution of Environmental Chemistry to the above mentioned toxic substances laws is to develop laboratory test methods, or mathematical correlations and models that predict the environmental fate of new chemical compounds. The third purpose of this Handbook is to help in the basic understanding and development of such test methods and models.

The last explicit purpose of the Handbook is to present, in concise form, the most important properties relating to environmental chemistry and hazard assessment for the most important series of chemical compounds.

At the moment three volumes of the Handbook are planned. Volume 1 deals with the natural environment and the biogeochemical cycles therein, including some background information such as energetics and ecology. Volume 2 is concerned with reactions and processes in the environment and deals with physical factors such as transport and adsorption, and chemical, photochemical and biochemical reactions in the environment, as well as some aspects of pharmacokinetics and metabolism within organisms. Volume 3 deals with anthropogenic compounds, their chemical backgrounds, production methods and information about their use, their environmental behaviour, analytical methodology and some important aspects of their toxic effects. The material for volume 1, 2 and 3 was each more than could easily be fitted into a single volume, and for this reason, as well as for the purpose of rapid publication of available manuscripts, all three volumes were divided in the parts A and B. Part A of all three volumes is now being published and the second part of each of these volumes should appear about six months thereafter. Publisher and editor hope to keep materials of the volumes one to three up to date and to extend coverage in the subject areas by publishing further parts in the future. Plans also exist for volumes dealing with different subject matter such as analysis, chemical technology and toxicology, and readers are encouraged to offer suggestions and advice as to future editions of "The Handbook of Environmental Chemistry".

Most chapters in the Handbook are written to a fairly advanced level and should be of interest to the graduate student and practising scientist. I also hope that the subject matter treated will be of interest to people outside chemistry and to scientists in industry as well as government and regulatory bodies. It would be very satisfying for me to see the books used as a basis for developing graduate courses in Environmental Chemistry.

Due to the breadth of the subject matter, it was not easy to edit this Handbook. Specialists had to be found in quite different areas of science who were willing to contribute a chapter within the prescribed schedule. It is with great satisfaction that I thank all 52 authors from 8 countries for their understanding and for devoting their time to this effort. Special thanks are due to Dr. F. Boschke of Springer for his advice and discussions throughout all stages of preparation of the Handbook. Mrs. A. Heinrich of Springer has significantly contributed to the technical development of the book through her conscientious and efficient work. Finally I like to thank my family, students and colleagues for being so patient with me during several critical phases of preparation for the Handbook, and to some colleagues and the secretaries for technical help.

I consider it a privilege to see my chosen subject grow. My interest in Environmental Chemistry dates back to my early college days in Vienna. I received significant impulses during my postdoctoral period at the University of California and my interest slowly developed during my time with the National Research

Council of Canada, before I could devote my full time of Environmental Chemistry, here in Amsterdam. I hope this Handbook may help deepen the interest of other scientists in this subject.

Amsterdam, May 1980   *O. Hutzinger*

Seventeen years have now passed since the appearance of the first volumes of the Handbook. Although the basic concept has remained the same some changes and adjustments were necessary.

Some years ago publishers and editor agreed to expand the Handbook by two new open-ended volume series: Air Pollution and Water Pollution. These broad topics could not be fitted easily into the headings of the first three volumes. All five volumes series are integrated through the choice of topics and by a system of cross referencing.

The outline of the Handbook is thus as follows:

1. The Natural Environment and the Biochemical Cycles,
2. Reactions and Processes,
3. Anthropogenic Compounds,
4. Air Pollution,
5. Water Pollution.

Rapid developments in Environmental Chemistry and the increasing breadth of the subject matter covered made it necessary to establish volume-editors. Each subject is not supervised by specialists in their respective fields.

A recent development is the 'Super Index', a subject index covering chapters of all published volumes, which will soon be available via the Springer Homepage http://www.springer.de or http://www.springer-ny.com or http://Link. springer.de.

With books in press and in preparation we have now published well over 30 volumes. Authors, volume-editors and editor-in-chief are rewarded by the broad acceptance of the 'Handbook' in the scientific community.

May 1997   *Otto Hutzinger*

Contents

Foreword

Photochemical reactions, i.e. reactions induced by UV or visible light, play a major role in the environment. Light is involved in a large number of reactions in the atmosphere, in natural waters, on soil and in living organisms. These reactions contribute to the synthesis of many organic substances or to their degradation and they participate in the chemical equilibrium of our planet. The chemical influence of light was reported several thousand years ago by the inhabitants of ancient Egypt, and some photochemical reactions have been known for several centuries, but the scientific development of photochemistry is relatively recent and is still progressing.

The aim of this volume is to describe the state of our knowledge about photochemical mechanisms and processes involved in the environment. Various aspects of environmental photochemistry are presented but it is evidently not possible, in just a few hundred pages, to give a complete overview of the influence of light.

On one hand, it seemed useful to report some recent results on the sunlight spectrum and the evaluation of absorption, refection and scattering phenomena on the activic fluxes arriving on the Earth. On the other hand, the absorption of light is a very fast process (the order of magnitude is $10^{-14} - 10^{-15}$ s) and in most cases several short-lived intermediates are involved in photochemical reactions before the formation of stable products. An attempt is made to present the recent techniques developed for studying these short-lived intermediates. Laser flash photolysis has been crucial for studying the mechanism of the phototransformation of phenol and its derivatives in aqueous solution.

Most of the other chapters deal with the photochemical reactions occurring in the atmosphere, clouds and surface waters. Photochemical reactions are the main way of eliminating organic substances in the atmosphere and they play a significant role in the degradation of slightly biodegradable compounds in fresh or marine surface waters. Special attention is focussed on the fate of organic pollutants such as PAHs, anthropogenic molecules including PCBs and pesticides which are distributed in the environment. The phototransformation of some biological substances in marine media is also presented since it plays a significant role in natural cycles and in the formation of humic substances.

The reactions involved in photochemical transformations may result from direct excitation of substances absorbing sunlight, or may be induced by various UV-absorbing species present in the environment which produce reactive intermediates such as hydrated electrons, singlet oxygen, superoxide ions, hydroxyl and peroxyl radicals. In some cases, sensitization by energy transfer may occur. These different aspects of photochemical transformations are presented. However, the photochemistry of ozone is not detailed here because it was the subject of an excellent review by R. P. Wayne in volume 2E of this series.

Another aspect of photochemistry presented here is the use of solar light for the detoxification of polluted water. Photocatalysis plays a minor role in natural processes but it is an efficient method which was intensively studied during the last decade for the elimination of organic pollutants. The revue presented in this volume is mainly focussed on mechanisms involved, but some practical applications of photocatalysis to solar treatment of real waste waters are also reported.

The important theme of biological photosynthesis is not presented here. It was not possible to treat it in this volume because the field is too large and has been the subject of many specialized books.

In brief, the present volume contributes to a better understanding of various photochemical processes which occur in our natural environment or are related to pollution.

P. Boule   Aubière, 1998

1 The Role of Solar Radiation in Atmospheric Chemistry

Sasha Madronich and Siri Flocke

Atmospheric Chemistry Division, National Center for Atmospheric Research, Boulder, Colorado 80307, USA. *E-mail: sasha@sasha.acd.ucar.edu*

Solar radiation at visible and ultraviolet wavelengths drives the chemistry of the atmosphere, by photo-dissociating relatively stable molecules into highly reactive radical fragments. Knowledge of photo-dissociation rate coefficients (J values) is crucial to understanding the behavior of global stratospheric and tropospheric ozone, the atmospheric lifetimes of gases such as carbon monoxide, methane, and non-methane hydrocarbons, and the formation of oxidants at urban and regional scales. J values depend on molecular parameters (absorption cross sections and photo-dissociation quantum yields) that are specific to the photo-reaction of interest, and on the availability of solar radiation at any specific location in the atmosphere. Advances in computer modeling of atmospheric radiative transfer now allow rapid calculation of J values for use in photo-chemistry models, and routinely include the effects of molecular absorbers and scatterers, clouds, aerosols, and surface reflections, for any location and time of the year. However, actual atmospheric conditions needed as input to the calculation are often not available. Direct measurements of J values, while in principle preferable, are technically difficult and limited in their temporal/spatial coverage, but generally support the theoretical calculations at least under optimal conditions (e.g., cloud free skies).

Keywords: solar radiation, spectral actinic flux, radiative transfer, photolysis rate coefficients, Earth-Sun geometry.

Contents

The Handbook of Environmental Chemistry Vol. 2 Part L
Environmental Photochemistry (ed. by P. Boule)
© Springer-Verlag Berlin Heidelberg 1999

List of Symbols and Abbreviations

A	surface albedo
a_1-a_6	Fourier expansion coefficients for declination
b_1-b_4	Fourier expansion coefficients for the equation of time
c_1-c_4	Fourier expansion coefficients for Earth-sun distance
d_n	day-of-year number
$E_\downarrow$	diffuse downwelling irradiance
$E_\uparrow$	diffuse upwelling irradiance
$E(r, \lambda)$	spectral irradiance at point r
EQT	equation of time
E_{sun}	irradiance of the direct solar beam
F_∞	solar irradiance at top of the atmosphere
$F_\downarrow$	diffuse downwelling actinic flux
$F_\uparrow$	diffuse upwelling actinic flux
$F(r, \lambda)$	spectral actinic flux at point r
F_{sun}	actinic flux of the direct solar beam
g	asymmetry factor
GMT	Greenwich mean time
$I_\uparrow$	diffuse upwelling radiance
$I_\uparrow$	diffuse downwelling radiance
$I(r, \theta, \varphi, \lambda)$	spectral radiance at position r, from direction θ, φ
$I(\tau, \theta, \varphi)$	radiance as a function of optical depth τ, from direction θ, φ
I_0	unattenuated collimated beam strength
$I_{diff}(r, \theta, \varphi, \lambda)$	diffuse spectral irradiance at position r, from direction θ, φ
I_N	collimated beam strength attenuated by sample
J	photolysis rate coefficient for a generic photodissociation process
J_{NO2}	photolysis rate coefficient for $NO_2 \rightarrow NO + O(^3P)$
J_{O1D}	photolysis rate coefficient for $O_3 \rightarrow O_2 + O(^1D)$
J_x	rate coefficient for a photodissociation process x
L	cell length
M	mass of liquid water column

N	total number of sample molecules within the cell
n_i	number of molecules/particles of constituent i
$P(\tau, \theta', \varphi'; \theta, \varphi)$	scattering phase function
r	radius
R_v	visual range
s	distance along an atmospheric path
s_1, s_2	arbitrary points in the atmosphere
s_x	sensitivity of a photodissociation process to changes in total ozone amount
t_n	local hour angle
V	cell volume
Ω	total overhead ozone column
β	aerosol extinction coefficient
δ	solar declination
$\phi(\lambda)$	molecular photodissociation quantum yield
φ	azimuth angle of incident light, measured from North
φ_o	solar azimuth angle
λ	wavelength of light
θ	zenith angle of incident light, measured from overhead
θ_o	solar zenith angle
θ_n	day angle (in radians)
ρ	cloud droplet mass density
$\sigma(\lambda)$	molecular absorption cross section
σ_i	total attenuation cross section of i molecules/particles
σ_{NO2}	molecular absorption cross section of NO_2
τ_{12}	optical depth between s_1 and s_2
τ_M	total absorption by mass M
ω_o	single scattering albedo

1
Introduction

Light from the sun (Fig. 1) supplies energy to our biosphere, atmosphere, land, and oceans. Many molecules present in the atmosphere are dissociated by the incoming photons, and the resulting reactive fragments initiate and drive atmospheric chemistry. Such photo-dissociation processes influence a broad range of environmental issues, from urban and regional ozone to acidity of precipitation and lifetimes of gases affecting global energy budgets.

Estimation of atmospheric photo-dissociation reaction rates is fundamental to the quantification, e.g., in models, of atmospheric composition, for both natural and polluted conditions. The problem is however complex, as the rates are different for different chemical reactions, and are in addition as highly variable as the ambient atmospheric light. Because of this variability, it is impractical to provide tabulations that could apply to all possible conditions. Instead, here the objectives are limited to a brief review of the fundamental concepts required to understand the rates of atmospheric photodissociation

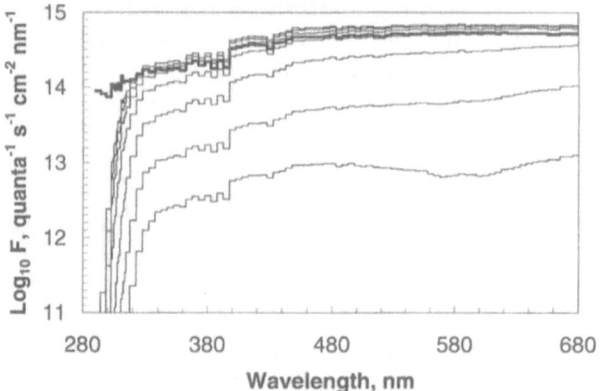

Fig. 1. Atmospheric spectral radiation. *Thick curve* shows solar irradiance incident at the top of the atmosphere, F_∞. *Thin curves* show the actinic flux F at the Earth's surface, progressively lower for solar zenith angles of 0°, 30°, 45°, 60°, 75°, 85°, and 90°

processes, and to providing some representative values which can be used for preliminary estimates. Sensitivities are examined to some atmospheric factors, such as ozone, clouds, and aerosols, that can modify the available photochemical radiation. When more accurate values specific to a given location and time are required, direct measurement or detailed modeling must be carried out.

2
Atmospheric Photo-Chemistry

Atmospheric photochemistry is largely driven by absorption of photons, especially at the more energetic ultraviolet (UV) wavelengths (UV-A 315–400 nm, UV-B 280–315 nm, UV-C 100–280 nm), by otherwise relatively unreactive major constituents of a simple atmosphere (e.g., N_2, O_2, H_2O) [1, 2]. Stratospheric ozone results from the photolysis process

$$O_2 + h\nu \; (\lambda < 240 \text{ nm}) \; \rightarrow \; 2O \qquad (1)$$

followed by

$$O + O_2 + M \; \rightarrow \; O_3 + M \qquad (2)$$

(M being any molecule). Ozone destruction also proceeds by photolysis-driven processes, and in the troposphere it controls the atmospheric oxidation capability (essentially the atmosphere's self-cleaning ability). One especially important aspect is the production of hydroxyl (OH) and hydroperoxyl (HO_2) radicals,

$$O_3 + h\nu \; (\lambda < 340 \text{ nm}) \; \rightarrow \; O_2 + O(^1D) \qquad (3)$$

followed by

$$O(^1D) + H_2O \; \rightarrow \; 2OH \qquad (4)$$

$$OH + O_3 \qquad \rightarrow \; HO_2 + O_2 \qquad (5)$$

$$HO_2 + O_3 \qquad \rightarrow \; OH + 2O_2 \qquad (6)$$

as well as the production of atmospheric peroxides, e.g.,

$$HO_2 + HO_2 \rightarrow H_2O_2 + O_2 \tag{7}$$

Photolysis controls the partitioning of the oxides of nitrogen, NO_x = nitrogen dioxide (NO_2)+nitric oxide (NO),

$$NO_2 + h\nu \ (\lambda < 420 \text{ nm}) \rightarrow NO + O \tag{8}$$

$$NO + O_3 \qquad\qquad\qquad \rightarrow NO_2 + O_2 \tag{9}$$

with NO_2 photodissociation also being a source of ozone (via the reaction at Eq. 2) that can lead to severe photochemical smog when catalyzed by carbon monoxide (CO) and hydrocarbons:

$$OH + CO \quad \rightarrow H + CO_2 \tag{10}$$

$$H + O_2 + M \rightarrow HO_2 + M \tag{11}$$

$$HO_2 + NO \quad \rightarrow NO_2 + OH \tag{12}$$

Numerous other photolysis reactions are also important, and with few exceptions the resulting photo-fragments tend to enhance atmospheric reactivity (e.g., HNO_3, H_2O_2, organic peroxides, nitrates, carbonyls).

Quantitative modeling of atmospheric chemistry requires that the rates of photodissociation be known as accurately as those of collisional reactions, with model-predicted concentrations strongly dependent on uncertainties inherent in their estimation (e.g., [3]). The required information is contained in the photolysis rate coefficient (alternatively termed photodissociation frequency or J value, here usually in units of s^{-1}), which is defined as a first-order contribution by the photolysis process to the local chemical rate equation, e.g., for nitrogen dioxide (reaction at Eq. 8), $d[NO_2]/dt = -J_{NO2}[NO_2]$ + other sources and sinks of NO_2 where t is time. A J value may be constructed by considering the sequential probabilities related to (1) photon availability at a given time and location, (2) the strength of absorption by the molecule, and (3) within the molecule, the fractional disposition of the absorbed energy into bond breaking vs other (e.g., quenching) possible routes. Recognizing the wavelength (λ) dependence of each,

$$J = \int F(\lambda)\, \sigma(\lambda)\, \phi(\lambda)\, d\lambda$$

where the actinic flux $F(\lambda)$ of course depends entirely on local optical conditions, while the absorption cross section $\sigma(\lambda)$ and quantum yield $\phi(\lambda)$ depend on the molecule of interest, and may be sensitive to various environmental conditions (e.g., temperature, pressure, humidity). In practice, the limits of spectral integration are reached when any of the factors in the integrand vanish. These three quantities give rise to a variety of behaviors. The characteristic spectra of the many atmospheric molecules lead to different lifetimes ranging from only a few seconds to months or longer. The light environment, too, varies with time and location, and is sensitive to scattered radiation from all directions (Sect. 4).

3
Molecular Absorption and Dissociation

Laboratory measurements of absorption cross sections, $\sigma(\lambda)$, and yields of dissociation per absorbed quantum, $\phi(\lambda)$, are available for many molecules of interest. Comprehensive, critically evaluated tabulations of such laboratory data are available (e.g., [4]). The basic measurement principle is to pass a collimated monochromatic beam through a vessel containing the molecules of interest. Measurements are made (with various degrees of difficulty) of the filled cell pathlength (L) and volume (V), total number of sample molecules (N), and beam strength with (I_N) or without sample (I_0). The transmission is governed by the Beer-Lambert law,

$$I_N/I_0 = \exp\left[-\sigma(\lambda) \, L \, N/V\right]$$

so that $\sigma(\lambda)$ is readily determined from relatively simple measurements.

Determination of the quantum yield is usually much more difficult, requiring absolute estimation (rather than ratios) of both the number of absorbed photons and the number of sample molecules dissociated. If several product channels are possible, the relative importance of the multiple pathways must also be quantified. Nonetheless, for many simple cases the quantum yields are large (close to unity) for photon energies well in excess of those of the dissociating bonds, and zero at much longer wavelengths – with the details of the fall-off not being important unless both $F(\lambda)$ and $\sigma(\lambda)$ happen to be large in that neighborhood. A spectral region of particular importance is the UV-B band where tropospheric radiation is strongly wavelength dependent, so that both quantum yields and absorption cross sections should be determined with high spectral resolution.

Uncertainties still exist in molecular cross sections and quantum yields. Differences among studies of 5–20% are common, and are occasionally larger. These put an absolute limit on the accuracy with which models can compute photodissociation rates.

4
Atmospheric Radiation

4.1
Directionality of Radiation

In contrast to laboratory collimated beam experiments, atmospheric radiation may be incident simultaneously from many different directions. Scattering, or redirecting, of solar photons leads to a complex angular redistribution of the sky radiation. Often the direct solar beam is only one component of the total radiation field, with large scattered contributions possible from clouds, aerosols, and air molecules, and surface reflections.

At any point r in the atmosphere, the spectral radiance $I(r, \theta, \varphi, \lambda)$ (sometimes called the intensity) describes the energy incident from a direction θ, φ, in unit time on a unit area perpendicular to the beam, per unit wavelength. Here θ and φ are the usual angular coordinates (θ = zenith normally measured from the overhead direction, and φ = azimuth normally measured from the North). Summing $I(r, \theta, \varphi, \lambda)$ over all directions gives several important radiometric quantities. The energy passing through a unit area of a fixed plane, or irradiance $E(r,\lambda)$ (sometimes called flux), is given by cosine-weighted integration

$$E(r, \lambda) = E_{sun} + \int_0^{2\pi} \int_0^{\pi/2} I_{diff}(r, \theta, \varphi, \lambda) \cos\theta \, \sin\theta \, d\theta \, d\varphi +$$

$$+ \int_0^{2\pi} \int_{\pi/2}^{\pi} I_{diff}(r, \theta, \varphi, \lambda) \cos\theta \, \sin\theta \, d\theta \, d\varphi$$

where E_{sun} is the irradiance of the direct solar beam, and the two separate integrals denote downwelling and upwelling diffuse (scattered) radiation, respectively (I_{diff}). The actinic flux, $F(r, \lambda)$, represents radiation incident into a volume element, and is given by integration of the radiance without bias to incoming direction,

$$F(r, \lambda) = \int_0^{2\pi} \int_0^{\pi} I(r, \theta, \varphi, \lambda) \sin\theta \, d\theta \, d\varphi =$$

$$= F_{sun} + \int_0^{2\pi} \int_0^{\pi/2} I_{diff}(r, \theta, \varphi, \lambda) \sin\theta \, d\theta \, d\varphi +$$

$$+ \int_0^{2\pi} \int_{\pi/2}^{\pi} I_{diff}(r, \theta, \varphi, \lambda) \sin\theta \, d\theta \, d\varphi$$

where the second equality again separates the contributions of the direct solar beam F_{sun}, and of the downwelling and upwelling diffuse radiation. The irradiance is a quantity commonly measured by radiometric instruments, but is not readily converted to actinic flux unless the angular distribution of the radiation is also known. The actinic flux is the quantity which must be used for the calculation of photochemical dissociation rate coefficients. Figure 2 illustrates the difference between these two quantities: The radiation crossing a unit area at angle θ is less (by $\cos\theta$) than that crossing a unit area normal to the beam. The latter applies to the photodissociation of atmospheric molecules, because these, being randomly oriented, present equal cross section to all directions.

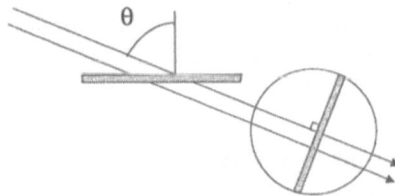

Fig. 2. Radiation incident at angle θ on a horizontal plane (the irradiance), and incident on a molecule which on average presents the same cross section to all directions (the actinic flux). Per unit target area, the actinic flux exceeds the irradiance by $1/\cos\theta$.

The distinction between irradiance and actinic flux may be further illustrated by considering the special case of hemispherically isotropic diffuse radiances, downwelling $I_\downarrow$ and upwelling $I^\uparrow$ and the direct solar beam incident at an angle θ_o. In this case the components of the actinic flux are given by

$$F_{sun} = E_{sun}/\cos\theta_o$$

$$F_\uparrow = 2\pi\, I^\uparrow = 2E_\uparrow$$

$$F_\downarrow = 2\pi\, I_\downarrow = 2E_\downarrow$$

Another example is a situation in which the upwelling radiation is due to reflection at a diffusing surface with albedo (or reflectivity) A, defined as the fraction of the incident energy (irradiance) that is reflected. If the reflected radiation is also assumed to be isotropic ($I^\uparrow$ independent of angle), it follows from the above equations that in the limit $E_\downarrow = 0$, $\theta_o = 0$, a bright diffusing surface (with $A \to 1$) can enhance the total actinic flux ($F_{sun} + F_\downarrow$) by up to a factor of three. In general, sky radiation is not isotropic and may have extremely complex angular distribution, for example near clouds.

4.2
Scattering

Molecular (Rayleigh) scattering leads to broadly varying angular distributions of the diffuse radiation (since air density is usually uniform). More complex distributions occur in the presence of cloud and other particles, especially if these are spatially inhomogeneous (e.g., broken clouds). Depending on the particle size (relative to λ), individual scattering events re-distribute the radiation from an incident direction (say $\theta = 0$, $\varphi = 0$) into a specific direction (θ, φ) with an angular probability phase function $P(\theta, \varphi)$. For small particle (e.g., air molecules) the Rayleigh limit gives

$$P(\theta, \varphi) = 3/4\,(1 + \cos^2\theta)$$

which is notably symmetric in the forward and back directions. For larger particles (e.g., haze and cloud) the scattering probability becomes pronouncedly peaked in the forward direction but with often-complex contributions in other directions (e.g., side and back). The phase functions of atmospheric particles are usually not well known, because they depend on a number of variables including size distribution, shape, and composition.

Useful estimates are nonetheless available for the first moment of the phase function, or asymmetry factor g,

$$g = \frac{1}{2} \int_0^{2\pi} \int_0^{\pi} P(\theta, \varphi)\, \cos\theta\, \sin\theta\, d\theta\, d\varphi$$

This at least distinguishes between relative forward/backward probabilities, with extreme values of +1 and –1 corresponding, respectively, to pure forward and backward scattering, and values of 0 for isotropic and Rayleigh scattering.

4.3
Absorption

Atmospheric absorbers include oxygen (Schumann-Runge and Herzberg structures, $\lambda < 240$ nm) and ozone (Hartley and Huggins, $\lambda < 340$ nm; Chappuis $\lambda > 360$ nm). A host of other gaseous absorbers (sulfur and nitrogen dioxides, organics) can attenuate the actinic flux in polluted conditions, and many other compounds can be detected in trace amounts but have negligible effects on optical transmission. Particles, liquid or solid, differ widely in their ability to absorb photons, e.g., soot in urban haze vs water droplets in pristine clouds.

The relative probability of absorption is parameterized by the single scattering albedo, defined as $\omega_o =$ scattered fraction = (scattering)/(absorption + scattering), and may be estimated from knowledge of the complex index of refraction of collected particles, or from detailed inverse analysis on the effects on the atmospheric radiation field.

4.4
Radiative Transfer

The atmosphere may be viewed as a reasonably smooth optical medium within which absorption and scattering can take place. The optical depth between any two points, s_1 and s_2, is defined as

$$\tau_{12} = \int_{s_1}^{s_2} \sum_i n_i \sigma_i \, ds$$

where s is the path between the points, and the summation extends over all constituents (e.g., molecules particles) each with total (absorption plus scattering) attenuation cross section σ_i and local density n_i. Along any such path, the propagation of radiation is given by the equation of radiative transfer,

$$\frac{dI(\tau, \theta, \varphi)}{d\tau} = -I(\tau, \theta, \varphi) + \frac{\omega_o}{4\pi} \int_0^{2\pi} \int_0^{\pi} I(\tau, \theta', \varphi') \, P(\tau, \theta', \varphi'; \theta, \varphi) \sin\theta' d\theta' d\varphi'$$

where ω_o is the effective single scattering albedo for the medium. The first term on the right hand side represents attenuation of the radiance along the path, while the second term accounts for the contributions to $I(\tau, \theta, \varphi)$ from scattering from all other directions. The dependence on wavelength has been omitted for simplicity.

Solution of the equation of radiative transfer is generally difficult [5–7]. Analytic solutions exist for only a few highly idealized situations. Two types of numerical solutions should be recognized. So-called exact methods (e.g., discrete ordinates, adding-doubling) are based on high order expansions and can be extremely intensive computationally, but give arbitrarily accurate results if the phase function is known and the geometric context is reasonably simple. Much faster computation of solutions can be achieved if the radiative transfer equation is first approximated, or equivalently by assuming that the diffuse radiances are given by simple angular functions. For example, the two-stream methods [8–10] approximate the downwelling and upwelling diffuse radiances, $I_\downarrow$ and $I_\uparrow$, as linear functions of $\cos\theta$. Comparisons with exact methods show that the two-stream and similar approximations often lead to errors in calculated actinic fluxes of 5–20% in the middle troposphere, with both higher and lower errors possible under some conditions [11]. But in view of the benefits of computational speed and simplicity, such accuracy can be acceptable when other input parameters (e.g., exact amount of cloud) are not well known anyway.

4.5
Earth-Sun Geometry

Atmospheric radiation is controlled in large part by the angular position of the sun in the sky. This position is specified by two angles, the solar zenith angle, θ_o, measured from the local vertical to the sun, and the solar azimuth angle φ_o, measured from a reference horizon direction (e.g., north). By symmetry arguments, it is readily seen that only θ_o affects the actinic flux, unless specific local factors are present (e.g., azimuth-dependent orographic obstructions of the horizon).

The solar zenith angle at latitude Φ and longitude Ψ may be computed from the expression

$$\cos\theta_o = \sin\delta \sin\Phi + \cos\delta \cos\Phi \cos t_h$$

where δ is the solar declination (the angle between the Earth's equatorial plane and the sun's direction, varying from around $-23.5°$ on 21 December to around $+23.5°$ on 21 June), and t_h is the local hour angle given by

$$t_h = \pi(GMT/12 - 1 + \Psi/180) + EQT$$

Here GMT is Greenwich mean time (readily computed from the local clock time and knowledge of the local time zone), and EQT is the so-called equation of time which accounts for variations in the apparent angular speed of the sun in the sky.

Methods to calculate EQT and δ are well established, but very accurate computations are quite complex (e.g., [12, 13]) due to the need to consider numerous orbital perturbations to the relative Earth-sun position. For many practical applications simpler expressions are sufficient, for example the Fourier expansions derived by fitting more accurate expressions [14]:

$$\delta = a_0 + a_1\cos\theta_n + a_2\sin\theta_n + a_3\cos 2\theta_n + a_4\sin 2\theta_n + a_5\cos 3\theta_n + a_6\sin 3\theta_n$$

and

$$EQT = b_0 + b_1\cos\theta_n + b_2\sin\theta_n + b_3\cos 2\theta_n + b_4\sin 2\theta_n$$

where

$$\theta_n = 2\pi d_n/365 \text{ (in radians)}$$

for day number d_n of the year ($d_n = 0$ for 1 January, 365 for 31 December). The coefficients a_0-a_6 and b_0-b_4 are given in Table 1. Use of these approximations yields θ_o with an accuracy of about 1°, which is sufficient for many practical

Table 1. Fourier expansion coefficients for parameterizations of solar declination $(a_0\text{-}a_6)$, equation of time $(b_0\text{-}b_4)$, and Earth-sun distance $(c_0\text{-}c_4)$. From [14]

i	a_i	b_i	c_i
0	0.006918	0.000075	1.000110
1	− 0.399912	0.001868	0.034221
2	0.070257	− 0.032077	0.001280
3	− 0.006758	− 0.014615	0.000719
4	0.000907	− 0.040849	0.000077
5	− 0.002697		
6	0.001480		

applications. Note that this formulation is the same for each year, not accounting for long-term drifts in orbital parameters, nor for leap years. Some simple limiting values of θ_o are noteworthy: $\theta_o = 0°$ for overhead sun, $\theta_o = 90°$ at sunrise and sunset. Also, local high-sun values (at solar noon, which may be slightly different than clock noon) of θ_o vary from $\Phi + 23.5°$ (winter solstice) to $\Phi - 23.5°$ (summer solstice), with $\theta_o = \Phi$ for both equinoxes.

A second orbital consideration, of lesser importance than the solar zenith angle but of significance to some specific applications, is the yearly cycle in the Earth-sun distance. This varies around the yearly averaged value ($R_o = 1$ astronomical unit, equal to around 1.496×10^{11} m), by around 3.5% peak-to-peak from a minimum distance in early January to a maximum distance in early July. Exact computations are again complex [12, 13] but for approximate calculation a simple Fourier expansion can again be used [14]:

$$(R_o/R_n)^2 = c_o + c_1\cos\theta_n + c_2\sin\theta_n + c_3\cos2\theta_n + c_4\sin2\theta_n$$

with the values of $c_0\text{-}c_4$ given in Table 1. The factor $(R_o/R_n)^2$ may be used directly to scale the incident solar radiation or the computed actinic fluxes and photolysis rate coefficients, because it affects all wavelengths equally. The effect on the incident radiation, while only ∼7% peak-to-peak, can be of some importance when examining inter-hemispheric differences. For example, incident solar radiation is highest during southern hemisphere summers, leading to slightly higher yearly-averaged atmospheric radiation levels compared to the northern hemisphere. As before, this simple formula does not account for year to year variations, e. g., leap years or secular orbital variations.

5
Reference Atmospheric Values

5.1
Model Description

Photodissociation rate coefficients (J values) can be calculated from a knowledge of the molecular cross section and quantum yield data for a target molecule, and from a calculation of the atmospheric actinic flux. Such cal-

Table 2. Spectral actinic flux at the Earth's surface, for cloud-free skies and different solar zenith angles. Also shown is the solar spectral irradiance incident at the top of the atmosphere, F_∞, at the average Earth-sun distance. Values given as Log_{10} of quanta s^{-1} cm^{-2} nm^{-1}

Wavelength Range, nm		Solar zenith angle, degrees							
		F_∞	0	30	45	60	75	85	90
289.9	294.1	13.95	9.26	8.52	7.35	5.93	5.41	4.98	4.06
294.1	298.5	13.91	11.27	10.82	10.09	8.73	7.81	7.35	6.55
298.5	302.5	13.87	12.38	12.09	11.63	10.70	9.36	8.81	8.13
302.5	303.5	14.04	13.01	12.79	12.44	11.72	10.36	9.67	9.03
303.5	304.5	13.99	13.06	12.86	12.54	11.87	10.53	9.79	9.17
304.5	305.5	14.03	13.25	13.08	12.79	12.19	10.92	10.09	9.47
305.5	306.5	13.96	13.28	13.11	12.84	12.29	11.07	10.18	9.56
306.5	307.5	14.02	13.43	13.28	13.04	12.53	11.40	10.43	9.81
307.5	308.5	14.05	13.55	13.41	13.19	12.73	11.67	10.64	10.01
308.5	309.5	13.97	13.53	13.40	13.19	12.76	11.76	10.70	10.06
309.5	310.5	13.96	13.60	13.48	13.29	12.90	11.98	10.89	10.22
310.5	311.5	14.15	13.83	13.71	13.53	13.16	12.28	11.18	10.50
311.5	312.5	14.06	13.79	13.69	13.52	13.17	12.35	11.26	10.54
312.5	313.5	14.08	13.85	13.75	13.60	13.27	12.50	11.42	10.68
313.5	314.5	14.08	13.89	13.80	13.65	13.33	12.60	11.54	10.78
314.5	317.5	14.08	13.95	13.87	13.73	13.45	12.80	11.81	10.98
317.5	322.5	14.10	14.06	13.99	13.87	13.64	13.09	12.23	11.36
322.5	327.5	14.17	14.19	14.13	14.03	13.83	13.36	12.65	11.82
327.5	332.5	14.24	14.28	14.23	14.14	13.95	13.52	12.87	12.10
332.5	337.5	14.20	14.27	14.22	14.13	13.96	13.54	12.93	12.22
337.5	342.5	14.24	14.32	14.27	14.19	14.02	13.62	13.03	12.34
342.5	347.5	14.23	14.32	14.27	14.19	14.03	13.64	13.04	12.37
347.5	352.5	14.25	14.34	14.29	14.22	14.06	13.67	13.07	12.40
352.5	357.5	14.26	14.35	14.31	14.24	14.08	13.70	13.10	12.43
357.5	362.5	14.23	14.32	14.28	14.21	14.06	13.68	13.08	12.41
362.5	367.5	14.32	14.42	14.38	14.31	14.16	13.79	13.19	12.51
367.5	372.5	14.35	14.45	14.41	14.34	14.20	13.83	13.22	12.55
372.5	377.5	14.29	14.39	14.35	14.29	14.15	13.78	13.17	12.50
377.5	382.5	14.34	14.45	14.41	14.35	14.21	13.85	13.24	12.56
382.5	387.5	14.25	14.36	14.32	14.26	14.13	13.77	13.15	12.48
387.5	392.5	14.35	14.46	14.42	14.37	14.24	13.89	13.27	12.59
392.5	397.5	14.26	14.37	14.34	14.28	14.16	13.81	13.19	12.51
397.5	402.5	14.51	14.62	14.59	14.53	14.41	14.07	13.44	12.76
402.5	407.5	14.53	14.64	14.61	14.55	14.43	14.10	13.47	12.78
407.5	412.5	14.55	14.66	14.63	14.57	14.46	14.13	13.49	12.81
412.5	417.5	14.57	14.68	14.65	14.60	14.48	14.16	13.52	12.83
417.5	422.5	14.56	14.67	14.64	14.59	14.48	14.17	13.52	12.83
422.5	427.5	14.56	14.67	14.64	14.60	14.49	14.17	13.53	12.84
427.5	432.5	14.51	14.63	14.60	14.55	14.44	14.14	13.49	12.80
432.5	437.5	14.59	14.70	14.67	14.63	14.52	14.22	13.57	12.87
437.5	442.5	14.60	14.72	14.69	14.64	14.54	14.24	13.59	12.89
442.5	447.5	14.64	14.75	14.73	14.68	14.58	14.29	13.64	12.93
447.5	452.5	14.67	14.79	14.76	14.72	14.62	14.33	13.68	12.97
452.5	457.5	14.67	14.78	14.76	14.71	14.62	14.33	13.68	12.97
457.5	462.5	14.67	14.79	14.77	14.72	14.63	14.35	13.70	12.97
462.5	467.5	14.67	14.79	14.77	14.72	14.63	14.36	13.70	12.98

Table 2 (Continued)

Wavelength Range, nm		Solar zenith angle, degrees							
		F_∞	0	30	45	60	75	85	90
467.5	472.5	14.67	14.79	14.77	14.73	14.63	14.36	13.71	12.98
472.5	477.5	14.68	14.80	14.78	14.74	14.65	14.38	13.73	12.99
477.5	482.5	14.70	14.81	14.79	14.75	14.66	14.40	13.74	12.99
482.5	487.5	14.66	14.78	14.76	14.72	14.63	14.37	13.71	12.95
487.5	492.5	14.68	14.80	14.77	14.73	14.65	14.39	13.73	12.97
492.5	497.5	14.69	14.81	14.79	14.75	14.66	14.41	13.75	12.99
497.5	502.5	14.68	14.79	14.77	14.74	14.65	14.40	13.74	12.96
502.5	507.5	14.69	14.81	14.78	14.75	14.66	14.41	13.75	12.95
507.5	512.5	14.70	14.81	14.79	14.75	14.67	14.42	13.76	12.96
512.5	517.5	14.68	14.79	14.77	14.73	14.65	14.40	13.75	12.94
517.5	522.5	14.67	14.79	14.77	14.73	14.65	14.41	13.75	12.93
522.5	527.5	14.69	14.80	14.78	14.75	14.67	14.43	13.76	12.93
527.5	532.5	14.71	14.82	14.80	14.76	14.68	14.44	13.77	12.92
532.5	537.5	14.71	14.82	14.80	14.76	14.68	14.44	13.77	12.91
537.5	542.5	14.70	14.81	14.79	14.76	14.68	14.44	13.77	12.90
542.5	547.5	14.71	14.82	14.80	14.77	14.69	14.45	13.78	12.90
547.5	552.5	14.71	14.82	14.80	14.77	14.69	14.45	13.79	12.89
552.5	557.5	14.72	14.83	14.81	14.77	14.70	14.46	13.79	12.89
557.5	562.5	14.71	14.82	14.80	14.76	14.69	14.45	13.78	12.85
562.5	567.5	14.72	14.82	14.80	14.77	14.69	14.46	13.78	12.83
567.5	572.5	14.72	14.82	14.80	14.77	14.69	14.46	13.78	12.81
572.5	577.5	14.73	14.83	14.81	14.78	14.70	14.47	13.79	12.82
577.5	582.5	14.73	14.83	14.81	14.78	14.70	14.47	13.80	12.83
582.5	587.5	14.73	14.84	14.82	14.78	14.71	14.49	13.82	12.85
587.5	592.5	14.72	14.82	14.80	14.77	14.70	14.47	13.81	12.84
592.5	597.5	14.73	14.83	14.81	14.78	14.71	14.48	13.82	12.84
597.5	602.5	14.73	14.83	14.81	14.77	14.70	14.48	13.82	12.82
602.5	607.5	14.73	14.83	14.81	14.78	14.71	14.49	13.83	12.83
607.5	612.5	14.73	14.83	14.81	14.78	14.71	14.49	13.84	12.85
612.5	617.5	14.72	14.82	14.80	14.77	14.70	14.49	13.85	12.87
617.5	622.5	14.73	14.83	14.81	14.78	14.72	14.51	13.88	12.91
622.5	627.5	14.72	14.83	14.81	14.78	14.71	14.51	13.89	12.92
627.5	632.5	14.72	14.83	14.81	14.78	14.72	14.52	13.91	12.94
632.5	637.5	14.72	14.83	14.81	14.79	14.72	14.53	13.92	12.97
637.5	642.5	14.72	14.83	14.81	14.79	14.72	14.53	13.94	12.99
642.5	647.1	14.72	14.83	14.81	14.79	14.73	14.54	13.95	13.01
647.1	655.0	14.72	14.83	14.81	14.78	14.72	14.54	13.96	13.02
655.0	665.0	14.70	14.82	14.80	14.77	14.72	14.54	13.97	13.04
665.0	657.0	14.71	14.83	14.81	14.79	14.73	14.55	14.01	13.08
675.0	685.0	14.71	14.82	14.81	14.78	14.73	14.56	14.02	13.10
685.0	695.0	14.70	14.82	14.80	14.78	14.73	14.56	14.04	13.12
695.0	705.0	14.70	14.81	14.80	14.77	14.72	14.56	14.05	13.13
705.0	715.0	14.70	14.81	14.80	14.77	14.72	14.57	14.06	13.15
715.0	725.0	14.69	14.80	14.79	14.77	14.72	14.56	14.07	13.16
725.0	735.0	14.69	14.80	14.79	14.77	14.72	14.57	14.08	13.17

culated *J* values are only approximations to the real ones, because information about the exact optical state of the atmosphere (e.g., cloud amount) is seldom available. Direct measurements of *J* values, discussed in Sect. 7, are preferable at least in principle, but in practice such measurements are difficult and specific to the conditions of the experiment. Here, model results are presented for one set of atmospheric conditions that falls within the range of common observations. Such conditions, while not necessarily typical of any specific location or time, provide a useful baseline for studies of sensitivity to various atmospheric conditions, given in Sect. 6.

The model used in these calculations is the Tropospheric Ultraviolet Visible (TUV) code [15]. Unless otherwise stated, standard conditions consist of cloud-free skies, vertical profiles of air, ozone, and temperature from the US Standard Atmosphere [16], 10% surface albedo, and a typical continental aerosol profile corresponding to 25 km visible range at ground level [17], scaled at other wavelengths with λ^{-1}. Extraterrestrial irradiances were taken from [18] for $\lambda < 350$ nm and from [19] at longer wavelengths. Atmospheric propagation was computed with an eight-stream discrete ordinates code [20] with a pseudospherical modification [21] for increased accuracy at low sun. The calculation includes absorption by atmospheric ozone with cross sections from [22], Rayleigh scattering using the cross section parameterization by [23], as well as aerosol scattering and absorption. The radiative transfer calculation was carried out at the center of each wavelength interval in Table 2. Absorption cross sections are largely in accord with recent critical evaluations [4] and are documented in [15].

5.2
Calculated Actinic Fluxes

The actinic flux at ground level is given in Table 2 (see also Fig. 1), for one set of atmospheric conditions. Note that even at ground level the actinic fluxes can be substantially larger than the incident solar irradiance at the top of the atmosphere, reflecting the contributions of scattered radiation. Previous tabulations (e.g., [24]) give values similar to those presented here, though based on some older data for the extraterrestrial solar flux and other inputs. Some tabulations (e.g., [25]) cannot be compared directly as they refer to irradiance on a horizontal surface, not actinic flux (see Sect. 4.1). It should be emphasized that the actinic fluxes given in Table 2 are strictly valid only for the atmosphere, and cannot be used directly in surface waters or soils. Additional consideration is required of numerous other factors, such as Fresnel reflections and refraction at air-water interfaces (including perturbations by waves and spray), absorption and scattering by dissolved molecules and suspended particulate matter (possibly biologic, e.g., phytoplankton), and, for the case soils, the occurrence of scattering and partial shading by soil roughness [26, 27, and references cited therein].

The necessity of careful radiative transfer calculations is also illustrated in Fig. 3, where the contribution of scattered radiation to the total actinic flux is shown for cloud-free conditions. At the surface, 40% or more of the total actinic

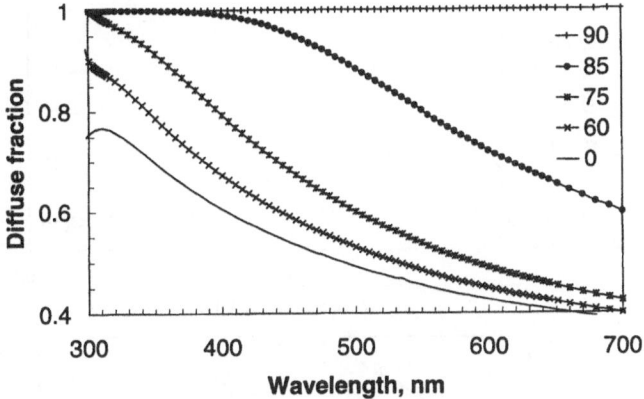

Fig. 3. Contribution of diffuse (scattered) radiation to the total spectral actinic flux at the surface. Calculations for cloud-free skies, 25 km ground-level visibility, 10 % surface albedo, with eight-stream pseudo-spherical discrete ordinates model. Solar zenith angles are given in the legend

flux stems from scattering, and is dominant in the ultraviolet range. At low sun, the diffuse fraction becomes larger at the surface (shown in the figure), but is less important in the upper atmosphere relative to the direct solar beam (not shown). Of course, clouds and heavier aerosol loading would further increase the scattered fraction.

5.3
Calculated Photolysis Rate Coefficients

Table 3 shows the J values for selected molecules of interest to atmospheric chemistry, computed with the actinic fluxes of Table 2. Note the wide range of photolysis lifetimes ($1/J$) at high sun: from a few seconds for NO_3 and a few minutes for NO_2, to several days or longer for other species. Figure 4 shows the altitude dependence of selected J values. Much of the increases with altitude are due to higher actinic fluxes, especially evident at stratospheric altitudes where ozone absorption is strongest. Some effects also arise from temperature and pressure profiles, as these variables affect the molecular properties. Notable are the contributions from the temperature dependence of ozone quantum yields and cross sections, and both temperature and pressure dependencies for aldehyde quantum yields. These vertical profiles of J values should be viewed as illustrative only, and can be quite different in the presence of clouds or other aerosol and gas distributions.

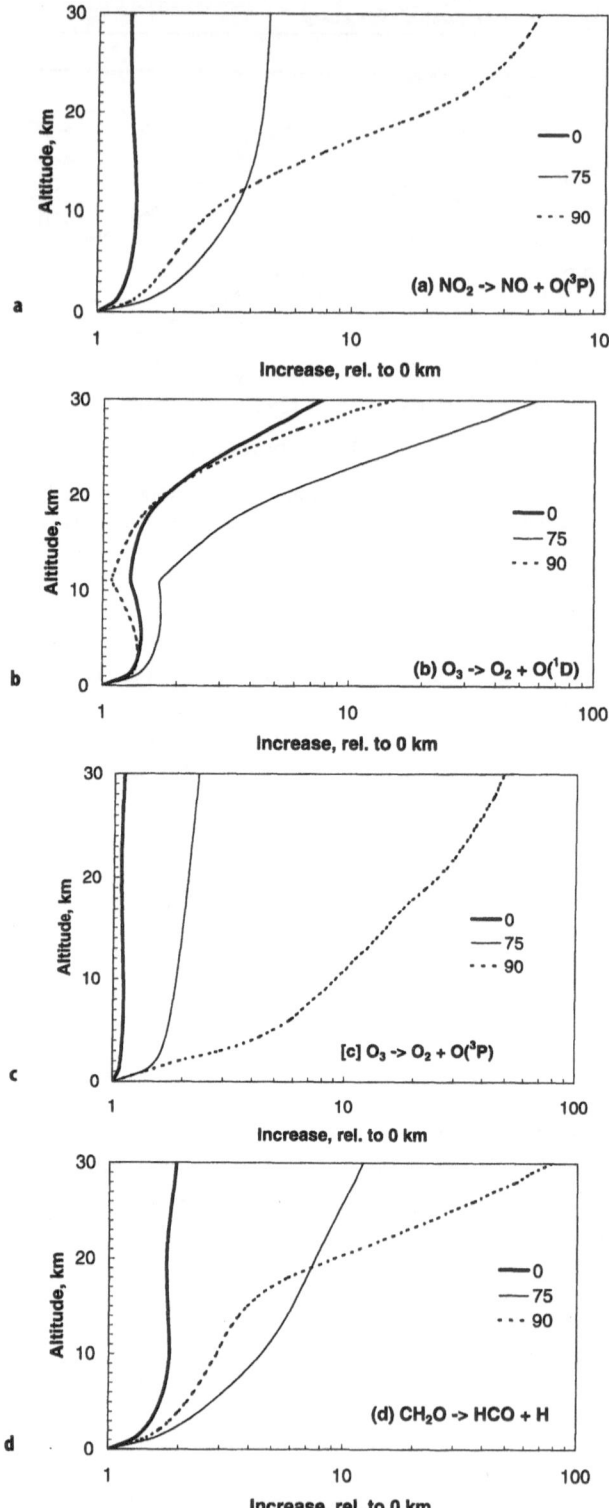

Fig. 4a–d. Altitude profiles of photodissociation coefficients, normalized to values at the surface (see Table 3), for solar zenith angles of 0° *(thick solid curve)*, 75° *(thin solid curve)*, and 90° *(dashed curve)*. Panel (a) for $NO_2 \rightarrow NO + O(^3P)$; (b) for $O_3 \rightarrow O_2 + O(^1D)$; (c) for $O_3 \rightarrow O_2 + O(^3P)$; (d) for $CH_2O \rightarrow HCO + H$

Table 3. Photolysis rate coefficients (J values) at the Earth's surface, for selected reactions at different solar zenith angles. Values given as $\text{Log}_{10}\ \text{s}^{-1}$

Reaction	Solar zenith angle, degrees						
	0	30	45	60	75	85	90
$O_3 \rightarrow O_2 + O(^1D)$	– 4.41	– 4.56	– 4.79	– 5.23	– 6.07	– 7.04	– 7.79
$O_3 \rightarrow O_2 + O(^3P)$	– 3.31	– 3.34	– 3.38	– 3.46	– 3.69	– 4.32	– 5.25
$NO_2 \rightarrow NO + O(^3P)$	– 2.00	– 2.05	– 2.11	– 2.26	– 2.63	– 3.24	– 3.93
$NO_3 \rightarrow NO + O_2$	– 1.60	– 1.62	– 1.65	– 1.72	1.94	– 2.58	– 3.56
$NO_3 \rightarrow NO_2 + O(^3P)$	– 0.71	– 0.73	– 0.76	– 0.84	– 1.07	– 1.73	– 2.64
$N_2O_5 \rightarrow NO_3 + NO_2$	– 4.29	– 4.36	– 4.47	– 4.68	– 5.15	– 5.81	– 6.53
$HNO_2 \rightarrow OH + NO$	– 2.66	– 2.70	– 2.77	– 2.93	– 3.31	– 3.92	– 4.60
$HNO_3 \rightarrow OH + NO_2$	– 6.13	– 6.23	– 6.39	– 6.69	– 7.29	– 8.08	– 8.85
$HNO_4 \rightarrow HO_2 + NO_2$	– 5.27	– 5.37	– 5.52	– 5.82	– 6.43	– 7.27	– 8.08
$H_2O_2 \rightarrow 2OH$	– 5.09	5.16	– 5.27	– 5.50	– 5.98	– 6.66	– 7.39
$CH_2O \rightarrow H + HCO$	– 4.46	– 4.54	– 4.66	– 4.91	– 5.45	– 6.21	– 7.02
$CH_2O \rightarrow H_2 + CO$	– 4.29	– 4.35	– 4.45	– 4.64	– 5.08	– 5.73	– 6.46
$CH_3OOH \rightarrow CH_3O + OH$	– 5.21	– 5.28	– 5.39	– 5.60	– 6.08	– 6.74	– 7.47
$CH_3ONO_2 \rightarrow CH_3O + NO_2$	– 5.97	– 6.08	– 6.24	– 6.54	– 7.15	– 7.95	– 8.74
$CH_3CHO \rightarrow CH_3 + CHO$	– 5.19	– 5.30	– 5.47	– 5.79	– 6.46	– 7.34	– 8.15
$CH_3COCH_3 \rightarrow$ products	– 5.90	– 6.00	– 6.15	– 6.43	– 7.02	– 7.81	– 8.61
$CH_3COO_2NO_2 \rightarrow$ products	– 6.07	– 6.16	– 6.29	– 6.55	– 7.07	– 7.79	– 8.55

6
Sensitivity to Atmospheric Conditions

6.1
Stratospheric Ozone

The tropospheric actinic flux can be reduced by stratospheric ozone absorption, especially at the energetic wavelengths below about 330 nm. Consequently, many photodissociation reactions in the troposphere depend sensitively on stratospheric ozone levels. Table 4 shows these sensitivity coefficients, defined as

$$s_x = - \frac{d \ln J_x}{d \ln \Omega}$$

or approximately as the percent increase in J_x for a 1% decrease in the overhead ozone column Ω. The largest sensitivity is found for the important reaction

$$O_3 + h\nu \ (\lambda < 340 \ \text{nm}) \rightarrow O_2 + O(^1D) \quad J_{O1D}$$

with smaller but non-negligible values for several other reactions. These sensitivities result from ozone absorption in the Huggins bands, with the exception of NO_3 which shows increasing sensitivity at low sun due to ozone's

Table 4. Sensitivity of photolysis rate coefficients at the Earth surface to changes in the total atmospheric ozone column. Values are defined as the percent increase in J of a specific reaction, for a 1 percent decrease in the ozone column

Reaction	Solar zenith angle, degrees						
	0	30	45	60	75	85	90
$O_3 \rightarrow O_2 + O(^1D)$	1.48	1.54	1.63	1.78	1.98	2.09	2.05
$O_3 \rightarrow O_2 + O(^3P)$	.08	.08	.08	.09	.12	.25	.48
$NO_2 \rightarrow NO + O(^3P)$	.02	.02	.02	.02	.02	.03	.04
$NO_3 \rightarrow NO + O_2$	.04	.05	.06	.08	.14	.32	.67
$NO_3 \rightarrow NO_2 + O(^3P)$	.03	.04	.04	.06	.11	.26	.50
$N_2O_5 \rightarrow NO_3 + NO_2$	.30	.30	.30	.29	.29	.32	.33
$HNO_2 \rightarrow OH + NO$	.01	.01	.02	.02	.02	.04	.05
$HNO_3 \rightarrow OH + NO_2$	.79	.80	.81	.82	.85	.88	.91
$HNO_4 \rightarrow HO_2 + NO_2$	.73	.74	.77	.81	.93	1.13	1.21
$H_2O_2 \rightarrow 2OH$	.34	.34	.34	.34	.35	.39	.42
$CH_2O \rightarrow H + HCO$	.40	.42	.44	.50	.62	.79	.90
$CH_2O \rightarrow H_2 + CO$	.13	.14	.15	.16	.21	.28	.34
$CH_3OOH \rightarrow CH_3O + OH$	.30	.30	.30	.30	.31	.35	.37
$CH_3ONO_2 \rightarrow CH_3O + NO_2$	.79	.80	.82	.85	.90	.97	1.01
$CH_3CHO \rightarrow CH_3 + CHO$	.83	.86	.91	1.00	1.18	1.37	1.40
$CH_3COCH_3 \rightarrow$ products	.64	.65	.68	.72	.82	.92	.99
$CH_3COO_2NO_2 \rightarrow$ products	.56	.55	.54	.54	.54	.57	.60

Chappuis absorption in the visible. The photodissociation of NO_2 is quite insensitive to stratospheric ozone changes.

The response of tropospheric chemistry to stratospheric ozone changes is of some interest. Long term reductions in stratospheric ozone have been reported in recent years [28], and it is likely that the enhanced UV penetration has led to increases in tropospheric reactivity. Increases in OH shorten the atmospheric lifetime of key tropospheric gases (e.g., [29–32]), and ultimately result in faster production of peroxides [32]. In high-NO_x environments (e.g., urban and rural polluted regions), higher J values are likely to result in additional tropospheric ozone production [29], while in low-NO_x regions ozone destruction is enhanced [29, 32]. Decreases in stratospheric transmission have also been observed, e.g., from sulfur dioxide and aerosols injected by the 1991 Mt. Pinatubo eruption, and have been correlated with temporary increases in tropospheric methane and carbon monoxide in the tropics, due to the reduced photolytic OH production rates [33].

6.2
Clouds

Clouds are highly variable in space and time. On the microscopic scale, cloud water may be present in either liquid or solid particles with complex size distributions, and with varying chemical composition depending on history of exposure to aerosols and various atmospheric gases. On the larger scale, in-

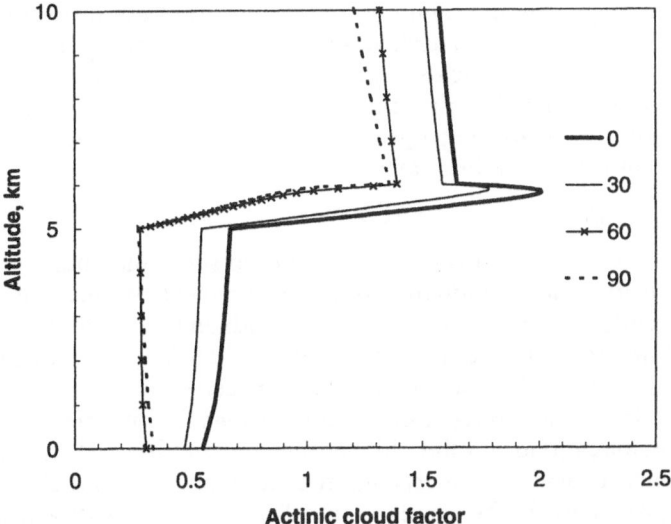

Fig. 5. Actinic flux at 400 nm in the presence of a uniform cloud between 5 and 6 km, normalized by cloud-free values. Cloud optical depth $\tau = 20$, asymmetry factor $g = 0.85$, single scattering albedo $\omega_0 = 0.9999$. Solar zenith angles are indicated in the legend

dividual clouds exhibit near-chaotic morphology, and multiple clouds are often present in both horizontal and vertical directions. Under such conditions, modeling of actinic flux and J values is extremely difficult, being limited more by lack of knowledge of detailed cloud optical parameters than by uncertainties in molecular data or radiative transfer methods. Nevertheless, several general aspects of the effect of clouds on photodissociation can be noted.

Figure 5 illustrate the vertical profile of the actinic flux in the presence of a horizontally uniform cloud. Above a cloud, actinic fluxes are always enhanced by the reflection provided by the cloud. Below cloud, actinic fluxes are usually (but not always) reduced relative to the cloud-free case. The situation inside a cloud is more complex, and both reductions and enhancements (especially just below cloud-top) are possible. Simple models and parameterizations have been developed (e.g., [34, 35]), and the general features have been confirmed by measurements [36, 37].

Cloud vertical optical depth τ, defined in the sense of Beer-Lambert attenuation of the direct solar beam, is approximately related to the liquid water content by

$$\tau = \text{(effective cross section of particles) (number of particles in column)} \sim$$
$$\sim 2\pi \text{ (effective radius, r)}^2 \text{ (mass of liquid water column, } M) \text{ (mass of each particle)}$$
$$\sim 3M/(2r\rho)$$

for spherical cloud droplets of mass density ρ, to be integrated over their size distribution. For smaller droplets, the effective cross sectional area $2\pi r^2$ should be replaced by the more complex Mie scattering functions.

Some perspective on the values of ω_o in clouds (but not in aerosols) may be gained by noting that rainwater is fairly transparent when collected in typical containers, e.g., several cubic centimeters. The total absorption by such a mass, say $\tau_M < 1$ for $M \sim 1$ g, is largely similar regardless of whether the mass is subdivided, while the scattering optical depth increases. For the above values subdivided into 10^{-5} m radius particles, ω_o is easily shown to be

$$\omega_o > 1/[2r\rho/(3M) + 1] \sim 0.998$$

Although this may seem very near unity, even such small numbers can have a substantial effect on the radiation for $\tau_M \sim 1$. Some enhancement of absorption can occur within a cloud due to the long pathlengths of multiply scattered photons, though of course quite differently for photons transmitted toward the ground than for those backscattered toward space.

With partial cloud cover, extremely complex vertical profiles of actinic flux become possible, and local values higher than for cloud-free skies are common. Simple models consider a superposition of fully overcast and cloud-free conditions, in proportion to the cloud fractional area coverage, with attention to the position of the sun relative to clouds [35, 38, 39]. More complex models, such as Monte Carlo calculations for three-dimensional cloud arrays of various aspect ratios [40, 41], show many interesting features, but for practical applications these too are hindered by insufficient knowledge of the properties of any specific real cloud.

6.3
Aerosols

Aerosol hazes are often more uniform than clouds, but present other difficulties because of their highly variable composition and size distributions, depending on aerosol type: mineral dust, soot, organic, sulfate, dry or wet, or otherwise heterogeneous mixtures. Absorption properties of various aerosols are still poorly characterized, especially in the UV spectral range, and multi-modal size distributions can lead to complex angular dependence of the phase functions and the resulting radiation field.

Visibility in the planetary boundary layer is often limited by the presence of aerosol hazes, and a simple relationship may be used to relate the visual range R_v (in km for 2% contrast at 550 nm) to the aerosol extinction β (in km^{-1}) $\sim 3.9/R_v$. While this strictly applies only to the horizontal direction, vertical mixing in the boundary layer (and knowledge of its height) can make visibility-based estimates useful also for solar actinic radiation. The vertical optical depth of the aerosol in the boundary layer is then estimated by multiplying the extinction β by the height of the boundary layer.

Figure 6 shows the effect of aerosols on J_{NO2} in the lower atmosphere, for overhead sun. In this illustration a considerable amount of aerosol is distributed uniformly through the first km, with extinction of 1.0 km^{-1} at 550 nm (3.9 km visible range). Non-absorbing aerosols ($\omega_o = 1$) cause an increase in J_{NO2} above the surface, relative to the aerosol-free case, while decreases are found in the presence of heavily absorbing aerosols ($\omega_o = 0.5$ and 0.75).

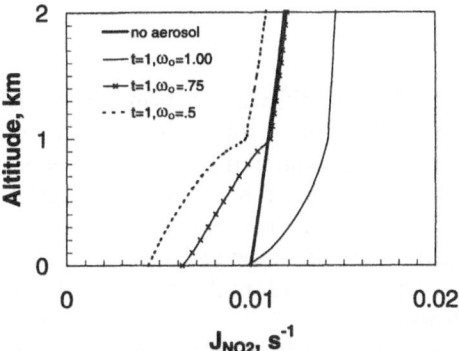

Fig. 6. J_{NO2} in the presence of aerosols in a 1 km boundary layer. Overhead sun, solid curve gives aerosol-free case, other curves for aerosol optical attenuation = 1.0 km^{-1} at 550 nm (3.9 km visible range) and single scattering albedo indicated in the legend

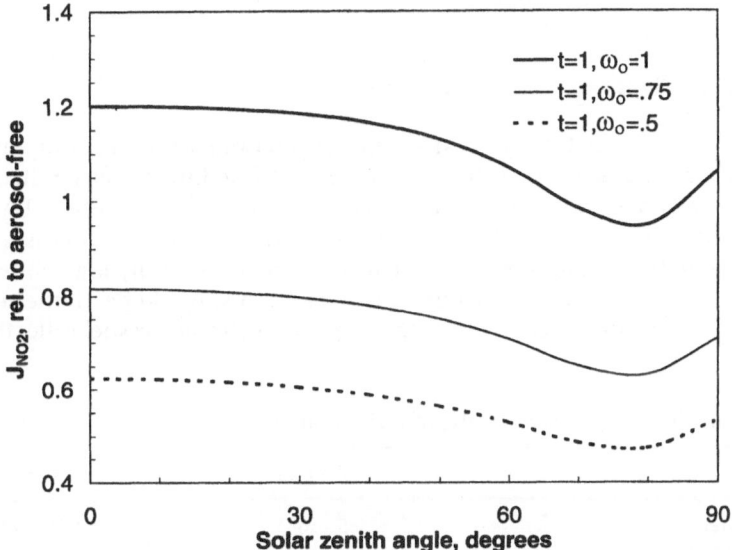

Fig. 7. J_{NO2} averaged vertically in a 1 km boundary layer containing aerosols with values of the aerosol optical depth and single scattering albedo indicated in the legend

Insofar as chemicals in the boundary layer can be mixed vertically, it is also of interest to consider the effect of aerosols on vertically averaged J_{NO2}, as shown in Fig. 7. With non-absorbing aerosols, enhancements are seen for θ_o smaller than about 60°, up to 20% for the case considered (3.9 km visible range). Absorbing aerosols reduce the average boundary layer J_{NO2}, but not as much as would be inferred from surface values (see Fig. 6).

6.4
Other Gaseous Absorbers

Tropospheric ozone can be a significant fraction of the total atmospheric ozone column ($\sim 1/10$) but is quite variable and depends on location and time. Urban and regional scale pollution contributes to tropospheric ozone formation. Absorption of tropospheric radiation can be substantial, especially at UV-B wavelengths, with reductions of 10–20% for J_{O1D} not being unusual. The tropospheric absorption can be enhanced by the strong Rayleigh scattering [42], which increases photon pathlengths through the ozone, relative to the direct beam (at high sun).

Absorption by other gases is usually not important, except in highly polluted regions where large amounts of NO_2, sulfur dioxide (SO_2), and various organic (especially carbonyl) gases may be present. For example, 100 ppb of NO_2 in a 1 km boundary layer yields an optical depth of about 0.1 in the UV-A region (with $\sigma_{NO2} \sim 5 \times 10^{-19}$ cm^2). SO_2 absorption occurs primarily in the UV-B region and therefore may affect J_{O1D}, but again large concentrations (several tens of ppb) are required for a significant effect on J values.

6.5
Surface Reflections

Solar radiation reflected by oceans, soils, vegetation, structure, and any other surfaces contributes to the total atmospheric actinic flux. Surface reflectivities are quite variable, and relatively little data is available, especially at UV wavelengths most important to photodissociation. A brief compilation is given in Table 5. Note that values for soils and vegetation are typically low, while nearly complete reflection may be encountered over fresh snow. Reflectivities of large bodies of water often exceed those expected from simple Fresnel reflection at a

Table 5. Ultraviolet reflectivity (albedo) of various surfaces

Surface	Albedo, %	Reference
Liquid water	5–10[a]	[43–45]
Clean dry snow	30–100	[43–46]
Dirty wet snow	20–95	[43, 45]
Ice	7–75	[43, 44]
Various vegetation	1–8	[43, 44, 46–49]
Black lava	1–3	[49, 50]
Various soils (sand, loam, silt, clay)	4–11	[47, 49, 50]
Limestone	8–12	[44]
Gypsum sand	16–30	[50]
White cement	17	[46]
Blacktop asphalt	4–11	[44, 47]
Black cloth	2	[46]

[a] Depending on solar zenith angle, wave roughness, sea-salt aerosol.

horizontal water-air interface, because of surface roughness (e.g., wave breaking) and scattering by sea-spray aerosol.

In many cases, particularly near the surface, it is useful to distinguish between local and regional albedo. The former refers to light reflected directly (e.g., by a nearby building) towards the target of interest, e.g., a photodissociating molecule or a measuring instrument. The latter refers to the extended area, several km^2 in size, surrounding the location of interest. On this larger scale, photons reflected upwards have a significant chance of being scattered again in the atmosphere, and thus will contribute to the total down-welling diffuse light. This accounts, for example, for the bright skies observed above snow-covered ground.

7
Actinometric Measurements

Numerous more or less direct measurements of photodissociation rate coefficients have been made using chemical actinometers [39, 49, 51–62]. Such measurements are generally difficult, requiring specialized instrumentation and careful consideration of numerous potential chemical and optical interferences, if accurate results are to be obtained.

Several chemical actinometers have been designed for the reaction

$$NO_2 + h\nu \ (\lambda < 420 \ nm) \ \rightarrow \ NO + O(^3P) \quad J_{NO2}$$

relying, for example, on the detection of product NO. The reaction proceeds in the confines of a UV-transparent quartz vessel exposed to ambient radiation. The measurements of J_{NO2} are generally in 10–30% agreement with theoretical models, and in most cases show the expected behavior with zenith angle and altitude, and to a lesser quantitative extent with aerosols and clouds.

Another important class of actinometers has been used for the reaction

$$O_3 + h\nu \ (\lambda < 340 \ nm) \ \rightarrow \ O_2 + O(^1D) \quad J_{O1D}$$

where the excited atoms react in the presence of added nitrous oxide,

$$O(^1D) + N_2O \ \rightarrow \ 2NO$$

followed again by detection of NO [63–68]. Figure 8 shows the results of one such study. The measurements are about 15% lower than the model predictions of absolute values, but, just as importantly, do show good agreement with the model on the dependence on the slant ozone column (the overhead vertical ozone column amount, divided by the cosine of the solar zenith angle). This level of agreement is acceptable considering the numerous sources of error possible in both measurements (e.g., flow calibrations, impurities and secondary reactions, optical perturbations) and in model inputs (e.g., extraterrestrial solar irradiance, ozone absorption and quantum yields) and radiative transfer calculations.

Radiometers filtered to mimic the photo-action spectrum ($\sigma\phi$) of J_{NO2} and J_{O1D} have been developed, and offer a simple surrogate measurement [68–71] Careful calibrations against chemical actinometers are required, and if the

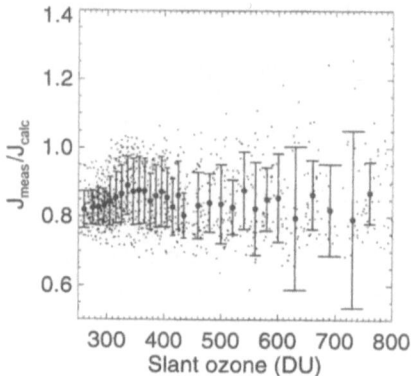

Fig. 8. J_{O1D} measurement/model comparisons at Mauna Loa. Small points are individual measurements for cloud-free skies, circles are averages with deviations. Adapted from [68]

spectral match between the filter and $\sigma\phi$ is imperfect, corrections must be applied when the spectrum of the ambient light changes under different solar zenith angles, ozone amounts, and various other atmospheric factors.

8
Conclusions

The uses of J values in atmospheric chemistry may be broadly divided into climatological and episodic. The first includes development of chemistry-transport models to study the impacts of future emission scenarios, as well as present and past atmospheric chemical states, and comparisons to long-term large-scale data records. Satellite data will be used increasingly, in combination with models, to provide information about the atmospheric light field. Temporal and spatial resolution is problematic with satellite measurements, and therefore such approaches are best suited to developing statistical measures (averages, variances) on extended time/space scales, rather than applied to any specific instant and location.

The episodic approach also has multiple objectives. Direct comparison between J measurements and models, for numerous episodes under many different conditions, is an essential catalyst for improvement of models (and perhaps measurements). There are, in addition to current research on atmospheric photo-dissociation, increasing efforts to monitor UV irradiances at the surface to assess the environmental impacts of stratospheric ozone reductions. Such measurements provide a major opportunity to evaluate atmospheric UV models and satellite-derived values. However it should be recognized that the estimation of the actinic flux is much more complex than that of the surface irradiance, both for the geometric differences discussed, and for the usual need to know its full vertical profile. Beyond this, direct J estimates obtained simultaneously with other physical and chemical measurements must show a degree of internal consistency, and are therefore fundamentally useful to test in detail our putative understanding of photochemical processing [72–74].

References

1. Finlayson-Pitts BJ, Pitts JN (1986) Atmospheric chemistry. Wiley-Interscience, New York
2. Graedel TE, Crutzen PJ (1993) Atmospheric change. WH Freeman, New York
3. Thompson AM, Stewart RW (1991) J Geophys Res 96:13,089
4. DeMore WP, Sander SP, Golden DM, Hamspon RF, Kurylo MJ, Howard CJ, Ravishankara AR, Kolb CE, Molina MJ (1997) Chemical kinetics and photochemical data for use in stratospheric modeling, Evaluation Number 12. JPL Publication 97-4 Pasadena.CA
5. Chandrasekhar S (1960) Radiative transfer. Dover, NY
6. Hansen JE, Travis LD (1974) Space Sci Rev 16:527
7. Goody RM, Yung YL (1989) Atmospheric radiation. Oxford University Press, New York
8. Joseph JH, Wiscombe WB, Weinman JA (1976) J Atmos Sci 33:2452
9. Meador WE, Weaver WR (1980) J Atmos Sci 37:630
10. Toon OB, McKay CP, Ackerman TP, Santhanam K (1989) J Geophys Res 94:16,287
11. Petropavlovskikh I (1996) PhD thesis. University of Brussels
12. Smart WM (1979) Textbook on spherical astronomy. Cambridge University Press, Cambridge
13. Duffet-Smith P (1988) Practical astronomy with your calculator. Cambridge University Press, Cambridge
14. Spencer JW (1971) Search 2:172
15. Madronich S, Flocke SJ, Zeng J, Petropavlovskikh I (1997) Tropospheric ultraviolet-visible model (TUV). National Center for Atmospheric Research NCAR Technical Note (in preparation)
16. US Standard Atmosphere (1976) National Oceanic and Atmospheric Administration (NOAA), National Aeronautics and Space Administration (NASA), United States Air Force, Washington DC
17. Elterman L (1968) UV, visible, and IR attenuation for altitudes to 50 km. Air Force Cambridge Research Laboratories (AFCRL) Report-68-0153, Cambridge, MA
18. Van Hoosier ME, Bartoe J-D, Brueckner GE, Printz DK (1987) Solar irradiance measurements 120-400 nm from Spacel Lab-2, IUGG Assembly, Vancouver
19. Neckel H, Labs D (1984) Solar Physics 90:205
20. Stamnes K, Tsay SC, Wiscombe W, Jayaweera K (1988) Appl Opt 27:2502
21. Zeng J, Petropavlovskikh I, Kunasz P, Madronich S (1997) Submitted to J Geophys Res
22. Molina LT, Molina MJ (1986) J Geophys Res 91:14,501
23. Nicolet M (1984) Planet Space Sci 32:1467
24. Demerjian KL, Schere KL, Peterson JT (1980) Adv Env Sci Tech 10:369
25. Frank R, Klöpffer W (1986) Spectral solar photon irradiance in central Europe and the adjacent North Sea, presented at the International Conference on Chemicals in the Environment, Lisbon.
26. Häder DP, Worrest RC, Kumar HD, Smith RC (1995) Ambio 24:174
27. Zepp RG, Callaghan TV, Erickson DJ (1995) Ambio 24:181
28. World Meteorological Organization (1994) Scientific assessment of ozone depletion. Global Ozone Research and Monitoring Project Report-37, Geneva
29. Liu SC, Tranier M (1988) J Atmos Chem 6:221
30. Madronich S, Granier C (1992) Geophys Res Lett 19:465
31. Tang X, Madronich S (1995) Ambio 24:188
32. Fuglestvedt JS, Jonson JE, Isaksen ISA (1995) Tellus 46B:172
33. Dlugokencky EJ, Dutton EG, Novelli PC, Tans PP, Masarie KA, Lantz KO, Madronich S (1996) Geophys Res Lett 23:2761
34. Madronich S (1987) J Geophys Res 92:9740
35. Chang JS, Brost RA, Isaksen ISA, Madronich S, Middleton P, Stockwell WR, Walcek CJ (1987) J Geophys. Res 92:14,681
36. van Weele M, Duynkerke PG (1993) J Atmos Chem 16:231
37. van Weele M, Vila-Guerau de Arrellano J, Kuik F (1995) Tellus 47B:353
38. Nack ML, Green AES (1974) Appl Opt 12:2405

39. Lantz KO, Shetter RE, Cantrell CA, Flocke SJ, Calvert JG, Madronich S (1996) J Geophys Res 101:14,613
40. Feigelson EM (1984) Radiation in a cloudy atmosphere. Reidel Boston
41. van Weele M (1996) PhD thesis. University of Utrecht, Utrecht
42. Brühl C, Crutzen PJ (1989) Geophys Res Lett 16:703
43. Kondratyev KY (1969) Radiation in the atmosphere. Academic Press New York
44. Blumthaler M, Ambach W (1988) Photchem Photobiol 48:85
45. Doda DD, Green AES (1980) Appl Opt 19:2140
46. Dickerson RR, Stedman DH, Delany AC (1982) J Geophys Res 87:45,933
47. Coulson KL, Reynolds DW (1971) J Appl Met 10:1285
48. McKenzie RL, Kotkamp M, Ireland W (1996) Geophys Res Lett 23:1757
49. Shetter RE, McDaniel AH, Cantrell CA, Madronich S, Calvert JG (1992) J Geophys Res 97:10,349
50. Doda DD, Green AES (1981) Appl Opt 20:636
51. Stedman DH, Niki H (1973) Environ Sci Technol 7:735
52. Jackson JO, Stedman DH, Smith RG, Hecker LH, Warner PO (1975) Rev Sci Instr 46:376
53. Harvey RB, Stedman DH, Chameides W (1977) J Air Pollut Control Assoc 27:663
54. Zafonte L, Rieger PL, Holmes JR (1977) Environ Sci Technol 11:483
55. Bahe FC, Marx WN, Schurath U, Roth EP (1979) Atmos Environ 13:1515
56. Dickerson RR, Stedman DH, Delany AC (1982) J Geophys Res 87:4933
57. Madronich S, Hastie DR, Ridley BA, Schiff HI (1983) J Atmos Chem 1:3
58. Parrish DD, Murphy PC Albritton DL Fehsenfeld FC (1983) Atmos Environ 17:1365
59. Madronich S, Hastie DR, Ridley BA, Schiff HI (1985) J Atmos Chem 1:151
60. Madronich S (1987) Atmos. Environ 21:569
61. Kelly P, Dickerson RR, Luke WT, Kok GL (1995) Geophys Res Lett 22:2621
62. Volz-Thomas A, Lerner A, Pätz HW, Schultz M, McKenna DS, Schmitt R, Madronich S, Röth EP (1996) J Geopys Res 101:18,613
63. Bahe FC, Schurath U, Becker KH (1980) Atmos Environ 14:711
64. Dickerson RR, Stedman DH, Chameides WL, Crutzen PJ, Fishman J (1979) Geophys Res Lett 6:833
65. Bairai ST, Stedman DH (1992) Geophys Res Lett 19:2047
66. Blackburn TE, Bairai ST, Stedman DH (1992) J Geophys Res 97:10,109
67. Müller M, Kraus A, Hofzumahaus A (1995) Geophys Res Lett 22:679
68. Shetter RE, Cantrell CA, Lantz KO, Flocke SJ, Orlando JJ, Tyndall GS, Gilpin TM, Madronich S, Calvert JG (1996) J Geophys Res 101:14,631
69. Junkerman W, Platt U, Volz-Thomas A (1989) J Atmos Chem 8:203
70. Brauers T, Hofzumahaus A (1992) J Atmos Chem 15:269
71. Hofzumahaus A, Brauers T, Platt U, Callies J (1992) J Atmos Chem 15:283
72. Davis DD, Chen G, Chameides W, Madronich S, Bradshaw J, Sandholm S, Rodgers M, Schendal J, Sachse G, Collins J, Wade L, Gregory G, Anderson B, Barrick J, Shipham M, Blake D (1993) J Geophys Res 98:23,501
73. Ridley BA, Madronich S, Chatfield RB, Walega JG, Shetter RE, Carroll MA, Montzka DD (1992) J Geophys Res 97:10,375
74. HauglustaineD, Madronich S, Ridley BA, Walega JG, Cantrell CA, Shetter RE, Hübler G (1996) J Geophys Res 101:14,681

2 Emission and Flash Techniques in Environmental Photochemistry

R. G. Brown

Centre for Photochemistry, Department of Chemistry, University of Central Lancashire, Preston, Lancashire, PR1 2HE, UK. *E-mail: R.G.BROWN@UCLAN.AC.UK*

This Chapter provides an introduction to the spectroscopic techniques used to study transient species produced during the photolysis of environmentally relevant chemicals. After a brief discussion of the photochemical processes which an excited state can undergo, the techniques of conventional (microsecond) flash photolysis and laser flash photolysis (nanosecond to femtosecond) are described, together with fluorescence lifetime methods. All the techniques are illustrated with appropriate examples.

Keywords: flash photolysis, femtochemistry, fluorescence lifetimes, time-correlated single photon counting.

Contents

1
Introduction

Interest in environmental photochemistry has exploded in the last twenty years due to legislative requirements and an increased public awareness with respect to environmental matters generally. It is now essential that we have as much information as possible about the fate of xenobiotic materials in the environment so that repetition of phenomena such as ozone depletion are avoided if at all possible. It is also clear, however, that mankind will continue to discharge

The Handbook of Environmental Chemistry Vol. 2 Part L
Environmental Photochemistry (ed. by P. Boule)
© Springer-Verlag Berlin Heidelberg 1999

xenobiotics into the environment as their total removal from industrial and household waste is ruinously expensive. It is therefore important to know whether materials which are "safe" in the form in which they are discharged remain safe as they undergo chemical and biochemical transformations in the environment. Direct photolysis or sensitisation can play a major role in these processes. Finally, there is considerable current interest in a range of methods with great potential for the destruction of potential pollutants. These involve UV photolysis either on its own or in the presence of oxidising agents (such as ozone and hydrogen peroxide) or semiconductor materials.

All of these areas of interest involve the input of energy in the form of photons of light, which leads to one or more chemical transformations via the formation of excited states and, in many cases, species such as radicals. In order to understand the chemistry or biochemistry which is initiated by the light energy, we need to study the absorption properties of the starting compounds and the nature and properties of the transient species (excited states and radicals) which are produced from them. The reader is introduced to the theory and techniques behind such measurements in this chapter and in the companion chapters by Zepp [1] and by Roof [2]. The latter provides excellent coverage of concepts such as light absorption, sensitisation, quantum yields of reaction and methods of characterising the products of photochemical reactions. In this chapter I therefore only briefly cover these ideas and concentrate on the methods which can be used to study the transient species produced in photochemical reactions.

2
Light Absorption by Molecules

The theory and concepts involved in the absorption of light energy by molecules is well covered in the chapters by Roof and Zepp and is only briefly summarised here. There is also a vast array of textbooks on the UV-visible spectroscopy of molecules (e.g. [3]). True environmental photochemistry at ground level (i.e. photochemistry under real environmental conditions) is driven by the absorption of photons of sunlight. The consequence of this is that any molecules which are going to be photochemically active are ones which absorb light at wavelengths of 300 nm or greater since this is the approximate lower limit to the sun's radiation which reaches the surface of the earth [1]. Such molecules will have to contain chromophores (functional groups or parts of the molecule) that undergo $\pi - \pi^*$ or $n - \pi^*$ transitions such as aromatic or heteroaromatic functions, conjugated polyenes and carbonyl groups. Most aliphatic compounds, alkenes and dienes will not be active since they absorb at much shorter wavelengths. Simple carbonyl compounds such as propanone (acetone) will only absorb weakly around 300 nm (due to the low absorption intentsity of $n - \pi^*$ transitions) – the energy of the carbonyl $\pi - \pi^*$ transition is of much higher energy as is apparent from Fig. 1 where the relative energies and orbital nature of the σ, π, n, π^* and σ^* states of a carbonyl group are shown. However, once a carbonyl is conjugated with other π systems, the absorption maximum (in nm) and absorption intensity both increase.

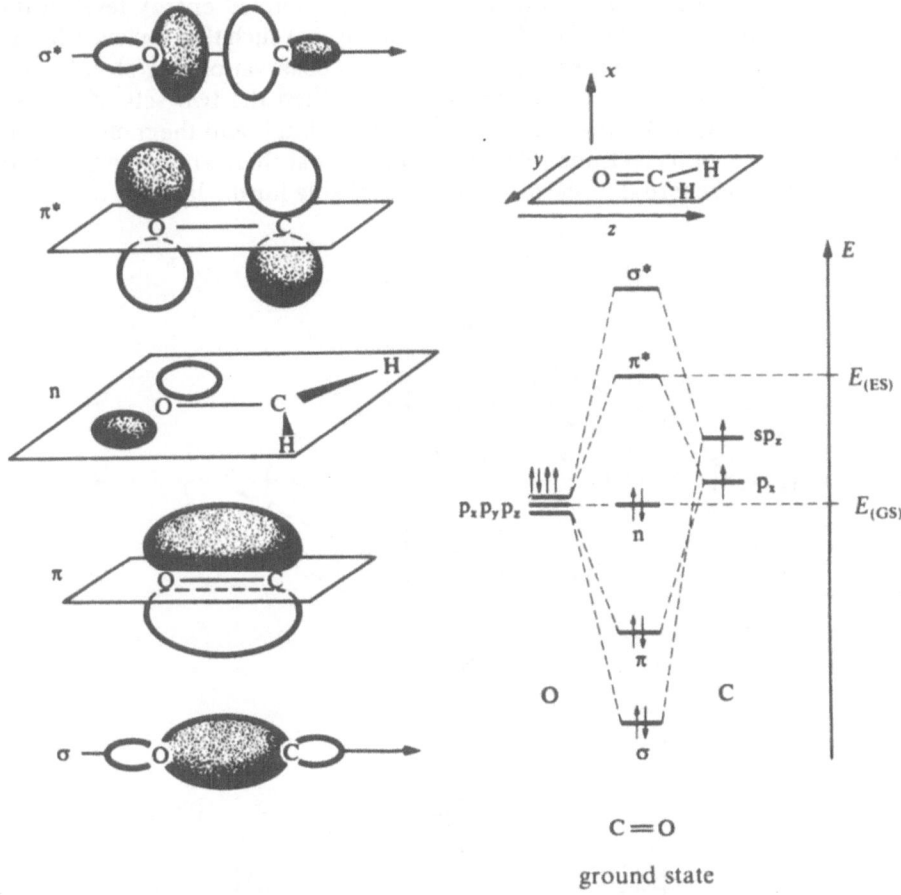

Fig. 1. The molecular orbitals of a carbonyl group. Reproduced with permission from [4]

The absorption intensity of a molecule is usually quantified in terms of its extinction coefficient (ε_λ) as defined by

$$\text{Absorbance (A)} = \log_{10} I_o/I_t = \varepsilon_\lambda \, cl \qquad (1)$$

where I_o is the intensity of a monochromatic light beam impinging on a sample containing the molecule of interest (at a concentration c mol dm^{-3}), I_t is the intensity of light transmitted by the sample and l is the path length of the light through the sample. Many absorption spectra are measured in cuvettes of 1 cm path length, but routine measurements with l between 1 mm and 20 cm are possible. Extinction coefficients of the order of 10–100 dm^3 mol^{-1} cm^{-1} are observed for n–π* transitions but much larger values (usually 1000–100000+) are observed for π–π* transitions [3]. The vast majority of absorption transitions originate from molecules which have a SINGLET state as their lowest energy state (the ground state). This is because virtually all

molecules have their electrons paired, i.e. every occupied energy level in the
molecule contains two electrons (with opposite spins) such that the overall spin
is zero. The best known exception to this general observation is molecular (di)
oxygen (O_2) which has a triplet ground state. When the two sets of oxygen
atomic orbitals are combined to form molecular orbitals and the molecular orbitals are filled with electrons (Fig. 2), it is found that there are two degenerate
(i.e. of the same energy) molecular orbitals available for the last two electrons.

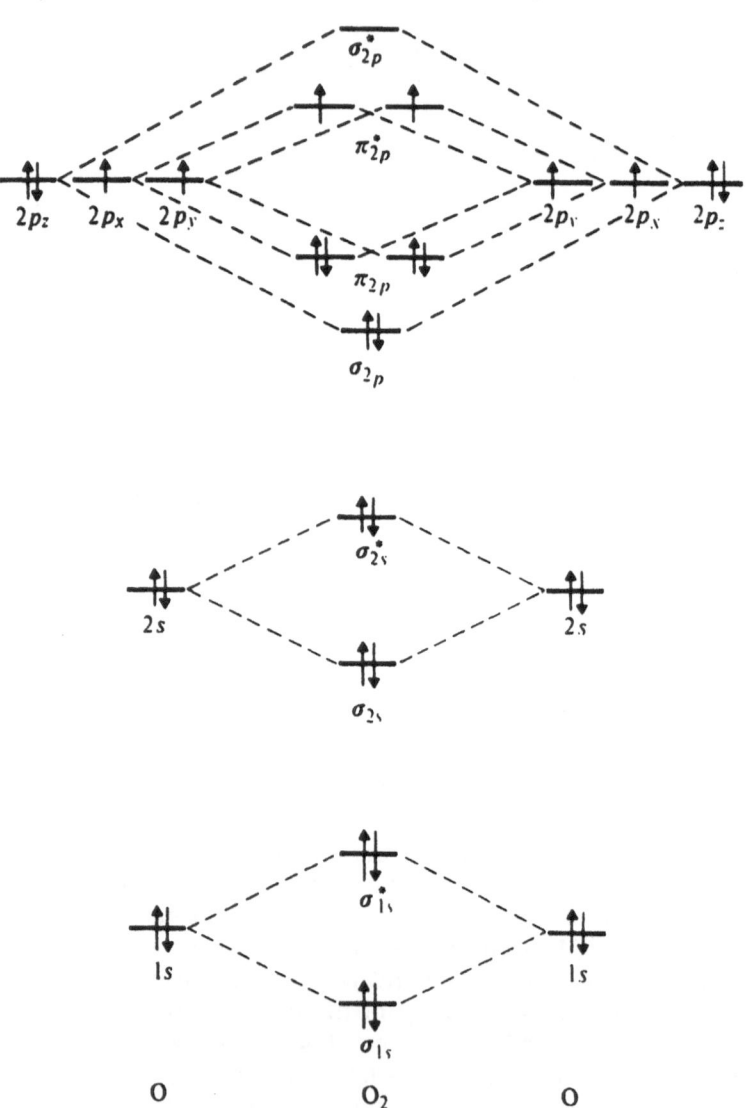

Fig. 2. The ground state of dioxygen. Reproduced with permission from [5]

The lowest energy arrangement is for each molecular orbital to contain one electron. This leads to three possible spin states $(+1, 0, -1)$ and the dioxygen ground state is therefore a TRIPLET state.

3
Fundamental Photochemical Concepts

The interest of the photochemist begins with the excited state which is formed by the interaction of the molecular ground state with a photon. The Grotthus-Draper law states that only the radiation absorbed by a molecule can produce photochemical change [3, 6] and the Stark-Einstein law indicates that each absorbed photon will produce one excited state [3, 6]. Since the vast majority of molecules have a singlet ground state (S_0), the excited state which is initially produced is also usually a singlet state. Absorption of light to form an excited spin state (e.g. a triplet state) different to that of the ground state is formally forbidden [6] but can be observed in the presence of a perturbation such as a high pressure of molecular oxygen [7] or a heavy atom such as xenon, bromine or iodine [8].

Any molecule will have a number of absorption bands in the UV-visible corresponding to transitions from the S_0 ground state to excited states S_1, S_2, S_3 etc. These bands will occur at shorter and shorter wavelength depending on the excited state which is produced. However, it is usually found that the higher excited states S_2, S_3 etc. rapidly undergo non-radiative decay (internal conversion) to produce the S_1 excited state. There are a few exceptions to this general rule such as azulene [9] and some thioketones [10] where the S_2 state is observable but these examples are quite rare.

Molecular photochemistry therefore almost invariably begins from the S_1 excited state and the processes which deplete this state are commonly depicted in a Jablonski diagram such as that shown in Fig. 3. If the S_1 excited state is produced in a vibrationally excited state (e.g. by absorption of a photon of energy higher than the 0–0 transition energy), in fluid solution at ambient temperature this extra vibrational energy is rapidly dissipated by collisions with solvent molecules – the process of vibrational relaxation (VR in Fig. 3). Subsequent processes will therefore originate from the zeroth vibrational level of S_1.

Following vibrational relaxation we can identify four different types of process which an excited state can undergo:

– radiative processes (fluorescence (F) and phosphorescence (P));
– non-radiative processes (internal conversion (IC) and intersystem crossing (ISC));
– photochemistry (i.e. a true chemical reaction is initiated);
– bimolecular interactions leading to quenching or photochemistry

The radiative processes, fluorescence (from S_1) and phosphorescence (from T_1), are directly observable as they involve the emission of a photon of light and are of particular relevance to the (environmental) photochemist since they provide an immediate way of observing the excited state. Both processes lead to the re-formation of the ground state and are classified as photophysical processes

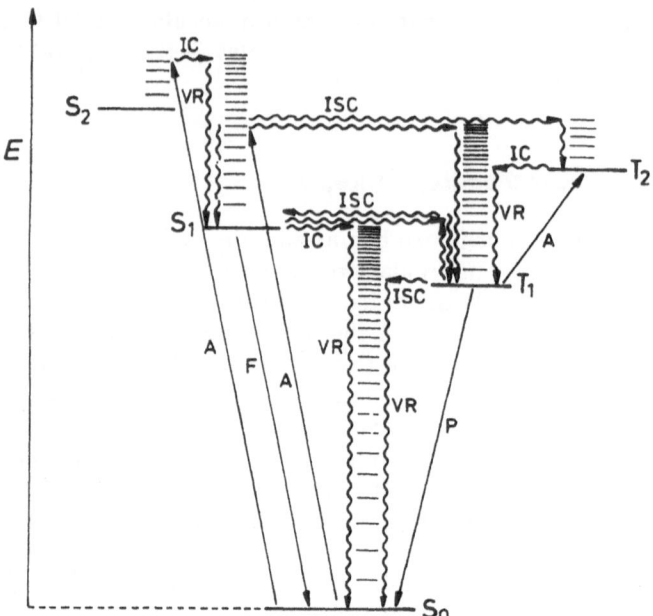

Fig. 3. A Jablonski diagram. Radiative processes (absorption (A), fluorescence (F) and phosphorescence (P)) are denoted by *straight arrows* and non-radiative processes (vibrational relaxation (VR), internal conversion (IC) and intersystem crossing (ISC)) by *wavy arrows*. Reproduced with permission from [3]

since the molecule has undergone no permanent chemical change in the cycle of light absorption followed by light emission. Similarly, the two non-radiative processes of internal conversion and intersystem crossing are also photophysical processes since they produce no overall chemical change. Both processes lead to the interconversion of an excited state into either another excited state of lower energy or the ground state without emission of a photon of light.

Internal conversion involves no change of spin (e.g. $S_2 \rightarrow S_1$, $S_1 \rightarrow S_o$, $T_2 \rightarrow T_1$) whereas intersystem crossing does involve a spin change (e.g. $S_1 \rightarrow T_2$, $T_1 \rightarrow S_o$). The latter is formally forbidden but becomes (weakly) allowed due to spin-orbit coupling [6]. Both internal conversion and intersystem crossing are believed to occur in two stages: the conversion of the upper state to a high vibrational level of the lower state which is equal in energy to the upper state, followed by vibrational relaxation.

Any excited state has the potential to undergo chemical reactions which are not possible in the ground state. These often involve an interaction with a second species such as a molecule of solvent. Alternatively, a bimolecular inter-action can occur which removes the excited state energy and regenerates S_o – so-called "quenching". Sensitisation is an example of the quenching of an excited state of one molecule to produce an excited state of a second (different) molecule.

All of the above processes may be characterised by a quantum yield (ϕ) defined as

$$\phi_f = \frac{\text{number of molecules undergoing the process/total}}{\text{total number of photons absorbed}} \qquad (2)$$

The consequence of the Stark-Einstein law is that the sum of the quantum yields for all the processes that a molecule undergoes is unity.

These processes will also be characterised by a particular rate constant (first or second order) and the properties of any excited state will be determined by the relative sizes of the individual rate constants. We are not able to measure these individual rate constants directly but can obtain information about them from other measurements. In particular, the lifetime (τ) of an excited state, defined as the time taken for the excited state concentration to decrease to $1/e$ of its initial concentration, is given for S_1 by

$$\tau = \frac{1}{k_f + k_{ic} + k_{isc} + k_{photochem} + k_Q[Q]} \qquad (3)$$

where the ks are the rate constants for all the various processes which S_1 can undergo.

If we combine lifetime information with quantum yields, we can calculate the individual rate constants. For example, if the fluorescence quantum yield (ϕ_f) of a compound is known, then

$$\phi_f = \frac{1}{k_f + k_{ic} + k_{isc} + k_{photochem} + k_Q[Q]} = k_f \tau \qquad (4)$$

The measurement of quantum yields for fluorescence (or phosphorescence) and photochemistry and the characterisation of bimolecular quenching are well established. However, it is much more difficult to determine quantum yields for the non-radiative processes, and these are often calculated by subtracting all the other quantum yields from unity.

In addition to their lifetimes and quantum yields, excited states and other transient species are also characterised by their spectroscopic properties – particularly their absorption and emission spectra – and these properties are frequently the means by which lifetime and quantum yield information is derived. Unfortunately, the lifetimes of transient species are often very short so that it is difficult to study them by steady-state methods since only small concentrations are present at any one time. Absorption methods are thus insufficiently sensitive although fluorescence and phosphorescence can still easily be observed. The alternative is to produce a much higher concentration of the transient by applying a relatively intense, short-lived flash of light to the sample and then using rapid absorption and emission spectroscopy to monitor the transient.

The techniques that are used to achieve these ends are the subject of the rest of this chapter.

4
Flash Photolysis Techniques

Flash photolysis is a method whereby a non-equilibrium situation can be created in a reaction system in a short interval of time. It provides a general means of preparing unstable intermediates in concentrations higher than can usually be obtained by other methods, which rarely provide non-equilibrium concentrations of free radicals or similar species high enough for direct observation. Along with pulse radiolysis, it is almost the only method available for the preparation and study of high concentrations of electronically excited molecules. It is applicable to gases, liquids, and solids, and to the whole available temperature range. In conjunction with physical methods of observation, the concentration of intermediates can be directly measured as a function of time and their physical and chemical properties determined [11].

Like the photochemical method itself, flash photolysis is of very general applicability and, with an appropriate choice of reacting system, most types of chemical change can be initiated and most classes of chemical intermediate can be prepared and studied. On the other hand, the innate specificity of photochemical reactions can be retained such that, by suitable choice of excitation wavelength, the primary process can often be very clearly defined.

In the flash photolysis technique a reactant is irradiated with an intense flash of infra-red, visible or ultraviolet light. The intensity must be sufficient to produce a measurable change in chemical composition, but of short duration compared with that of the ensuing reactions which are to be studied. Soon after the introduction of the technique, in 1949, following the pioneering work of Norrish and Porter for which they were awarded the Nobel Prize for Chemistry in 1967, the time resolution, as determined by the duration of the electric discharge flash lamp, was improved to a few microseconds and there was little change in this capability until about 1967 when the pulsed laser began to be used. The laser has, over the last thirty years, transformed the time resolution of the flash photolysis technique by reducing the times which can be resolved by a factor of a billion and by providing light sources which are both monochromatic and spatially coherent. It should be stressed, however, that for the study of the great variety of chemical changes which occur in the time range between one microsecond and one second, the conventional flash photolysis technique using pulsed electric discharge lamps is less expensive and entirely satisfactory for most purposes.

4.1
Conventional Flash Photolysis

A typical conventional flash photolysis arrangement consists of three parts:

- a photolysis flash lamp for producing a short pulse of light of very high intensity along with associated energy storage units – usually capacitors – and equipment for charging and initiating the discharge (triggering);
- a reaction vessel;
- some arrangement for physical detection of the transients.

The ideal source for use in producing or monitoring absorbing species in flash photolysis is one which combines the highest intensity of visible or ultraviolet light and the shortest duration. Furthermore, the lamp must be reproducible from flash to flash in both spectral output and intensity and must produce an adequate number of flashes before deteriorating. In practice, increasing the energy of the flash increases the duration so that, for a particular application, it is necessary to balance these two factors. Although these requirements may apply to both the excitation and monitoring light sources, it is the characteristics of the former which determine the time resolution of the experimental apparatus.

Light flashes for microsecond work and longer are almost invariably produced by the discharge of a capacitor through a quartz tube filled with a moderate pressure of a rare gas (usually xenon) [11]. Lamps for flash photolysis applications usually have energies in the range 10–3000 J. Higher energies than this are rarely required. At the higher energies, where conversion of electrical energy into light energy is efficient (1–50%), a 10-fold reduction in flash duration requires a 100-fold reduction in energy. Although it is possible to dissipate large energies in carefully designed circuits to give short flash durations, the time resolution in conventional flash photolysis is normally determined by the tail on the flash profile. At low energies, it is very difficult to obtain both short and intense light pulses, particularly since conversion efficiencies are of the order of 10^{-4}%.

Once the characteristics of the excitation source have been decided upon, the transient species which are produced by the excitation flash may be monitored by flash spectroscopy or by kinetic spectrophotometry. The first of these uses a second, short duration flash (the spectroflash) at some predetermined time after the excitation flash to record the absorption spectrum of the transient(s) that have been produced. Kinetic spectrophotometry utilises a continuous monitoring beam which measures the change in absorbance at a chosen wavelength for a period of time after the excitation flash and thus provides information about the kinetics of the species absorbing at the chosen wavelength.

A typical experimental set-up for flash spectroscopy is shown in Fig. 4. In conventional flash photolysis, a cylindrical reaction vessel is usually used with the excitation lamps being placed co-linear with it and the monitoring beam (whether a continuous beam or a spectroflash) passing through the length of the vessel. The delay between the excitation flash (photoflash) and the spectroflash is determined electronically and will usually be variable from microseconds to seconds. Once the spectroflash has passed through the reaction vessel, it is spectrally dispersed and the spectroflash intensity as a function of wavelength measured with, for example, a diode array and recorded on computer. These have now replaced the photographic plates used in earlier work [11]. Calculation of absorbance values using Beer's Law (Eq. 1) requires the acquisition of reference intensities – this simply involves operation of the spectroflash without a preceding photoflash. Once these standard calculations have been undertaken, the spectral information will be presented in the usual way as absorbance vs wavelength.

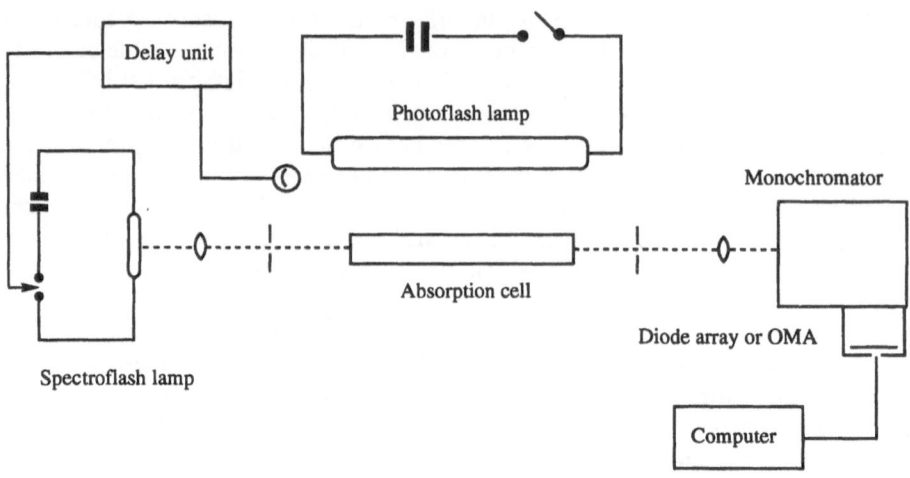

Fig. 4. Experimental layout for conventional flash photolysis by the flash spectroscopic method

Unless a previous study of the system has been carried out, spectroscopic investigations may reveal positions and approximate magnitudes of transient absorptions. Transients produced by the photolysis flash may have absorption bands which are as broad as 200 nm (in the case of large polycyclic molecular triplet states) or as narrow as 0.01 nm (in the case of free radicals such as CH, NH, etc.) [11, 12].

The principal information which can be derived from flash spectroscopy is as follows.

- The extent of permanent reaction: comparison of the spectra before and after a photolysis flash may show a decrease in absorption of the reactant or additional absorption due to permanent products. If these are extensive it may be necessary to resort to the tedious procedure of replacing the reaction mixture after each flash. Fortunately, in practice, it is found that many reactions can be studied without causing permanent change in one flash. Many are "relaxation" systems, i.e. reactions in which the dark reaction following photolysis regenerates the original reactants. In others it is often the case that the extinction coefficients of the intermediates are so high that they may be readily studied when their concentration is very small compared with that of the reactants so that the extent of permanent removal of reactants can be kept relatively small.
- The spectra of transient intermediates: absorption spectra which are not present in the records before or after the experiment, but only at relatively short delay times, must be attributed to transient changes in composition. Similarly, if the absorption by the parent substance is less at short delay times than before or after the experiment, it follows that part of the reactant has been converted temporarily into some other substance. The identification of the intermediates responsible for any new absorption spectra often calls for much ingenuity and often a thorough knowledge of the chemistry if the spectra are not already known.

– Lifetimes and kinetics of transient intermediates: these are more conveniently studied by kinetic spectrophotometry (see later) but preliminary
information from flash spectroscopy is often valuable. If more than one intermediate is present, or if the intermediate has a fine structure, as have most
simple radicals in the gas phase, the flash spectroscopic method becomes
almost essential. Even some larger radicals in solution, e. g. benzyl, which has
a single narrow band [13], would almost certainly be missed without a
preliminary survey by flash spectroscopy.

One of the most difficult problems associated with the spectroscopic investigations of transient species is that neither their concentration nor their extinction
coefficients are known, so that, if the reaction order of the intermediate is
higher than first, absolute rate constants are also unobtainable. The absolute
concentration of a transient species whose extinction coefficient is unknown
may, in principle be determined by one of two methods:

(a) product analysis, by conventional methods;
(b) spectroscopic determination of the concentration of all other intermediates
    at the same time as the absorption of the transient species is determined.

Both methods imply a complete knowledge of the reaction mechanism: it is
therefore necessary to carry out a rather complete investigation of the kinetics
in relative terms before the final determination of absolute concentrations, and
each system presents its own problems.

Nowadays, it is quite rare to find flash spectroscopy in use on the microsecond (or slower) timescale. Transient spectra are more often measured by
recording kinetic traces (see following section) at a series of wavelengths and
constructing the spectrum at a given time delay from them. This is possible
given the reliability and reproducibility of current flashlamps and nanosecond
laser systems, but can be experimentally tedious if the sample undergoes
significant photolysis after a single excitation flash, e. g. 4-chlorophenol in
water [14].

The equipment for kinetic spectrophotometry is shown in Fig. 5. The
spectroflash lamp used in flash spectroscopy is replaced with a continuous
white light source such as a xenon lamp or quartz halogen lamp. Following its
transit through the sample cell, the monitoring light beam passes into a
monochromator and the intensity of the beam at the wavelength chosen by the
monochromator is measured with a photomultiplier whose output as a
function of time is recorded using an oscilloscope, transient digitizer or directly
by a computer.

The advantages of the spectrophometric method for detailed kinetic studies
have already been indicated. Even before the transient spectra have been assigned it may be necessary to resort to kinetic spectrophotometry in order that
the mechanism of the change can be elucidated as preliminary to the assignment of the spectra. If weak absorptions are present, preliminary kinetic
scanning at 10 nm followed by 5 nm intervals at wavelengths longer than the
first singlet absorption band should detect most transient species with lifetimes
longer than the resolution time of the apparatus. If there are a number of

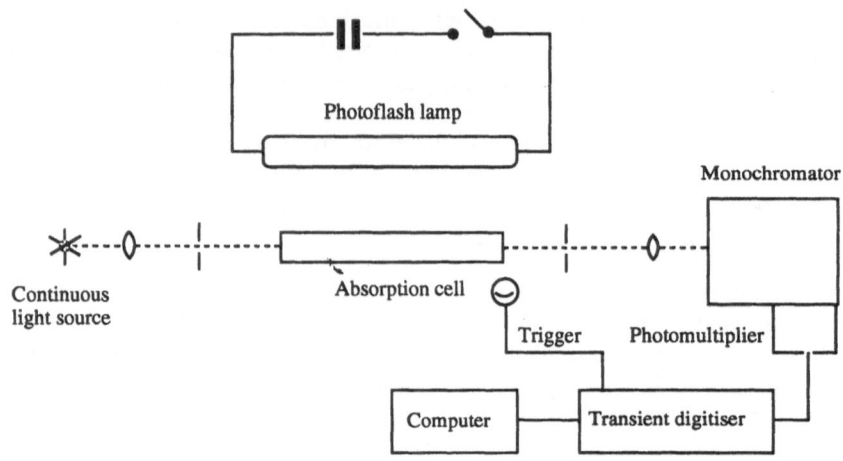

Fig. 5. Experimental layout for conventional flash photolysis by kinetic spectrophotometry

species, with overlapping spectra, whose concentrations vary, then it will be necessary to investigate the kinetics at a number of wavelengths as far apart as possible. For species with lifetimes comparable to that of the flash, care must be taken to allow for scattered light, which is wavelength dependent.

When measuring the kinetics of transients it is necessary not only to measure the transmitted light intensity through the sample at times following the flash, but also to measure the output from the photodetector with the monitoring light beam both "on" and "off" when the sample has not been excited. The latter measurements allow calculation of the light intensity transmitted by the sample in the absence of any transients (I_o) and, when combined with the first set of measurements, the transmitted light intensity as a function of time when transient species are present ($I_t(t)$). It is then possible to calculate the absorbance of the transient as a function of time ($A(t)$) using Eq. (1). All spectroscopic studies must utilise some relation between absorbance and the product $\varepsilon_\lambda cl$ for the transient species and since the relation cannot usually be directly tested on the spectrum of a transient product it is necessary to assume the validity of the Beer-Lambert law (Eq. 1). Reaction orders and rate constants may then follow.

Although all the early applications of flash photolysis were to the reactions of simple molecules in the gas phase [11], by far the greatest activity is now in organic, and to a lesser extent inorganic, reactions to solution. Since the spectra of complex molecules in solution rarely show fine structure, the kinetic spectrophotometric method is almost invariably used. The kinetics of triplet state decay in gases and liquids cannot be studied by luminescence methods, as can the decay of the excited singlet, and the flash method has been particularly valuable in this case. The effects of solvent viscosity, of triplet energy, of triplet concentration and of quenching molecules (either those intentionally added or present as impurities) are now well understood [6]. The radiationless conversion and intersystem crossing processes have received detailed study in solution, and some study in the gas phase. Physico-chemical properties of the excited states, such as acidity constants, can now be determined almost as readily as those of the ground state.

There can be few reactions in photochemistry in which an investigation by flash photolysis and kinetic spectrophotometry does not assist in the elucidation of the mechanism and in providing new quantitative data on rate constants of the intermediate reactions involved. In principle one should observe the first excited state reached by absorption, lower states reached by radiationless conversion, isomers, free radicals and all other intermediate stages in the photochemical transformation. This ideal state of affairs is rarely attained; some of the intermediates may have lifetimes shorter than the resolving times of the instrumentation, some of the intermediates may have weak absorptions or overlapping spectra. Nevertheless, in some investigations as many as five transient intermediates have been observed and followed kinetically in the transformation of a single substance [15,16]. At the present time, most of the spectra of the intermediates likely to be encountered in the study of the new reaction are probably unknown and the most efficient way of deciding whether flash photolysis is likely to contribute useful information to a problem is to carry out a rapid preliminary search of the whole spectral region, and over all available times. If new transient spectra are observed (and in our experience this is usually the case), they can be used to derive new quantitative, kinetic information about the reaction.

An excellent illustration of the information that can be obtained using conventional flash photolysis is the classic work of Porter and Wright [17, 18] and Norrish and co-workers [19, 20] on the gas phase chemistry and photochemistry of oxides of chlorine. Much of this chemistry is now being revisited given the current interest in atmospheric chlorine chemistry. Porter and Wright [17, 18] were able to demonstrate the formation of ClO by flashing a mixture of chlorine and oxygen. The species absorbs as a series of sharp lines in the ultraviolet and Porter and Wright were able to eliminate the obvious reaction for the formation of ClO, namely

$$Cl + O_2 \rightarrow ClO + O \qquad (5)$$

and concluded that it was formed by the reaction

$$2Cl + O_2 \rightarrow 2ClO \qquad (6)$$

possibly involving the formation of a ClOO species as an intermediate.

The decay of ClO was observed to be second order i.e.

$$2ClO \rightarrow Cl_2 + O_2 \qquad (7)$$

again with the possibility of an intermediate species; Cl_2O_2 in this case. Subsequent work by Lipscomb et al. [19] enabled the second order rate constant at 293 K to be evaluated as $6.2 \times 10^7 \, dm^3 \, mol^{-1} \, s^{-1}$ which agrees well with the value of $4.8 \times 10^7 \, dm^3 \, mol^{-1} \, s^{-1}$ derived from Porter and Wright's measurements [17, 18]. This chemistry was expanded to include compounds such as chlorine monoxide, Cl_2O [20].

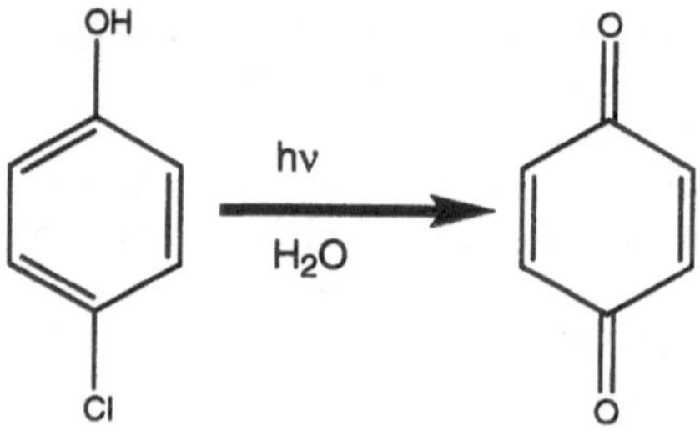

Scheme 1. The primary photoreaction of 4-chlorophenol (1) in aerated water

Since this early work, conventional flash photolysis has been used for studies on an enormous range of organic and inorganic materials and systems [6, 11]. One example of a novel use of the technique was recently reported by Lipczynska-Kochany and Bolton [21, 22]. They employed a microsecond xenon flash coupled with HPLC detection to the analysis of the aqueous photo-chemistry of 4-chlorophenol. The photochemistry of this system, like many aqueous systems, is complex and leads to the formation of a variety of photo-products. Steady-state photolysis [23–26] only provides information about changes in the concentrations of the various products with time and it is often impossible to decide which of the products are primary photoproducts and which result from secondary photolysis. Excitation with a single flash of short time duration (microseconds) such that the primary photoproduct(s) are not formed until after the flash has ended means that secondary photolysis is avoided. When coupled to sensitive HPLC detection, such that low percentage conversions can be analysed, it is then possible to identify primary and secondary photoproducts.

In the case of 4-chlorophenol (1) in aerated water, the primary photoproduct was found to be p-benzoquinone (2) (Scheme 1) [21, 22]. Other photoproducts such as p-hydroquinone and 2-hydroxy-p-benzoquinone were thus attributed to secondary photolysis of the primary photoproduct. These conclusions have been confirmed by other steady state and flash studies [24–27] and the technique has been applied to other systems including 3- and 4-bromophenol [28–30].

4.2
Nanosecond Laser Flash Photolysis

Although the conventional technique still has much to commend it, most laboratories now tend to use nanosecond laser flash photolysis because of the 1000-fold improvement in time resolution that this affords. The replacement of xenon flash lamps with, e.g., a neodymium/YAG laser limits the excitation wavelengths that are available but this is not usually problematic. Most of the concepts and techniques that have been discussed in relation to conventional

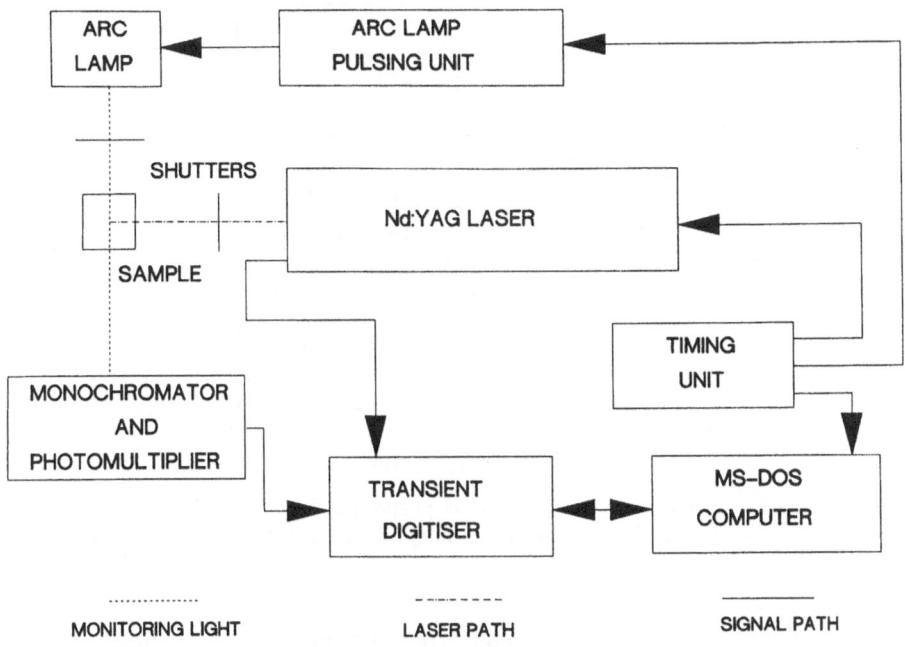

Fig. 6. Experimental layout for nanosecond laser flash photolysis

flash photolysis apply equally well to nanosecond laser flash photolysis; the only major difference is the change of excitation source with the associated improvement in time resolution.

A typical nanosecond laser flash photolysis system is shown in Fig. 6. A Q-switched Nd:YAG laser produces pulses of 1064 nm light with pulse widths (FWHM) of the order of 10–20 ns. The fundamental can be frequency-doubled, -tripled or -quadrupled to give excitation pulses at 532, 355 or 266 nm respectively. These wavelengths give good coverage of the near UV and visible regions of the spectrum but the Nd:YAG laser can be used to pump a dye laser or be replaced by, e.g., a ruby laser (694/347 nm) if required. As the Nd:YAG laser produces a continuous stream of nanosecond pulses (usually at a frequency of a few tens of Hertz), it is essential to insert a shutter between the laser and the sample so that premature excitation does not occur.

The formation of transient species is monitored with a white light source passed through the sample (usually in a 1 cm square cuvette) at right angles to the laser beam. An arc lamp such as a high pressure xenon lamp will usually be used to provide the monitoring beam, especially as it is possible to "pulse" the lamp (i.e. enhance the lamp intensity by a factor of 10–100 for a few tens of microseconds) and thus greatly improve the sensitivity of the monitoring beam. Again, a shutter is interposed between the arc lamp and sample to protect the latter from unnecessary photolysis. Once the monitoring beam has passed through the sample, it is passed through a monochromator and the intensity of the beam at the chosen wavelength is then detected with a photomultiplier

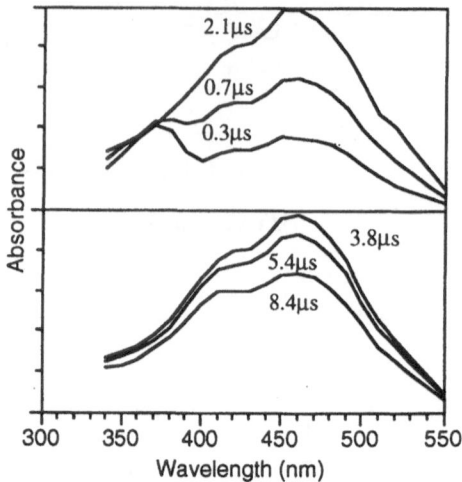

Fig. 7. Transient absorption spectra of 4-chlorophenol (10^{-3} mol dm^{-3}) in aerated water at various delay times. Absorbance range for each spectrum, 0–0.1. Reproduced with permission from [14]

as in conventional flash photolysis. The photomultiplier output is either recorded on a transient digitizer (a digital oscilloscope, or a transient recorder) and then stored on computer as shown in Fig. 6 or is fed direct to an analogue-digital converter board in the computer. Finally, the apparatus requires a timing unit to co-ordinate the firing of the laser with opening the shutters, pulsing the arc lamp (if required) and triggering data collection. This will now routinely all be carried out under computer control once the apparatus is set-up and optimised.

The photochemistry of aqueous 4-chlorophenol again provides an excellent illustration of the use of nanosecond flash photolysis. Experimental work by Grabner et al. [27] and Durand et al. [14] shows very clearly that two transient species are formed when an aerated aqueous solution of **1** is excited at 266 nm (frequency-quadrupled Nd:YAG). The initial absorption peaks around 370 nm (Fig. 7) which is then transformed in 1–2 μs to a second, broad absorption band in the 350–550 nm region. The first transient is produced immediately (i.e. in less than a few tens of nanoseconds) whereas the second transient (monitored at 470 nm) shows a clean "grow-in" followed by a decay (Fig. 8). The kinetics of this species show acceptable fits to a biexponential function:

$$\text{Absorbance} = A_1 \exp(-k_1 t) + A_2 \exp(-k_2 t) \tag{8}$$

where $A_1 = -A_2$, $k_1 = (1.0 \pm 0.1) \times 10^6 \, \text{s}^{-1}$ and $k_2 = (8.5 \pm 1.0) \times 10^4 \, \text{s}^{-1}$ [14]. Below 400 nm, where both transient species absorb, the kinetic traces can be fitted using the same rate constants but now $A_1 \neq -A_2$. The spectral and kinetic data leads to the assignment of the two transient species as 4-oxo-cyclohexa-2,5-dienylidene (**3**) and *p*-benzoquinone-*O*-oxide (**4**) respectively (Scheme 2). The spectra of these species, both of which are known [31, 32], are in good agreement with the spectra of the two transients that are observed.

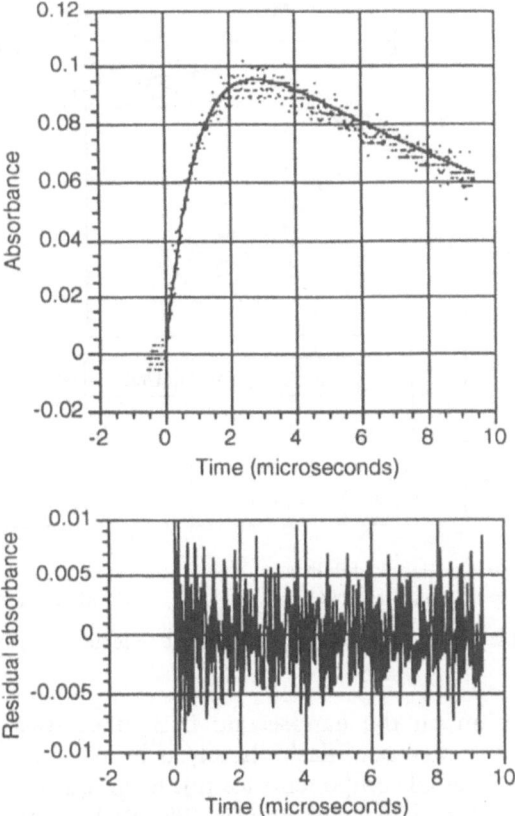

Fig. 8. Transient absorption of 4-chlorophenol in aerated, aqueous solution monitored at 470 nm, together with biexponential fitted curve and residuals. Reproduced with permission from [14]

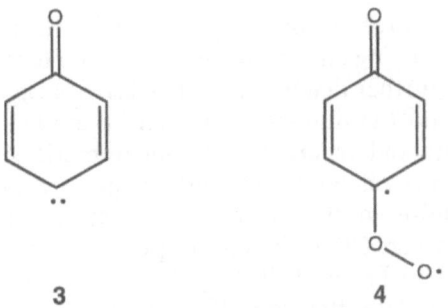

Scheme 2. The structures of 4-oxo-cyclohexa-2,5-dienylidene (3) and *p*-benzoquinone-*O*-oxide (4)

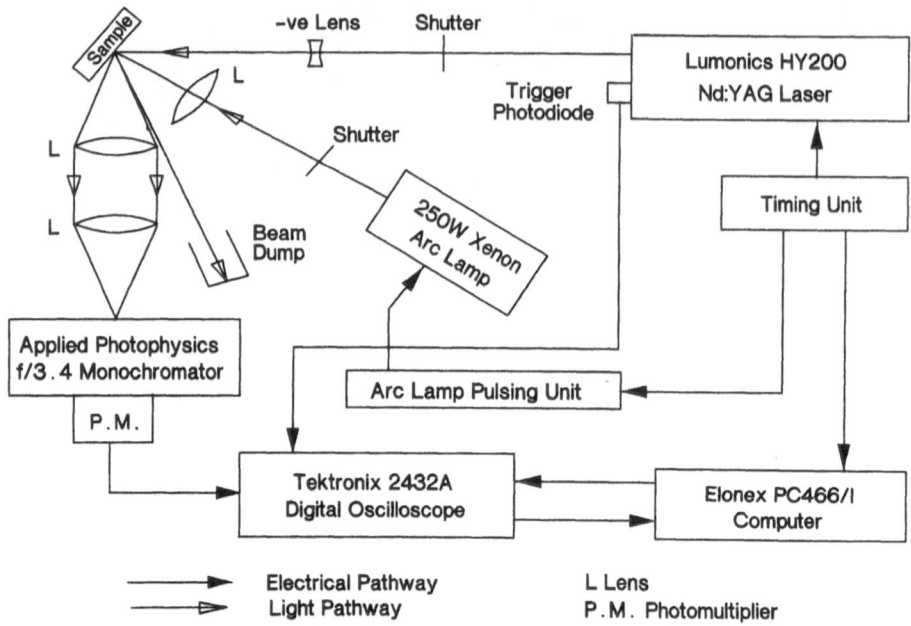

Fig. 9. Experimental layout for nanosecond laser diffuse reflectance flash photolysis

A recent variation on the nanosecond flash photolysis technique is the invention of diffuse reflectance flash photolysis by Wilkinson and co-workers [33, 34]. The experimental components for this technique are identical to those required for nanosecond flash photolysis (Fig. 9) but instead of using absorption measurements to monitor the production of transient species they are detected by diffuse reflectance. This technique is therefore highly advantageous for studying strongly scattering or opaque samples and is particularly useful for studying photochemistry on surfaces.

4.3
Picosecond and Femtosecond Laser Flash Photolysis

The generation of picosecond or sub-picosecond laser pulses relies upon the technique of mode-locking and the reader is referred to the excellent summary by Fleming [35] for further details. Solid state lasers such as neodymium-glass, neodymium:YAG and titanium-sapphire can be mode-locked as well as dye lasers. The resulting pulses are *not* monochromatic since the uncertainty principle places limits on the wavelength range of a pulse of a given time duration. As the pulse width decreases, the wavelength range of the pulse increases and vice versa. Often the output pulses from these systems are sufficiently short-lived to be used directly. However, if further compression is required, passing the pulses through a fibre optic increases the spectral range of the pulse [36] which can then be temporally compressed using a pair of matched gratings as described by Treacy [37]. Using these techniques it is possible to achieve output pulses of a few fs FWHM, but maintaining a reliable and reproducible output from such systems is not trivial. The same was true of picosecond lasers until a few years ago but these are now much easier to maintain and use.

Once sub-nanosecond time resolution is required, the techniques used to measure the spectra and kinetics of transients have to be different to those outlined above. The response times of conventional photodetectors are inadequate for real-time monitoring of sub-nanosecond changes of light intensity and flash photolysis measurements are almost invariably carried out by pump-probe techniques. The sample is excited with one laser pulse (the pump) and the effects of the excitation are monitored by a second, much weaker probe pulse. The latter may be a coloured, single-wavelength pulse or a white light pulse, but the probe pulse is always derived from the pump pulse so that the temporal relationship between the two is well-defined. A typical experimental set-up is shown in Fig. 10.

Here, femtosecond pulses of 120–150 fs FWHM at approximately 600 nm are produced by amplifying the output of a dye laser. The resulting pulse train is then split so that approximately 20% of the light is focused into a cell of water to produce a white light continuum (the probe) and the other 80% is frequency

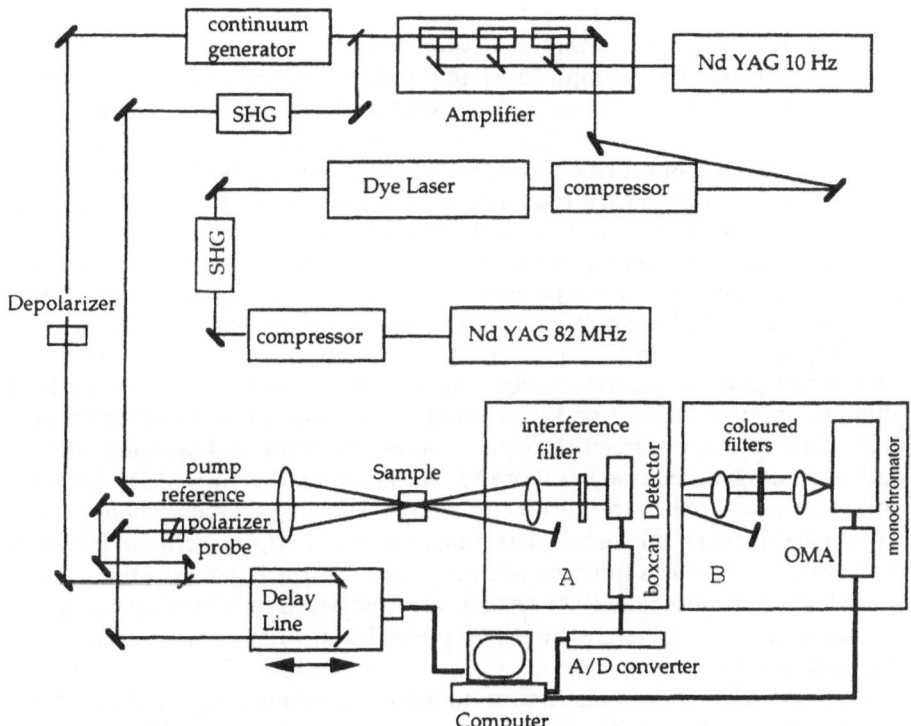

Fig. 10. Experimental layout for femtosecond pump-probe transient absorption measurements. Reproduced with permission from [38]

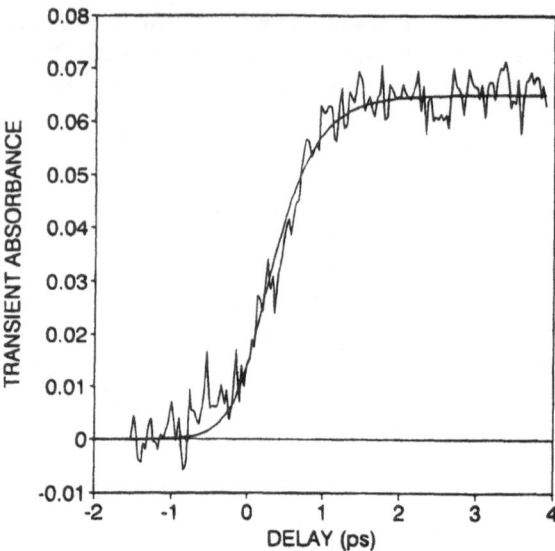

Fig. 11. Time evolution of the transient absorption of pyrene in solution (10^{-3} mol/l in *n*-octane). The pump is set at 290 nm and the probe at 460 nm. Reproduced with permission from [38]

doubled in a single harmonic generator (SHG) crystal to produce 5–15 µJ pulses of UV light (the pump). The pump pulses are used to excite the sample and the effect of the excitation is monitored by determining the change in intensity of the probe beam which is delayed temporally with respect to the pump by an optical delay line. Prior to the probe beam entering the delay line, a beam splitter is used to produce a reference beam which passes through the sample before the pump pulse arrives and thereby generates an I_o value for absorbance measurements. The probe pulse then arrives at some time (0–1 ns, typically) after the pump and interrogates the sample for the effect of the pump excitation. The resulting data consists of either an absorbance value at a given wavelength and a given time delay between the pump and probe if the white-light probe pulse is passed through an interference filter and detected by a photodiode, or an absorption spectrum of the transients present at the set time delay if the probe is passed through a monochromator and detected with an optical multichannel analyser (OMA). Variation of the time delay between pump and probe pulses then leads to a data set showing how absorbance varies with time (Fig. 11) if the former detection system is used, or a series of transient absorption spectra as a function of time (Fig. 12) if the latter detection system is used. Most femtosecond laser systems are now based on a titanium : sapphire laser system for the generation of the fs pulses but the detection techniques are still as detailed above.

There are a number of alternative methods for monitoring the effect of the pump pulse on the sample. Zewail has published a large number of superb reports on femtosecond photochemistry where they use the probe pulse to ionise the sample and detect photoproducts in a time-of-flight mass spectro-

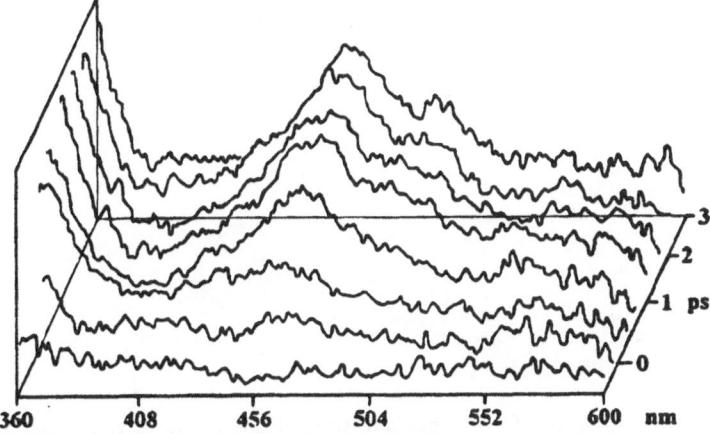

Fig. 12. Evolution in the first 3 ps of the transient absorption spectrum of pyrene in solution between 360 and 600 nm. Reproduced with permission from [38]

meter [39–41]. If the sample fluoresces, the fluorescence intensity as a function of time may be determined by the up-conversion technique [35]. Here the fluorescence (at frequency ω_1) is mixed in a non-linear crystal with a fs probe pulse (at frequency ω_2) to generate light at frequency ω_3 $(=\omega_1+\omega_2)$ which is detected by a photodiode. If the intensity of the probe pulse is constant, the intensity of the up-converted light as a function of pump-probe delay time is directly related to the fluorescence intensity. It is necessary to use this technique to obtain fs or low ps time-resolution. The up-converted signal is only produced when both ω_1 and ω_2 are present so the ω_2 probe pulse acts as a time gate for the fluorescence. If the fluorescence signal varies more slowly (say on a few tens of ps timescale or greater) it may be possible to use a streak camera or the time-correlated single photon counting technique (see later) to detect the fluorescence.

To illustrate the sort of kinetic and mechanistic information that can be obtained from ultrafast flash spectroscopy, we return to some of the chlorine/oxygen chemistry mentioned earlier. The photodissociation of chlorine dioxide (ClO_2 or $OClO$) in the near-UV is of interest due to its possible role in stratospheric ozone depletion. The principal reaction channel for this molecule is

$$OClO \xrightarrow{\ h\nu\ } ClO + O \tag{9}$$

but a minor channel

$$OClO \xrightarrow{\ h\nu\ } Cl + O_2 \tag{10}$$

also exists and the products of both channels are species which are active in atmospheric odd oxygen reactions.

The ultrafast dynamics of OClO fragmentation have been studied by Baumert et al. using femtosecond excitation (80 fs FWHM) to the A^2A_2 state and subsequent multiphoton ionisation (MPI) time of flight (TOF) detection to

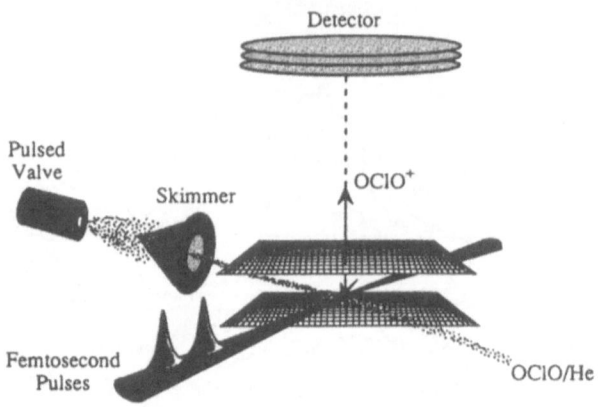

Fig. 13. Experimental layout for femtosecond time-of-flight mass spectrometry of gaseous OClO). Reproduced with permission from [42]

monitor the concentrations of the species involved in the reaction [42]. The experiment is shown diagramatically in Fig. 13. A supersonic molecular beam of OClO mixed with helium is excited by a pair of femtosecond pulses which are delayed with respect to one another. The first fs pulse produces the A^2A_2 OClO excited state whilst the second fs pulse causes ionisation of the OClO A^2A_2 excited state (or one of the reaction products) by multiphoton absorption. These ionised species are then detected by a TOF mass spectrometer. The intensities of the ion signals (see, e.g., Fig. 14) are related to the concentrations

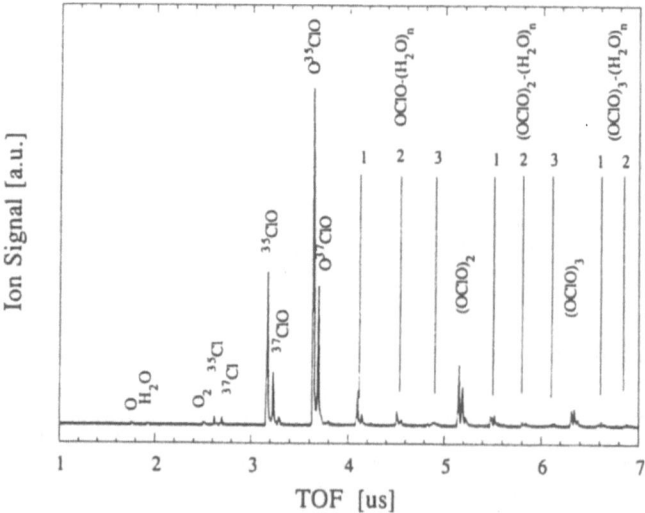

Fig. 14. A TOF spectrum taken in a femtosecond pump-probe arrangement. Note that the prominent masses are the OClO parent and the ClO product. Reproduced with permission from [42]

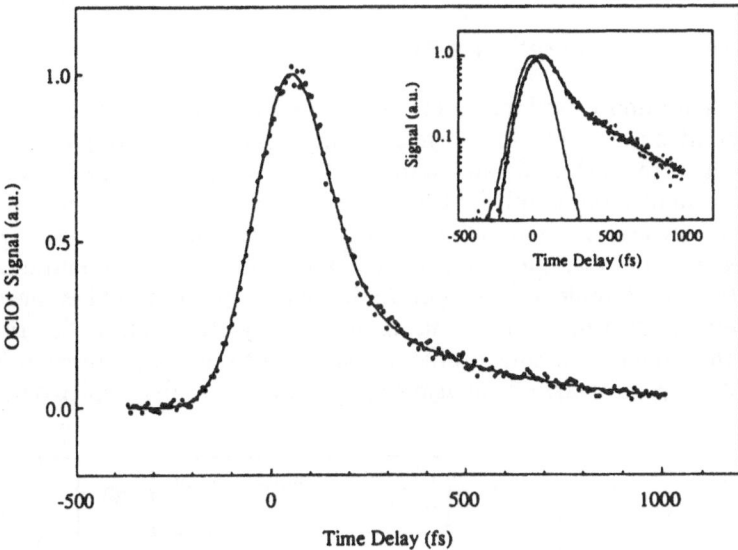

Fig. 15. A typical femtosecond transient taken on the $O^{35}ClO$ parent mass. The *solid line* is a double exponential fit to the data including convolution with the cross correlation. The *insert* shows the decay on a logarithmic signal axis, in which the biexponential character is evident. Reproduced with permission from [42]

of the unionised species in the molecular beam. If the experiment is performed for different time delays between the pairs of pulses, the resulting data slows how the concentrations of the various species vary with time. Figure 15 shows how the concentration of the $O^{35}ClO$ parent varies with the time between the two pulses. The decay is actually biexponential (see inset where the intensity is plotted logarithmically) with decay times of 50 and 400 fs. These reflect a mechanism whereby fragmentation occurs not only from the initially excited A^2A_2 state (OClO*) but also from lower energy states (denoted by [OClO]$^{\neq}$ in Scheme 3).

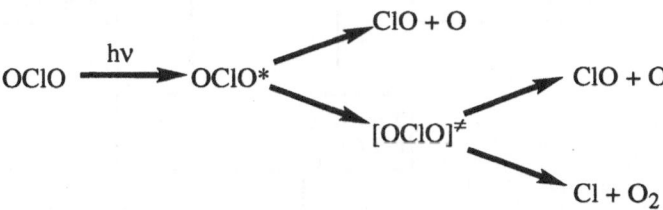

Scheme 3. The reactions of OClO following UV excitation

5
Measurement of Fluorescence Kinetics

The most commonly used technique for the measurement of fluorescence decay characteristics is that of time-correlated single photon counting (TCSPC). For full details of the technique and associated technologies the reader is referred to the excellent monograph by O'Connor and Phillips [43]. There are also a number of reviews devoted to the technique and its applications [44–47].

The essence of TCSPC is that if the time taken for a photon of fluorescence to be emitted by a sample following excitation with a flash of light is measured a large number of times, the resulting probability distribution describes the fluorescence intensity vs time profile for the sample under investigation. The experiment is typically carried out using equipment such as that depicted in Fig. 16.

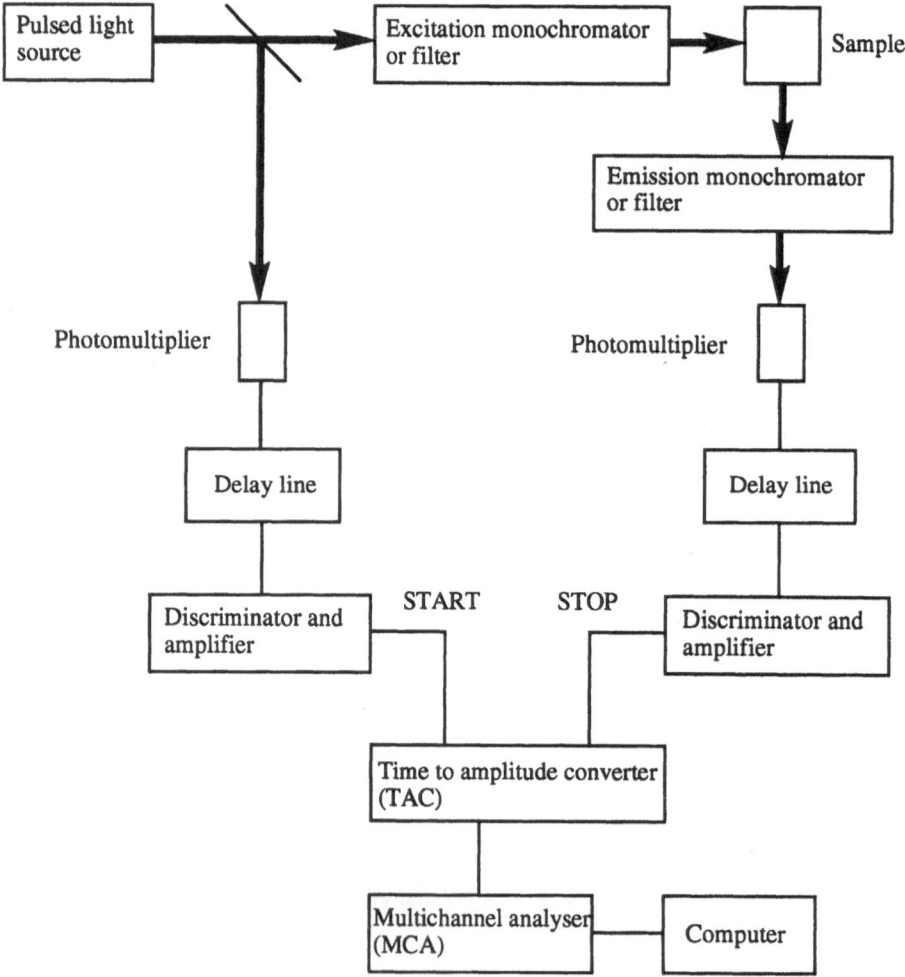

Fig. 16. A typical experimental system for time-correlated single photon counting

The sample is excited with a flash of light from an appropriate light source and a trigger pulse (the START pulse), generated from the light flash itself or from the instrumentation producing it, is used to initiate a linear voltage ramp in a time-to-amplitude converter (TAC).

Fluorescence from the sample is detected by a photomultiplier and the output from the photomultiplier (the STOP pulse) is also fed to the TAC. The two delay lines allow the relative temporal positions of the START and STOP pulses to be varied. This is essential given that the two pulses are generated at two different points in the experimental set-up and therefore may not both arrive at the TAC within the time-span of the voltage ramp and prove it has been initiated. However, once the latter has been assured, the delay times are kept constant for the measurement of a given fluorescence decay profile (and its associated instrumental response profile). The pulse from the photomultiplier STOPS the voltage ramp in the TAC; the voltage reached is thus a measure of the time elapsed between the arrival of the START and STOP pulses at the TAC.

The TAC output is usually stored in a multichannel analyser (MCA). This consists of two parts – an analogue to digital converter which converts the TAC voltage to a digital number and a memory device to store the resulting numbers. The MCA memory is divided into a number of channels (usually $2^8 - 2^{12}$ channels) each of which corresponds to a range of voltages generated by the TAC and hence to a range of elapsed times between the START and STOP pulses. For example, if the memory were set up such that each channel corresponded to a time range of 0.1 ns, then the first channel would correspond to START/STOP time differences between 0 and 0.1 ns, the second channel to time differences between 0.1 and 0.2 ns etc. Every time a TAC voltage arrives at the MCA the number in the channel corresponding to that voltage is increased by one. As the experiment is repeated a large number of times, the fluorescence decay profile is gradually built up.

It is essential that a maximum of one fluorescence photon per excitation pulse is detected by the photomultiplier to prevent 'pile-up' errors [48], i.e. a bias toward fluorescence photons which are generated soon after the excitation pulse (since any second or later photons are not detected). In practice, fluorescence photons tend to be detected at a level of the order of 1% or less of the excitation rate to not only avoid "pile-up" but also because samples are often fairly weakly fluorescent and the TAC/ADC combination cannot handle a higher rate. For example, in a typical experiment using the synchrotron radiation source (SRS) at the CLRC Daresbury Laboratory (Warrington, UK) as the excitation source [49], fluorescence photons are accepted at a maximum rate of 30 000 counts per second (cps) based on the 3.125 MHz repetition rate of the SRS in single bunch mode. At this count rate, an instrumental response profile will normally be acquired in a few seconds and a fluorescence decay profile in a few minutes for 10 000 counts in the most intense channel. In practice, if a decay profile displays single exponential kinetics, acceptable precision can be obtained with 10 000 counts in the most intense channel [50].

If the excitation pulse is very short-lived with respect to the fluorescence decay, then the decay profile can be immediately analysed using standard kinetic models. Normally, however, the excitation pulse has a lifetime which is finite with respect to the fluorescence decay in which case it is necessary to take account of the width of the excitation pulse. There are two facets to this; measurement of the excitation pulse profile and subsequent convolution of this profile with a chosen kinetic model to see if the model adequately reproduces the observed fluorescence decay profile.

The excitation pulse profile (or instrumental response) is measured by replacing the fluorescent sample with a scattering material such as Ludox (a colloidal silica suspension in water) and tuning both monochromators or filters to the same wavelength. The photomultiplier then detects scattered excitation light and the equipment builds up a profile of its intensity as a function of time when the same conditions regarding count rate as for a fluorescence decay profile are obeyed.

Once an instrumental response profile [$I(t)$] has been acquired, it is combined with the measured fluorescence decay profile [$F(t)$] to yield the true decay of the sample $D(t)$. These three functions are related through the integral

$$F(t) = \int_0^t I(t') \, D(t - t') \, dt' \tag{11}$$

In practice, the form of $D(t)$ is obtained by making an informed guess as to the type of expression which is appropriate for it and optimising any adjustable parameters in the expression to give the best agreement between the integral on the RHS of Eq. (13) and $F(t)$. For example, the fluorescence decay of naphthalimide 5 (Scheme 4) in 2-propanol has been analysed using both monoexponential (Eq. 12) and biexponential (Eq. 13) decay profiles:

$$D(t) = A \, e^{-t/\tau} \tag{12}$$

$$D(t) = A_1 e^{-t/\tau_1} + A_2 e^{-t/\tau_2} \tag{13}$$

In the case of the monoexponential function, there are two adjustable parameters (A and τ) which may be varied to search for a good fit between calculation and experiment. In most software packages which undertake this

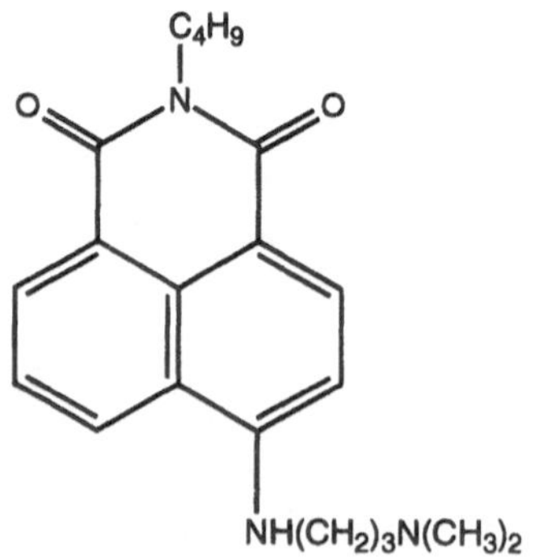

5

Scheme 4. Structure of 4-(3-(N,N-dimethylamino)propylamino)-1,8-naphthalimide (5)

task, the user is required to provide initial guesses for A and τ (and other parameters such as background counts). The software then undertakes a series of iterations where the values of A and τ are varied in a systematic way until the best value of a parameter such as the reduced chi-square (χ^2) is obtained. There are a number of different approaches to undertaking this iteration [43, 51] but the most widely used is probably non-linear least squares curve fitting. This is the basis of the software that we use [49] together with a Marquardt algorithm [52] to speed up attainment of the best χ^2.

The result of applying this software to the fluorescence decay of 5 in 2-propanol is shown in Figs. 17 and 18 which also contain the instrumental response

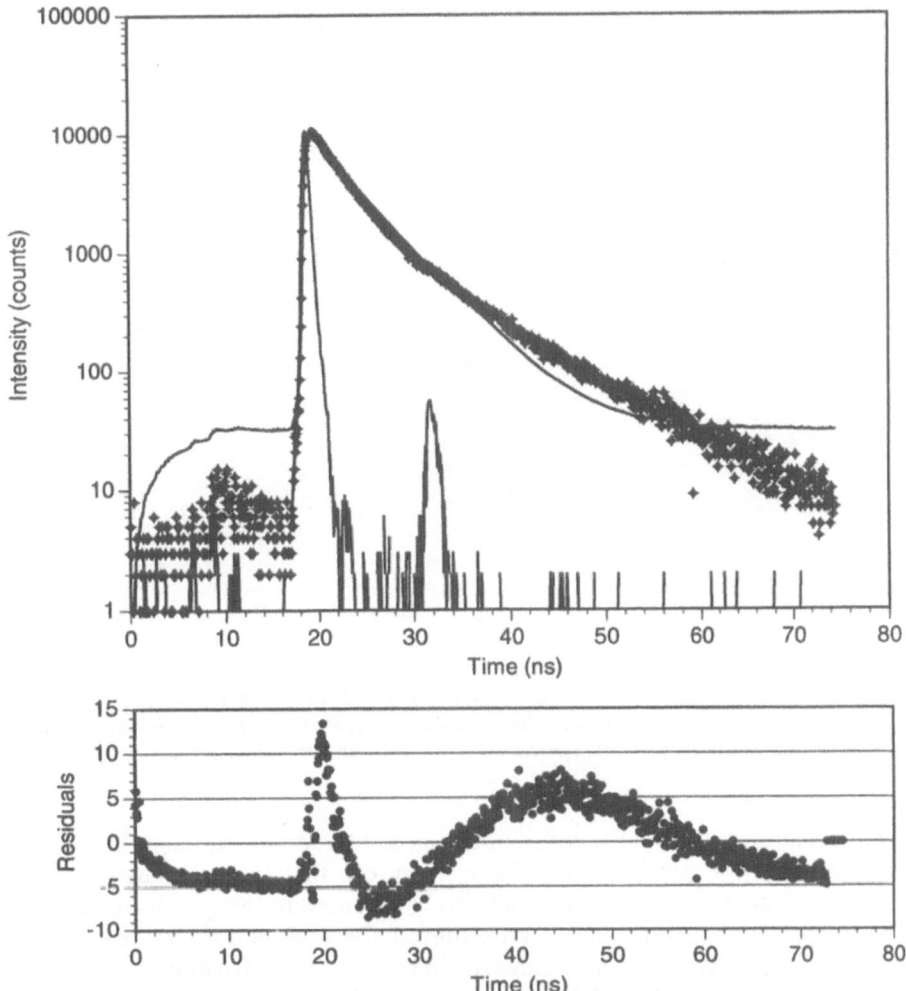

Fig. 17. Fluorescence decay profile for compound 5 in 2-propanol recorded at the CLRC Daresbury Laboratory using the SRS. The instrumental response profile, calculated fit for a monoexponential decay ($\chi^2 = 17.4$) and the resulting residuals are also shown

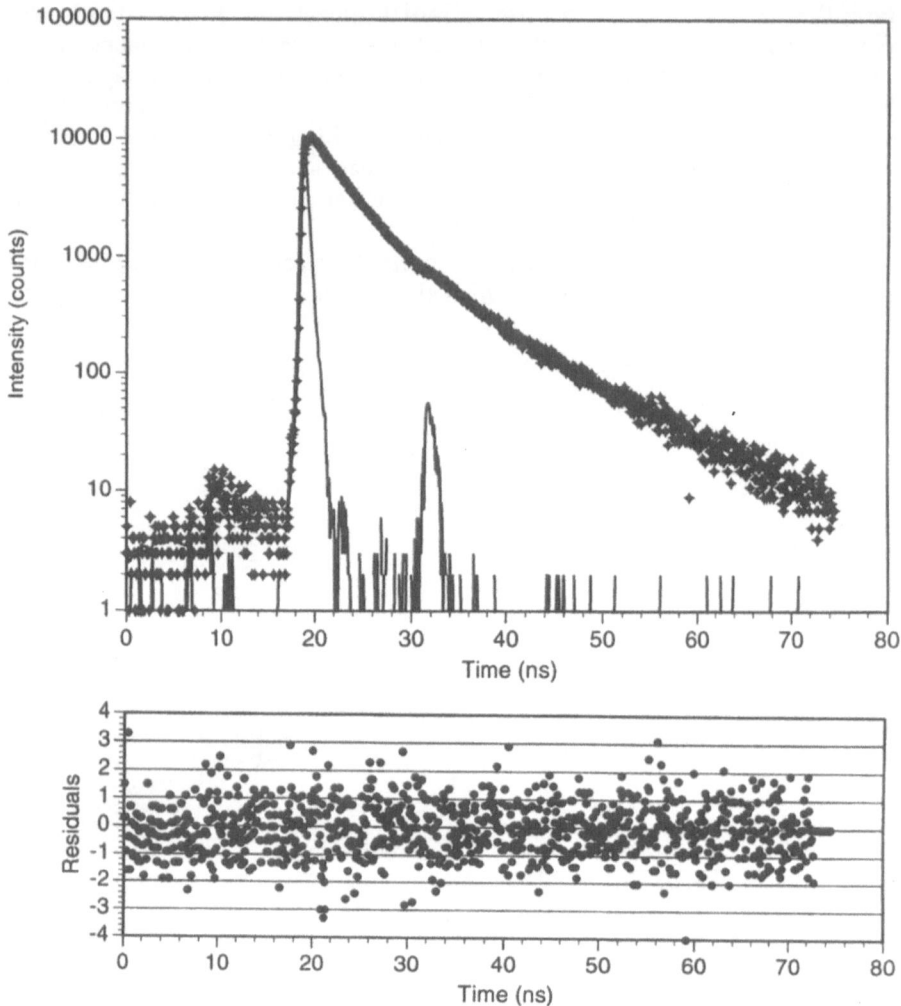

Fig. 18. Fluorescence decay profile for compound 5 in 2-propanol recorded at the CLRC Daresbury Laboratory using the SRS. The instrumental response profile, calculated fit for a biexponential decay ($\tau_1 = 3.2$ ns, $\tau_2 = 9.5$ ns, $\chi^2 = 0.996$) and the resulting residuals are also shown

profile. Figure 17 shows the result of trying to fit a mono-exponential decay to the experimental profile. It can be seen by eye that the calculated and experimental decays do not match and this is mirrored in the χ^2 value (17.4) and the residuals. If a good fit has been achieved, the χ^2 should lie in the region of 1.0 (usually 0.8–1.2) and the residuals will be randomly distributed about zero. The monoexponential model clearly fails both these tests. By contrast, fitting a biexponential function to the decay is much more successful (Fig. 18) reflecting the fact that this system forms an intramolecular exciplex [53] where biexponential kinetics would be expected [54].

O'Connor and Phillips outline a number of other functions and parameters which may be used to judge the goodness of fit [43]. Apart from χ^2 and the residuals, the principal ones which this author has used himself and seen others use are auto correlation of the weighted residuals [55] and the Durbin-Watson parameter [56].

The choice of function used in any attempt to fit experimental fluorescence decays must, of course, be based on the photo-chemistry/physics that the excited state undergoes. A function which gives a good mathematical fit but which makes no physical sense is clearly pointless. An alternative approach to the data analysis is therefore that of "target analysis" or "compartmental analysis" whereby the fit to an actual physical model is tested and real physical parameters such as rate constants are extracted from the data [57, 58]. This technique really comes into its own for complex systems – for simpler systems such as first order kinetics or exciplex formation, there is little to choose between the two approaches.

The increased computing power that has become available to researchers in recent years has enabled the development of "global analysis" [59, 60]. This method of data analysis involves the simultaneous fitting of a number of decay curves for a sample or a series of related samples. For example, a number of measurements may have been made on a sample at different excitation and emission wavelengths. The series of decay curves thus obtained would be expected to show related decay parameters and can therefore be fitted using the same rate constants. Alternatively, in a quenching experiment, the fluorescence kinetics will be expected to vary in a known manner as the quencher concentration is varied and this systematic variation can be built into a global analysis of the decay curves to yield the quenching rate constant as one of the output parameters.

The principal alternative to TCSPC for the measurement of fluorescence decay profiles is the phase-shift technique [61–63]. A typical experimental layout is shown in Fig. 19. This employs a continuous light source whose intensity is modulated (frequency f). The fluorescence from the system is therefore modulated at the same frequency but with a phase delay (ϕ) and a modulation depth (m) which are both related to the fluorescence kinetics. For example, for first order kinetics where the fluorophore lifetime is τ,

$$\tan \phi = f\tau \tag{14}$$

$$m = 1/\sqrt{(1 + f^2\tau^2)} \tag{15}$$

The excitation source may be an arc lamp or CW laser whose output is modulated with, e.g., a Pockels cell. This allows frequencies in the MHz range to be achieved and thus nanosecond lifetimes to be measured. Shorter lifetimes require picosecond pulsed lasers [64]. For all phase-shift instruments it is very important that the modulation frequency is adjustable since the relationships between ϕ, m and the kinetic parameters can become quite complex for systems exhibiting complicated kinetics [63]. It is therefore necessary to make a series of measurements of ϕ and m vs f in order to obtain the kinetic parameters.

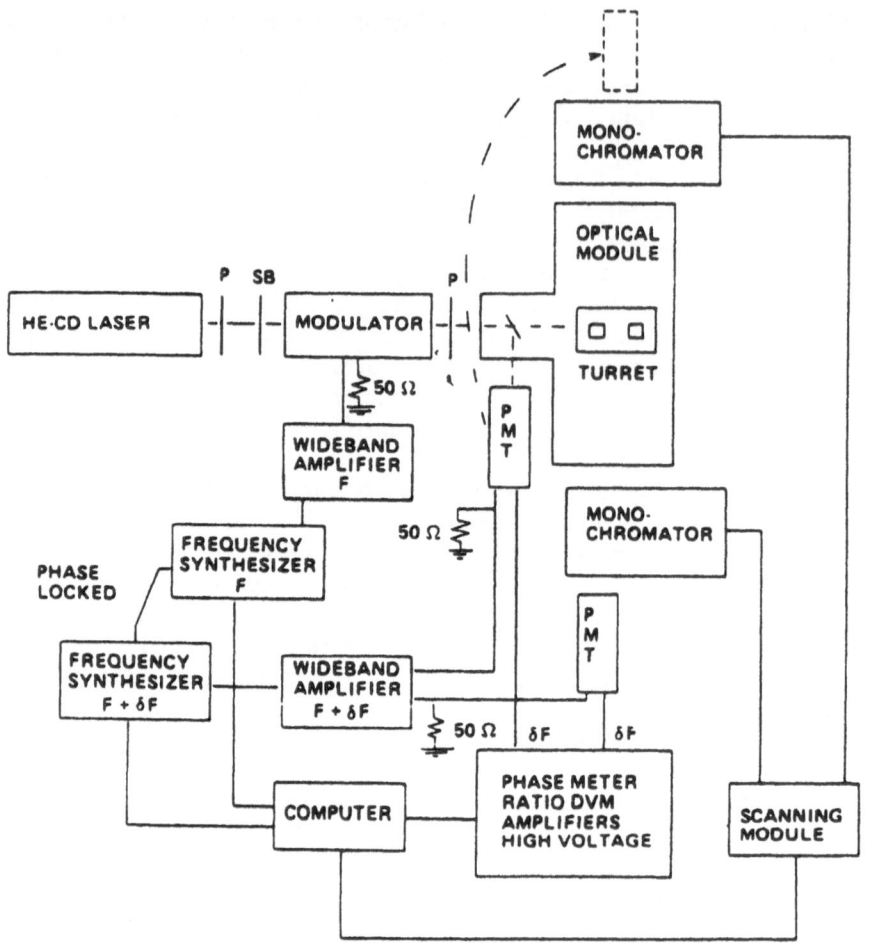

Fig. 19. Schematic representation of a variable- frequency phase-modulation fluorometer. P, polarizer; SB, Soleil-Babinet compensator; F, frequency; δF, cross-correlation frequency; PMT, photomultiplier tube. Reproduced with permission from [63]

The phase shift technique has been developed by Lakowicz and Gratton and co-workers to a level where its capabilities are similar to those of TCSPC. It is especially useful in systems where rapid measurements are required, either because of sample fragility or the need to make a lot of measurements on one sample, because, in the simplest case, a single determination of φ is all that is required for the measurement of the lifetime of a monoexponential decay. It has therefore found considerable application in fluorescence lifetime imaging microscopy [65], although recent developments in Birch's group are also opening up the use of TCSPC to this field [66].

Both the TCSPC and phase shift techniques are also applicable to the measurement of time-resolved emission spectra, either measured directly [43] or determined from a set of decays at different emission wavelengths [67], and

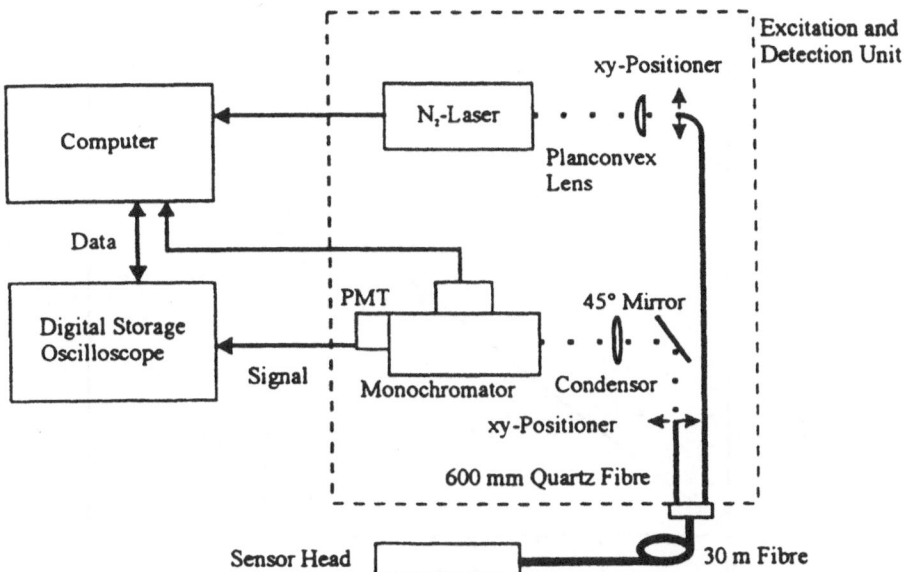

Fig. 20. Experimental layout of a remote sensor system for polycyclic aromatic hydrocarbons based on time-resolved fluorescence. Reproduced with permission from [69]

time-resolved fluorescence anisotropies [67]. Readers requiring standard compounds to characterise equipment are referred to the paper by Lampert et al. [68].

The TCSPC and modulation techniques are in widespread use to assist researchers in the characterisation of the photochemistry, photophysics and photobiology of a vast array of different systems. They provide direct access to kinetic parameters and are thus invaluable in arriving at an understanding of the behaviour of the excited states of interest. It is also possible to use lifetime/kinetic differences between different fluorophores in order to separate them when they are present together in the same sample. Such measurements in environmental photochemistry are exemplified by the following two examples.

Kotzick et al. have reported the use of time-resolved fluorescence for monitoring the concentrations of polycyclic aromatic hydrocarbons (PAHs) in the field [69]. Measurement of PAHs in groundwater is complicated by their low concentrations and their overlapping fluorescence spectra (both with each other and with humic material present in the water). However, the longer lived fluorescence from a PAH such as pyrene can be discriminated from the emission from other fluorophores and measurements of ground water pyrene levels can be made in situ using the equipment shown in Fig. 20.

The response of the system correlates extremely well with the results of HPLC determinations of pyrene from the same water samples (Fig. 21). The fluorescence measurements are much faster to make than the HPLC ones and can also be made in situ. However, the monitoring of compounds with shorter fluorescence lifetimes will be difficult/impossible.

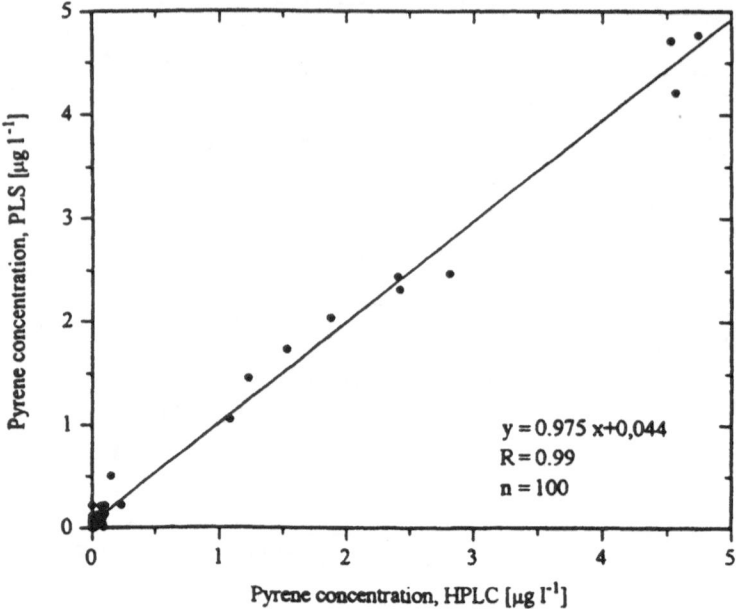

Fig. 21. Performance of the time-resolved sensor for polycyclic aromatic hydrocarbons compared with HPLC detection. Reproduced with permission from [69]

Time-resolved fluorescence has also proved to be a valuable tool in monitoring photosynthetic performance [70, 71], especially for systems which are under stress [72]. It is clear that the fluorescence decay profile corresponding to emission from chlorophylls in a photosynthetic organism is very sensitive to environmental perturbations such as nitrogen depletion [73] and to damage [74, 75]. For example, the fluorescence decay profiles of spruce needles are composed of a sum of three exponential components (Fig. 22) [75]. The three components exhibit lifetimes of 100–150 ps, 400–600 ps and 3.5–5.0 ns, with the intensity being virtually all contained within the two shorter-lived components to approximately equal degrees. When these needles are exposed to ozone, the contribution of the longer-lived component to the overall decay increases markedly, even when there is no visible sign of ozone damage and can approach 10% of the overall intensity [75]. Time-resolved fluorescence thus provides an excellent early warning of damage to, or disruption of, the function of photosynthetic systems.

Acknowledgements. I thank Dr. David Worrall of the University of Loughborough for providing me with Figs. 6 and 9.

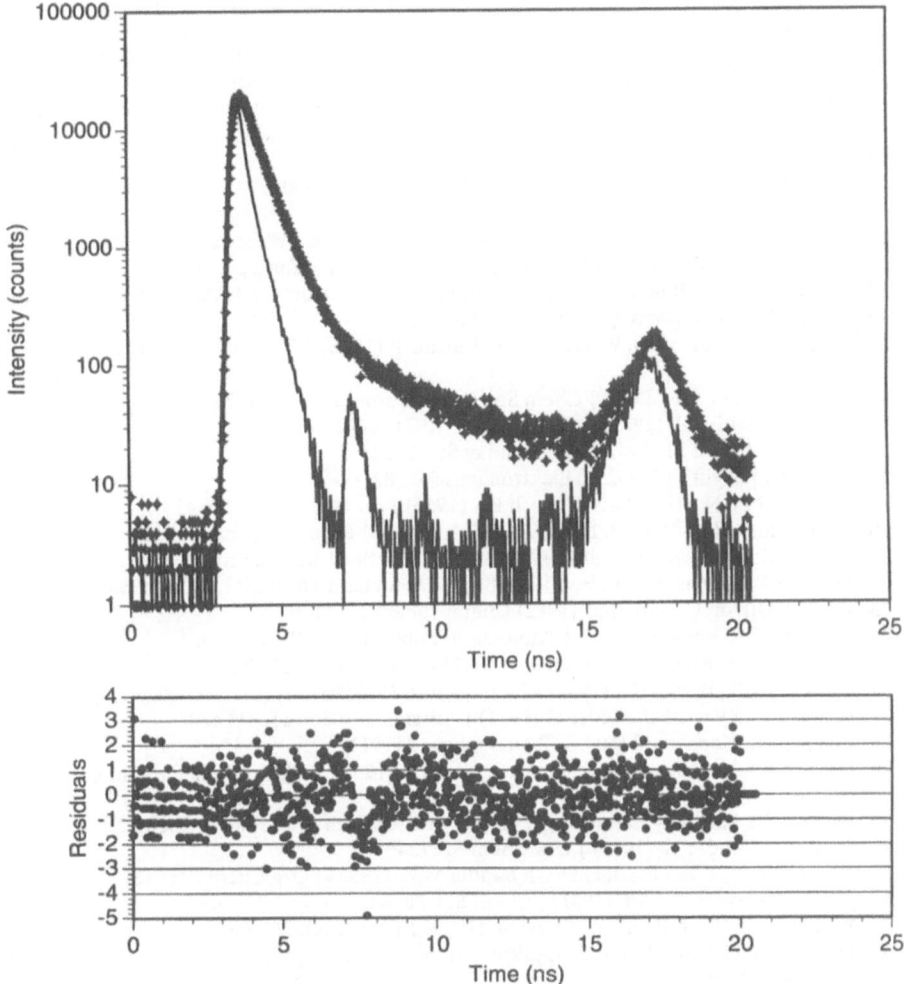

Fig. 22. Fluorescence decay profile and residuals plot for control spruce needles fitted with three exponential components. Calculated parameters $a_1 = 0.136$, $\tau_1 = 0.134$ ns, $a_2 = 0.019$, $\tau_2 = 0.571$ ns, $a_3 = 0.0003$, $\tau_3 = 4.904$ ns. The instrumental response profile is also shown. Reproduced with permission from [75]

References

1. Zepp RG (1982) Experimental approaches to environmental photochemistry. In: Hutzinger O (ed) Handbook of environmental chemistry, vol 2B. Springer, Berlin Heidelberg New York, p 19
2. Roof AAM (1982) Basic principles of environmental photochemistry. In: Hutzinger O (ed) Handbook of environmental chemistry, vol 2B. Springer, Berlin Heidelberg New York, p 1

3. Klessinger M, Michl J (1994) Excited states and photochemistry of organic molecules. VCH, New York
4. Braun AM, Maurette M-T, Oliveros E (1991) Photochemical technology. Wiley, Chichester
5. Sharpe AG (1992) Inorganic chemistry, 3rd edn. Longman, Harlow
6. Gilbert A, Baggott J (1991) Essentials of molecular photochemistry. Blackwell, London
7. Evans DF (1957) J Chem Soc 1957:3885
8. Grabowska A (1963) Spectrochim Acta 19:307
9. Beer M, Longuet-Higgins HC (1955) J Chem Phys 23:1390
10. Maciejewski A, Steer RP (1993) Chem Rev 93:67
11. Porter G (1963) Flash photolysis. In: Friess SLA, Lewis ES, Weissberger A (eds) Technique of organic chemistry, vol VIII, part II. Interscience, New York, p 1055
12. Herzberg G (1950) Molecular spectra and molecular structure 1. Van Nostrand, New York
13. Porter G, Windsor MW (1957) Nature 180:187
14. Durand AP, Brown RG, Worrall D, Wilkinson F (1996) J Photochem Photobiol A: Chem 96:35
15. Porter G, Tchir MF (1970) J Chem Soc Chem Commun 1970:1372
16. Porter G, Tchir MF (1971) J Chem Soc (A) 1971:3772
17. Porter G, Wright FJ (1952) Disc Faraday Soc 14:23
18. Porter G, Wright FJ (1952) Z Elektrochem 56:782
19. Lipscomb FJ, Norrish RGW, Thrush BA (1956) Proc Roy Soc A 233:455
20. Edgecombe FHC, Norrish RGW, Thrush BA (1957) Proc Roy Soc A 243:24
21. Lipczynska-Kochany E, Bolton JR (1990) J Chem Soc Chem Commun 1990:1596
22. Lipczynska-Kochany E, Bolton JR (1991) J Photochem Photobiol A: Chem 58:315
23. Boule P, Guyon C, Lemaire J (1982) Chemosphere 11:1179
24. Oudjehani K, Boule P (1992) J Photochem Photobiol A: Chem 68:363
25. Durand A-PY, Brattan D, Brown RG (1992) Chemosphere 25:783
26. Durand A-PY, Brown RG (1995) Chemosphere 31:3595
27. Grabner G, Richard C, Köhler G (1994) J Amer Chem Soc 116:11470
28. Lipczynska-Kochany E (1992) Chemosphere 24:911
29. Lipczynska-Kochany E, Kochany J, Bolton JR (1992) J Photochem Photobiol A: Chem 62:229
30. Lipczynska-Kochany E, Kochany J (1993) J Photochem Photobiol A Chem: 73:23
31. Bucher G, Sander W (1992) J Org Chem 57:1346
32. Arnold BR, Scaiano JC, Bucher GF, Sander WW (1992) J Org Chem 57:6469
33. Kessler RW, Wilkinson F (1981) J Chem Soc Faraday Trans I 77:309
34. Wilkinson F, Worrall DR, Williams SL (1995) J Phys Chem 99:6689
35. Fleming G (1986) Chemical applications of ultrafast spectroscopy. Oxford University Press, New York
36. Shank CV, Fork RL, Yen R, Stolen RH, Tomlinson WJ (1982) Appl Phys Lett 40:761
37. Treacy E (1969) IEEE J Quant Electron QE-5:454
38. Foggi P, Peltini L, Sànta I, Righini R, Califano S (1995) J Phys Chem 99:7439
39. Zewail AH (1993) Femtosecond reaction dynamics. In: Wiersma DA (ed) Proceedings of the colloquium Femtosecond Reaction Dynamics', Amsterdam, 17–19 May 1993. Royal Netherlands Academy of Arts and Sciences, Amsterdam
40. Zewail AH (1995) Femtochemistry: concepts and applications. In: Manz J, Wöste L (eds) Femtosecond chemistry. VCH, Weinheim
41. Polanyi JC, Zewail AH (1995) Acc Chem Res 28:119
42. Baumert T, Herek JL, Zewail AH (1993) J Chem Phys 99:4430
43. O'Connor DV, Phillips D (1984) Time-correlated single photon counting. Academic Press, London
44. Binkert Th, Tschanz HP, Zinsli PE (1972) J Luminesc 5:187
45. Knight AEW, Selinger BK (1973) Austr J Chem 26:1
46. Demas JN (1988) Time-resolved and phase-resolved emission spectroscopy. In: Schulman SG (ed) Molecular luminescence spectroscopy, part 2: methods and applications. Wiley-Interscience, New York

47. Malliaris A (1992) Appl Spectrosc Rev 27:51
48. Coates PD (1968) J Phys E Ser 2 1:878
49. Sparrow R, Brown RG, Evans EH, Shaw D (1986) J Chem Soc Faraday Trans II 82:2249
50. Yguerabide J (1972) Methods Enzymol 26:498
51. Eaton DF (1990) Pure & Appl Chem 62:1631
52. Bevington PB (1969) Data reduction and error analysis for the physical sciences. McGraw-Hill, New York
53. Mitchell K, Brown RG, Yuan D, Chang S-C, Utecht RE, Lewis DE (1998) I Photochem Photobiol A: Chem 115:157
54. Lewis C, Ware WR (1973) Molec Photochem 5:261
55. Grinvald A, Steinberg IZ (1974) Analyt Biochem 59:583
56. Durbin J, Watson GS (1950) Biometrika 37:409; Durbin J, Watson GS (1951) Biometrika 38:159
57. Beechem JM, Ameloot M, Brand L (1985) Analyt Instrum 14:379
58. Ameloot M, Beechem JM, Brand L (1986) Biophys Chem 23:155
59. Knutson JR, Beechem JM, Brand L (1983) Chem Phys Lett 102:501
60. Beechem JM, Ameloot M, Brand L (1985) Chem Phys Lett 120:466
61. Spencer RD, Weber G (1969) Ann NY Acad Sci 158:361
62. Merkelo H, Hammond JR, Hartmann SR, Derzko ZI (1970) J Luminesc 1/2:502
63. Lakowicz JR, Maliwal BP (1985) Biophys Chem 21:61
64. Berndt KW, Gryczynski I, Lakowicz JR (1991) Analyt. Biochem 192:131
65. Lakowicz JR, Koen PA, Szmacinski H, Gryczynski I, Kúsba J (1994) J Fluoresc 4:117
66. McLoskey D, Birch DJS, Sanderson A, Suhling K, Welch E, Hicks PJ (1996) Rev Sci Instrum 67:2228
67. Holzwarth AR (1995) Methods Enzymol 246:334
68. Lampert RA, Chewter LA, Phillips D, O'Connor DV, Roberts AJ, Meech SR (1983) Analyt Chem 55:68
69. Kotzick R, Haaszio S, Niessner R (1995) SPIE 2504:107
70. Evans EH, Brown RG (1994) J Photochem Photobiol B: Biol 22:95
71. Joshi MK, Mohanty P (1995) J Sci Ind Res 54:155
72. Lichtenthaler HK, Rinderle U (1988) CRC Crit Rev Analyt Chem 19:529
73. Evans EH, Allen MM, Brown RG (1991) Brit Phycol J 26:85
74. Schneckenburger H, Schmidt W (1992) Radiat Environ Biophys 31:73
75. Evans EH, Brown RG, Wellburn AR (1992) New Phytol 122:501

3 Atmospheric Degradation of Anthropogenic Molecules

T. J. Wallington[1] and O. J. Nielsen[2]

[1] Ford Motor Company, Ford Research Laboratory, Mail Drop SRL-3083, Dearborn, Michigan 48121–2053, USA. *E-mail: twalling@ford.com*
[2] Atmospheric Chemistry Division, Plant Biology and Biogeochemistry Department, Risø National Laboratory, DK-4000 Roskilde, Denmark. *E-mail: ole.john.nielsen@risoe.dk*

This chapter provides an overview of the atmospheric photochemical degradation of important anthropogenic molecules. Starting with an introduction to atmospheric chemistry, the concept of atmospheric lifetime is presented and the degradation mechanisms of alkanes, aromatics, nitrogen oxides, SO_2, chlorofluorocarbons, halons, hydrofluorocarbons, and hydrochlorofluorocarbons are discussed. Areas of uncertainty are highlighted.

Keywords: Atmospheric chemistry, Atmospheric lifetimes, CFCs (chlorofluorocarbons), Nitrogen oxides, Urban smog, SO_2.

Contents

The Handbook of Environmental Chemistry Vol. 2 Part L
Environmental Photochemistry (ed. by P. Boule)
© Springer-Verlag Berlin Heidelberg 1999

1
Introduction

Human activities result in the release of a large quantity and variety of chemical compounds into the atmosphere. These compounds are degraded in a complex series of reactions. The atmosphere is a highly oxidizing environment and degradation proceeds via a sequence of reactions in which the pollutants are oxidized in successive steps resulting in increasingly polar and less volatile products. Eventually the pollutant is either completely oxidized, e.g. the conversion of methane into CO_2 and H_2O, or it is converted into partially oxidized species which are removed via wet and/or dry deposition to the earth's surface. While the degradation reactions are beneficial because they remove pollutants from the air, they can have unwanted side effects. The degradation products and intermediates can lead directly, or indirectly, to environmental impacts on local, regional, or global scales. Local scale air pollution is responsible for the photochemical smog present in many large scale metropolitan areas. On a time scale of hours in the presence of sunlight, atmospheric chemical reactions convert reactive hydrocarbons and nitrogen oxides into a mixture of oxidants such as ozone and peroxyacetyl nitrate (PAN – $CH_3C(O)O_2NO_2$) that is generically known as urban smog. The deleterious effects of smog are typically experienced 10–100 km from the pollution sources. Regional scale problems are perhaps best exemplified by the phenomenon of acid precipitation. The atmospheric oxidation of SO_2 into sulfuric acid (H_2SO_4) occurs on a time scale of several days and acidic precipitation occurs typically 500–1000 km downwind of the pollution source. Notable examples include the acidic precipitation experienced in Northern Scandinavia originating from emissions in central Europe, acidic precipitation across Japan resulting from emissions from the Asian mainland, and acidic rain in the North Eastern U.S. from emissions in the Ohio river valley. Stratospheric ozone depletion is an example of a global scale problem. Stratospheric ozone loss is caused by the release of chlorofluorocarbons (e.g. CF_2Cl_2) and Halons (e.g. CF_3Br and CF_2ClBr). CFCs and Halons have no significant loss mechanisms in the lower atmosphere and within several years of their release they become uniformly distributed in the lower atmosphere on a global scale. CFCs and Halons are transported into the stratosphere as part of the natural air circulation. In the stratosphere the CFCs encounter harsh solar UV irradiation ($\lambda < 250$ nm) which is blocked from the lower atmosphere by absorption by the ozone layer. UV irradiation of CFCs and Halons releases Cl and Br atoms which then participate in ozone destruction reactions.

To assess the environmental impact of the release of a compound into the atmosphere there are several issues that need to be considered. The first step is to determine its atmospheric lifetime. As discussed above the atmospheric lifetime determines the geographical extent of the possible direct environmental impact. To calculate the atmospheric lifetime of a chemical compound we need information on the kinetics of its reaction with key atmospheric trace species such as OH and NO_3 radicals and O_3, the rate of its photolysis, its solubility and hence propensity towards wet deposition, and finally its rate of dry deposition.

If the compound is long lived (>10 years) and released in substantial quantities it may contribute to greenhouse warming and it would be appropriate to measure its infrared spectrum and calculate its potential effect on the radiative balance of the atmosphere. Once the lifetime of the pollutant has been established the next step is to determine the degradation products and intermediates of its atmospheric oxidation and assess whether they pose any environmental threat.

The recent search for environmentally acceptable CFC replacements provides a good example of how detailed studies of the atmospheric degradation mechanism of a class of chemical compounds lead to an understanding of their environmental impact. Hydrofluorocarbons (e.g. CF_3CFH_2, also known as HFC-134a) and hydrochlorofluorocarbons (e.g. CF_3CCl_2H, also known as HCFC-123) are important classes of chemical compounds which are used as replacements for CFCs in a variety of applications. When HFCs and HCFCs were first proposed as CFC replacements very little was known about their atmospheric chemistry. Initially there was speculation that fluorine-containing free radical species generated during the atmospheric degradation of HFCs and HCFCs (CF_3O_x, FCO_x, and FO_x) might adversely impact stratospheric ozone [1] and that the atmospheric degradation of HFCs and HCFCs might form toxic products which could accumulate in the environment [2]. To assess the environmental acceptability of HFCs and HCFCs the chemical industry sponsored a number of experimental and theoretical studies of their atmospheric chemistry. As a result we now have a good understanding of the atmospheric degradation mechanisms of such compounds. In fact, it can be argued that of all the different classes of chemical compounds emitted into the atmosphere we understand the chemistry of HFCs and HCFCs the best! It is now clear that initial concerns regarding the environmental impact of such compounds were unfounded. The recent work establishing the atmospheric degradation mechanism of HFCs and HCFCs provides a notable example of the successful application of fundamental research to solve a practical environmental problem.

In this chapter we will examine the atmospheric degradation mechanisms of the following important classes of anthropogenic molecules: alkanes, alkenes, aromatics, nitrogen oxides, SO_2, CFCs and Halons, and finally HFCs and HCFCs. Our intent is not to give an exhaustive account of the photochemical oxidation of every man-made chemical species but rather to present examples of the degradation mechanisms of a few representative members of each class of pollutant. First, we need to consider the general features of atmospheric chemistry.

2
General Atmospheric Chemistry

The atmosphere is a giant inhomogeneous photochemical reactor in which temperature, pressure, radiation flux, and composition vary widely. All of the energy for atmospheric chemistry comes from solar radiation which is absorbed by various components of the atmosphere. Figure 1 shows the direct

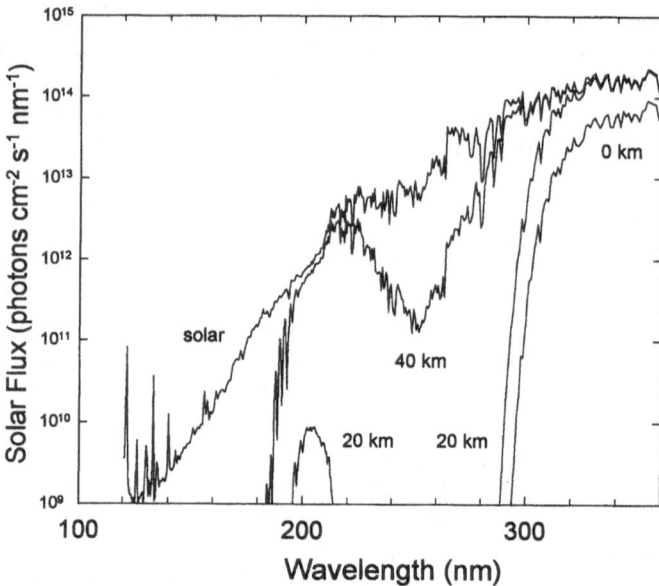

Fig. 1. Direct solar flux over the wavelength range 120–360 nm at the top of the atmosphere (labeled solar) and at 40, 20, and 0 km [83]

solar flux over the wavelength range 120–360 nm at the top of the atmosphere (labeled solar), and at 40, 20, and 0 km altitude. Molecular oxygen is responsible for the absorption of UV radiation of wavelengths less than 200 nm at altitudes above 40 km. Absorption by ozone in the stratosphere shields the earth's surface from UV radiation of wavelengths less than 300 nm.

Gas-phase, solution phase, and heterogeneous reactions all play important roles in atmospheric chemistry. The mean atmospheric composition is given in Table 1. N_2, O_2, and Ar comprise 99.9% of the atmosphere and, for all practical purposes, the relative proportion of these gases is constant in the lower 100 km of the atmosphere. We are concerned here with the fate of pollutants such as CO, volatile organic compounds, halocarbons, sulfur compounds, and nitrogen oxides which are present in trace amounts and whose concentrations vary significantly both spatially and temporally.

In discussions of atmospheric processes it is useful to divide the atmosphere into different regions. Temperature profiles provide the most convenient basis for this division. Figure 2 shows the altitude profile of atmospheric temperature and pressure up to 50 km [3]. The first 10–15 km of the atmosphere is characterized by a temperature profile in which colder air overlays warmer air. This situation is caused by the fact that the predominant heat source for this region of the atmosphere is the warm surface of the earth. The temperature profile of the troposphere is inherently unstable and results in strong vertical convective mixing from which the region derives its name (tropos is Greek for folding). Intense tropical thunderstorm systems can transport molecules from close to the earth's surface to the top of the troposphere within a few minutes. However,

Table 1. Average composition of dry air

Gas	Average Mixing Ratio	
	%	(parts per million)
N_2	78.1	(781,000)
O_2	20.9	(209,000)
Ar	0.9	(9,000)
CO_2	0.036	(360)
Ne	0.0018	(18)
He	0.00052	(5.2)
CH_4	0.00017	(1.7)
H_2	0.00006	(0.6)
N_2O	0.00003	(0.3)
CO	0.00001	(0.1)
O_3	0.000008	(0.08)
Non methane hydrocarbons	0.000001	(0.01)
Halocarbons	0.0000002	(0.002)
Sulfur compounds	0.00000008	(0.0008)
Nitrogen oxides (NO_y)	0.00000005	(0.0005)

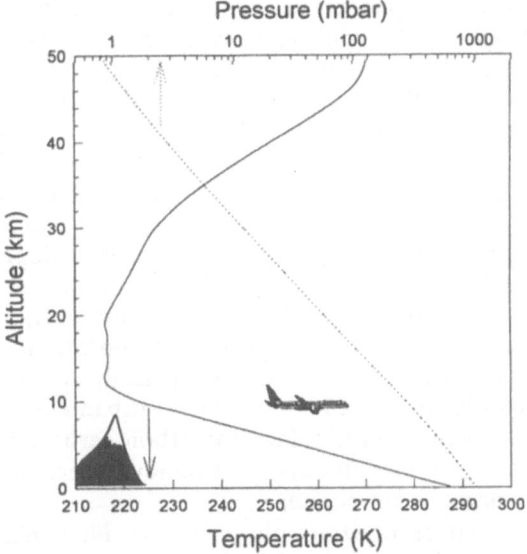

Fig. 2. Temperature and pressure profile of the atmosphere up to 50 km taken from the US Standard Atmosphere [3]

more typical mixing times are of the order of days to weeks. More than 90% of the mass of the atmosphere is located in the troposphere and it is here that the vast majority of anthropogenic molecules are degraded. In contrast to the troposphere, the stratosphere which lies at approximately 15–50 km altitude is heated principally from above by the absorption of solar UV radiation. In the stratosphere, warm air lies on top of cooler air which is an inherently stable situation resulting in a layered structure which gives the region its name (stratos is Greek for layered). Vertical mixing in the stratosphere proceeds slowly, typically on a time scale of several years. The region which marks the boundary between the two different regions is called the tropopause. In this chapter we will be concerned mainly with the photochemistry that occurs in the troposphere.

The driving force for most of the chemistry that occurs in the atmosphere is the formation of hydroxyl (OH) radicals via photolysis of ozone to form $O(^1D)$ atoms which react with water vapor.

$$O_3 + h\nu \ (\lambda < 320 \ nm) \ \rightarrow \ O(^1D) + O_2(^1\Delta_g) \tag{1}$$

$$O(^1D) + H_2O \ \rightarrow \ 2OH \tag{2}$$

The flux of UV light, O_3, and H_2O vapor combine to give a potent source of OH radicals. OH radicals react with almost everything emitted into the atmosphere. The atmospheric lifetimes of many pollutants are determined by their reactivity towards OH radicals. The generation of OH radicals is the primary mechanism by which the atmosphere cleanses itself. Only compounds such as CFCs, Halons, and N_2O which are inert towards OH radical attack survive transport through the troposphere into the stratosphere where they can damage the ozone layer.

The dominant loss of OH radicals is reaction with CO and organic compounds such as CH_4, both reactions produce peroxy radicals. Peroxy radicals play a key role in atmospheric chemistry. They are intimately involved in the formation and destruction of ozone and in the photooxidation of all organic compounds in the atmosphere [4]. The lifetime of OH radicals with respect to reactions (3) and (5) is of the order of a second and in the day time a steady state condition is established. The OH radical concentration in the atmosphere varies with location, time of day, season, and meteorological conditions, a reasonable global 24 hour global average is $(0.5 - 1.0) \times 10^6 \ cm^{-3}$ [5-7].

Although only 10% of atmospheric ozone resides in the troposphere (0-15 km altitude) it has a profound impact on tropospheric chemistry. Ozone concentrations in the troposphere vary from typically 20-40 ppb for a remote pristine site to 100-200 ppb in a highly polluted urban environment. Ozone is a reactive molecule which readily adds to carbon-carbon double bonds [8]. Reaction with ozone provides an important removal mechanism for many unsaturated reactive organic compounds.

There are two sources of tropospheric ozone. First, transport from the stratosphere in meteorological events known as "tropospheric folding" in which a layer of stratospheric air is entrained in tropospheric air flow and mixed into the troposphere. Second, peroxy radical reactions which oxidize NO to NO_2. For example, in the OH radical initiated oxidation of CO:

$$OH + CO \ \rightarrow \ H + CO_2 \tag{3}$$

$$H + O_2 + M \ \rightarrow \ HO_2 + M \tag{4}$$

$$HO_2 + NO \ \rightarrow \ OH + NO_2 \tag{5}$$

$$NO_2 + h\nu \ (\lambda < 420 \ nm) \rightarrow NO + O \tag{6}$$

$$O + O_2 + M \rightarrow O_3 + M \tag{7}$$

$$net: CO + 2O_2 \rightarrow CO_2 + O_3$$

and in the oxidation of CH_4:

$$OH + CH_4 \rightarrow H_2O + CH_3 \tag{8}$$

$$CH_3 + O_2 + M \rightarrow CH_3O_2 + M \tag{9}$$

$$CH_3O_2 + NO \rightarrow CH_3O + NO_2 \tag{10}$$

$$CH_3O + O_2 \rightarrow HCHO + HO_2 \tag{11}$$

$$HO_2 + NO \rightarrow OH + NO_2 \tag{5}$$

$$2 \times (NO_2 + h\nu \ (\lambda < 420 \ nm) \rightarrow NO + O \ ) \tag{6}$$

$$2 \times (O + O_2 + M \rightarrow O_3 + M) \tag{7}$$

$$net: CH_4 + 4O_2 \rightarrow H_2O + 2O_3 + HCHO$$

The formaldehyde formed in the oxidation of CH_4 can react with OH or photolyze leading to further NO_2 formation and ozone production:

$$HCHO + OH \rightarrow H_2O + HCO \tag{12}$$

$$HCHO + h\nu \rightarrow H + HCO \tag{13}$$

$$HCO + O_2 \rightarrow HO_2 + CO \tag{14}$$

$$H + O_2 + M \rightarrow HO_2 + M \tag{15}$$

$$2 \times (HO_2 + NO \rightarrow OH + NO_2) \tag{5}$$

Under certain conditions the atmospheric oxidation of one CH_4 molecule can lead to the formation of four molecules of ozone. Recent modeling calculations suggest that on a global scale, transport from the stratosphere accounts for approximately 10% of tropospheric ozone while peroxy radical chemistry is the source of the remaining 90% [9].

At night, the nitrate radical, NO_3, takes over from the OH radical as the dominant radical species which initiates the degradation reactions. The nitrate radical is formed by the reaction of NO_2 with O_3. NO_3 radicals absorb strongly in the visible and are photolyzed rapidly during the day. Hence, NO_3 radicals are present in significant quantities only at night. The NO_3 radical is much less reactive than OH radicals but nevertheless is an important oxidant species for many organic compounds. NO_3 radical concentrations can reach levels of the order of $10^9 \ cm^{-3}$ in polluted nighttime air [10]. NO_3 radicals react with a variety of organic species via H atom abstraction and/or addition mechanisms [11–15].

$$NO_3 + HCHO \rightarrow HNO_3 + HCO \tag{16}$$

$$HCO + O_2 \rightarrow HO_2 + CO \tag{14}$$

$$NO_3 + CH_3CH=CH_2 + M \rightarrow CH_3 C(\cdot)HCH_2ONO_2 + M \tag{17}$$

3
Atmospheric Lifetimes

The concept of atmospheric lifetime is useful in discussions of the atmospheric degradation of anthropogenic molecules [5]. It can be defined in several ways. Most simply put it can be expressed as the turnover time which is the atmospheric burden of a given species divided by its rate of emission, assuming a constant emission rate and steady state condition. Alternatively, it can be stated as the reciprocal of the pseudo first order rate constant (k') for its removal:

$$Atmospheric\ Lifetime\ (\tau) = \left(\frac{1}{k'}\right)$$

Most pollutants are lost from the atmosphere by several routes. For example while 90% of CO is removed from the atmosphere via reaction with OH radicals, about 10% is removed via microbial activity in soils [16]. Similarly, the bulk of CH_4 oxidation occurs via OH radical attack however microbial action and reactions (18) and (19) also contribute significantly [5].

$$O(^1D) + CH_4 \rightarrow OH + CH_3 \tag{18}$$

$$Cl + CH_4 \rightarrow HCl + CH_3 \tag{19}$$

In such cases the atmospheric lifetime is given by:

$$Atmospheric\ Lifetime\ (\tau) = \frac{1}{\sum\limits_i k_i'} = \frac{1}{\sum\limits_i \frac{1}{\tau_i}}$$

where k_i' and τ_i and the pseudo first order loss rate and lifetime with respect to process i. For example if a compound has a lifetime of 10 years with respect to one loss mechanism and 20 years with respect to another loss mechanism then the overall lifetime will be 7 years. In practice it is often difficult to assign a unique value of atmospheric lifetime of a given species. Difficulties arise because of several complicating factors.

Firstly, the concept of a single atmospheric lifetime is only applicable to compounds which are persistent enough to become uniformly mixed throughout the atmosphere. The CFCs provide a good example of such long lived species. Highly reactive compounds like alkenes have lifetimes which are dependent on the exact location where they are emitted. Alkenes emitted into air masses with low levels of ozone will survive considerably longer than if emitted into polluted air masses with high ozone levels.

Secondly, in cases where the pollutant is released in extremely large amounts, for example CO and CH_4, the lifetime can actually change with the rate of emission. Thus, large scale emission of CH_4 will reduce the OH radical concentration and lead to an increase in the atmospheric lifetime of CH_4 and indeed of all other species which are removed via reaction with OH.

Thirdly, for some species (most notably CO_2) there are removal processes in which the species equilibrates with large reservoirs. Atmospheric CO_2 equilibrates with CO_2 dissolved in the oceans and with the terrestrial biota within approximately 4 years [17]. However, the majority of the CO_2 in these reservoirs is returned to the atmosphere within a few years. It is only the relatively small fraction of CO_2 that is transferred to the deep ocean that can be considered to be permanently lost from the atmosphere. Loss of CO_2 from the atmosphere cannot be represented by a simple exponential decay but is instead is a complex function [18, 19]. As a guide the atmospheric lifetime of CO_2 is approximately 50–200 years [17].

Fourthly, the starting point for lifetime estimations is often laboratory generated kinetic data for reaction of the compound of interest with OH radicals. The bimolecular rate constants measured in laboratory kinetic experiments need to be converted into a pseudo first order rate constant for loss of the compound, k'. In principal this conversion is simple, i.e. the bimolecular rate constant merely has to be multiplied by the OH concentration ([OH]). In practice there are difficulties associated with the choice of an appropriate value of [OH]. At present we can not measure the global OH concentration field directly. The OH radical concentration varies widely with location, season, and meteorological conditions. To account for such variations requires use of sophisticated 3 D computer models of the atmosphere. In such models the OH concentration field is computed using measured or estimated concentration fields of the precursor molecules and photon flux data. The resulting OH field is then tuned such that it correctly predicts the lifetime of methyl chloroform (CH_3CCl_3) with respect to OH radical attack. From measurements of the atmospheric turnover time of CH_3CCl_3 (4.9 years) [20], its lifetime with respect to loss in the stratosphere (45 years), and its lifetime with respect to loss in the oceans (85 years) the tropospheric lifetime of CH_3CCl_3 with respect to OH radical attack has been inferred to be 5.9 years [17]. Methyl chloroform is the calibration molecule of choice because it has a long history of precise atmospheric measurements, it has no natural sources, its industrial production is well documented, and because the kinetics of reaction (20) are well established, $k_{20} = 1.8 \times 10^{-12} \exp(-1500/T)$ cm^3 molecule^{-1} s^{-1} [21].

$$OH + CH_3CCl_3 \rightarrow \text{products} \qquad (20)$$

The use of 3 D computer models to calculate atmospheric lifetimes is a rather cumbersome approach and access to such models is limited. A simpler technique to estimate the tropospheric lifetime of compound with respect to OH attack is to scale the tropospheric lifetime of CCl_3CH_3 by the rate constant ratio $k(OH + CCl_3CH_3)/k(OH + X)$.

$$OH + X \rightarrow \text{products} \qquad (21)$$

If reactions (20) and (21) have the same temperature dependence then the rate constant ratio can be evaluated at any given temperature. If not, then the temperature chosen for the comparison is important. Prather and Spivakovsky [22] have shown that 277 K is the optimal temperature for comparison. Hence, a reasonable estimate of the atmospheric lifetime of a compound with respect to OH radical attack in the troposphere can be made using:

$$\tau \, (compound) = \frac{k \, (OH + CCl_3CH_3)}{k \, (OH + compound)} \times 5.9 \; years$$

Over the past 10–20 years the kinetic data base for reaction of OH radicals with atmospheric pollutants has improved dramatically to a point where uncertainties in k_{OH} are typically in the range 10–20%. Critically evaluated data for OH radical reactions of atmospheric importance are available in several excellent reviews [8, 15, 21, 23, 24]. Likewise there are also extensive data bases available for NO_3 [8, 13] and O_3 reactions [8].

Table 2 lists atmospheric lifetimes for a range of anthropogenic molecules. As seen from Table 2 the lifetimes of pollutant molecules range from minutes or hours

Table 2. Atmospheric lifetimes for selected compounds

Compound	Atmospheric Lifetime
CO_2	50–200 years[a] [17]
CO	50 days[b]
CH_4 – methane	12.2 years [17]
C_2H_6 – ethane	60 days[b]
C_3H_6 – propane	12 days[b]
C_8H_{18} – iso-octane[b]	3.5 days[b]
C_2H_2 – acetylene	14 days[b]
C_2H_4 -ethene	1.4 days[b]
C_3H_6 – propene	11 h[b]
C_6H_6 – benzene	10 days[b]
$C_6H_5CH_3$ – toluene	1.9 days[b]
$C_6H_4(CH_3)_2$ – m-xylene	12 h
N_2O – nitrous oxide	120 years [17]
CF_2Cl_2 – CFC-12	105 years [103]
$CFCl_3$ – CFC-11	50 years [103]
CF_3CCl_2H – HCFC-123	1.4 years [103]
CF_3CFH_2 – HFC-134a	14 years [103]
CF_3Br – Halon 1301	65 years [103]
CF_3H – Fluoroform	250 years [103]
CF_4 – carbon tetrafluoride	50,000 years [17]
SF_6 – sulfur hexafluoride	3,200 years [17]
SO_2	12 days[b]
CF_3I – trifluoroiodomethane	<1 day [104]
HCHO	6 h[d]

[a] No single lifetime can be given, see text for details.
[b] Estimated using [OH] $=1\times10^6$ cm^{-3} and the following rate data k (OH + CO) $= 2.4 \times 10^{-13}$ [21], k(OH + C_2H_2) $= 8.1 \times 10^{-13}$ [24], k(OH + C_2H_6) $= 1.8 \times 10^{-13}$ [21], k(OH + C_3H_8) $= 9.2 \times 10^{-13}$ [21], k(OH + C_6H_6) $= 1.2 \times 10^{-12}$ [24], k(OH + $C_6H_5CH_3$) $= 6.0 \times 10^{-12}$ [15], k(OH + $C_6H_4(CH_3)_2$) $= 2.4 \times 10^{-11}$, k(OH + i-octane) $= 3.3 \times 10^{-12}$ [24], k(OH + C_2H_4) $= 8.5 \times 10^{-12}$ [24], k(OH + C_3H_6) $= 2.6 \times 10^{-11}$ [24], k(OH + SO_2) $= 9.6 \times 10^{-13}$ [66].
[c] i-octane $= 2,2,4$-trimethyl pentane, used to define octane scale of motor fuels.
[d] Approximate lifetime with respect to photolysis, see page 507 of ref. [105].

for photolabile compounds such as CF_3I and HCHO to millennia for perfluoro compounds such as CF_4 and SF_6. Currently there is some debate as to whether atmospheric lifetime should be used as a measure of environmental acceptability. It has been argued that it is prudent to restrict the manufacture of compounds which have excessively long atmospheric lifetimes [5]. The crux of the argument is that because their degradation is so slow any unforeseen adverse environmental impacts of such species would, for all practical purposes, be permanent.

4
Degradation of Alkanes

More than 130 different types of alkanes have been identified in the atmosphere [25] and they comprise a major fraction of the organic pollutants found in urban atmospheres (see Table 3). They are released in large amounts in activities connected with the extraction, refining, distribution, and combustion of fossil fuels. Alkanes are also released during the combustion of organic matter and by microbiological processes associated with the decay of organic matter. Alkanes are relatively unreactive species (the old chemical name for this class of compounds "paraffins" is derived from Latin words parum (little) and affinis (having affinity)). Alkanes do not react appreciably with ozone, are not subject to photolysis in the lower atmosphere, and react only slowly with NO_3 radicals. The atmospheric oxidation of alkanes is initiated by reaction with OH radicals. This reaction produces an alkyl radical which rapidly (within a μs) adds O_2 to give an alkyl peroxy radical (RO_2).

$$OH + RH \rightarrow R + H_2O \tag{22}$$

$$R + O_2 + M \rightarrow RO_2 + M \tag{23}$$

Table 3. The 32 most prevalent non-methane organic compounds (NMOC) in urban air [41]

Species	% NMOC[a]	Species	% NMOC[a]
2-methyl-butane	7.8	o-xylene	1.5
n-butane	7.2	3-methyl-pentane	1.5
toluene	6.4	n-hexane	1.3
propane	4.6	2-methyl-hexane	1.3
ethane	3.9	2,2,4-trimethyl-pentane	1.2
m, p-xylene	3.4	methylcyclopentane	1.0
2-methyl-propane	3.2	ethylbenzene	1.0
n-pentane	3.1	formaldehyde	1.0
acetylene	3.0	acetaldehyde	1.0
ethene	2.7	m-ethyltoluene	0.9
1-2-4-trimethylbenzene	2.3	propene	0.9
4-methyl-nonane	2.1	2-methyl-propene	0.9
2-methyl-pentane	2.1	c-2-pentene	0.8
benzene	2.0	3-methyl-hexane	0.8
n-decane	1.9	2,3,3-trimethyl-1-butene	0.8
acetylene	1.5	n-nonane	0.7

[a] % defined in terms of ppbC, i.e. the % of carbon atoms accounted for by each species.

In the atmosphere peroxy radicals react with NO, NO_2, HO_2 radicals and other peroxy radicals ($R'O_2$). The importance of these reactions is dictated by the abundances of NO, NO_2, and HO_2 radicals and by the rates of the reactions of RO_2 radicals with these species. In the troposphere the concentrations of NO, NO_2, and HO_2 vary widely, however, for the present purposes reasonable average concentrations are approximately $(2.5-10) \times 10^8 \, cm^{-3}$ [72]. Under atmospheric conditions, typical rate constants for the reactions of RO_2 radicals with NO, NO_2, and HO_2 radicals lie in the ranges $(8-20) \times 10^{-12}$, $(5-10) \times 10^{-12}$, and $(5-15) \times 10^{-12} \, cm^3$ molecule^{-1} s^{-1}, respectively [4]. Hence, on average these reactions are of comparable importance in the atmospheric fate of RO_2 radicals. On a local scale one reaction may dominate because of variation in the concentrations of NO_x (NO and NO_2) and HO_2 radicals. Thus, in remote marine locations with low NO_x levels, reaction of RO_2 radicals with HO_2 will be much more important than in urban air masses with high NO_x concentrations. The reactivity of peroxy radicals towards other peroxy radicals varies over many orders of magnitude depending on the nature of the "R" group. Consequently, it is not possible to provide a simple accounting of the importance of peroxy self and cross reactions compared to the other possible loss mechanisms of RO_2 radicals. It has been shown by computer modeling studies that cross and self reactions of peroxy radical play an important role in the atmospheric degradation mechanism of organic compounds released into environments containing low levels of NO_x, for example in the Amazon Basin [26]. The atmospheric lifetime of RO_2 radicals is approximately 1 minute. These reactions are illustrated for the case of methane in Fig. 3.

The reaction of peroxy radicals with HO_2 radicals can proceed via two reaction channels to give either organic hydroperoxides or carbonyl products.

$$RO_2 + HO_2 \rightarrow ROOH + O_2 \qquad\qquad (24a)$$

$$RO_2 + HO_2 \rightarrow R'CHO + H_2O + O_2 \qquad\qquad (24b)$$

The reaction of unsubstituted alkyl peroxy radicals like CH_3O_2 and $C_2H_5O_2$ with HO_2 leads to the formation of hydroperoxides in essentially 100% yield [27,28]. In contrast, substituted peroxy radicals (e.g. CH_2FO_2, CH_2ClO_2, $HOCH_2O_2$, and $CH_3OCH_2O_2$) react with HO_2 via two pathways to give hydroperoxide and carbonyl products [29–32]. The factors which determine the branching ratio k_{24a}/k_{24b} for any given RO_2 radical are unknown and further research is needed in this area. The net effect of reaction (24) is to convert two reactive radical species into a relatively unreactive hydroperoxide or carbonyl compound. Reaction (24) slows down the free radical driven photochemical oxidation reactions and reduces the formation of ozone. As indicated in Fig. 3 the hydroperoxides have an atmospheric lifetime of the order of a few days and undergo both photolysis and reaction with OH radicals resulting in the regeneration of RO_x radicals.

It is well established that the reaction of peroxy radicals with NO_2 proceeds by a simple association mechanism in which an alkyl peroxy nitrate is formed. The exothermicity associated with formation of the RO_2-NO_2 bond resides

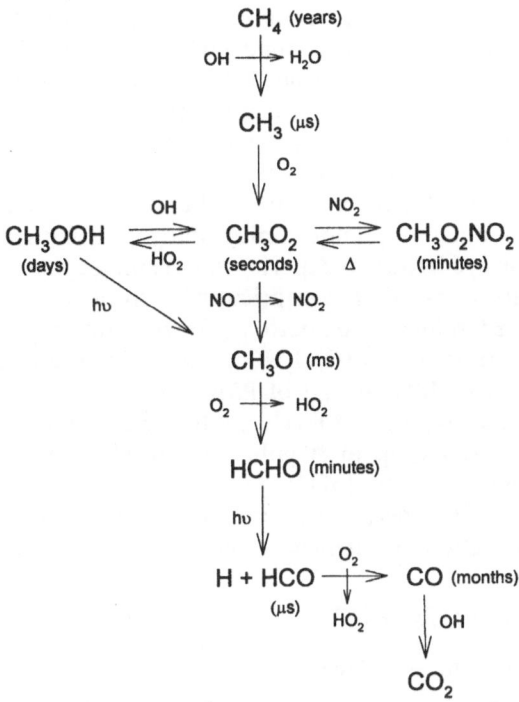

Fig. 3. Atmospheric degradation mechanism of methane showing the central role played by the methyl peroxy radical (CH_3O_2). Values in parentheses are order of magnitude estimates for the lifetimes of the various species

initially in the RO_2NO_2 molecule. Collisional deactivation is required to remove this excess energy.

$$RO_2 + NO_2 + M \rightarrow RO_2NO_2 + M \qquad (25)$$

The RO_2-NO_2 bond in alkyl peroxy nitrates is relatively weak (22–25 kcal mol^{-1}) and the peroxy nitrate decomposes rapidly to reform the reactants [33]. For example, in 760 Torr of air the lifetimes of $CH_3O_2NO_2$ at room temperature and –10°C are 0.6 and 51 s while for $C_2H_5O_2NO_2$ the corresponding lifetimes are 0.3 and 25 s [33].

$$RO_2NO_2 + M \rightarrow RO_2 + NO_2 + M \qquad (-25)$$

Hence, the net effect of the reaction of alkyl peroxy radicals with NO_2 is to sequester carbon in the form of peroxy nitrates for a short period of time. Peroxy acyl nitrates ($RC(O)O_2NO_2$) such as $CH_3C(O)O_2NO_2$, $C_2H_5C(O)O_2NO_2$ and $C_6H_5C(O)O_2NO_2$ have RO_2-NO_2 bonds which are somewhat stronger than their alkyl counterparts (28 kcal mol^{-1} for $CH_3C(O)O_2NO_2$). This slight increase in bond strength has a dramatic effect on their atmospheric chemistry. The lifetime of $CH_3C(O)O_2NO_2$ at room temperature is 0.6 h while at –10°C it is 13 days. The lifetimes of peroxy acyl nitrates with respect to thermal decomposition are sufficient at the low temperatures characteristic of the upper troposphere to allow long range transport of NO_x, and yet short enough at the warmer temperatures of the lower troposphere to release NO_2 which can then participate in photochemical reactions [34–36]. While peroxy acyl nitrates are not formed as primary products in the atmospheric oxidation of alkanes, they are important secondary products. For example, ethane is oxidized to give acetaldehyde which can then react to give PAN.

Peroxy acetyl nitrate (PAN) is the most abundant peroxy acyl nitrate and is an important component of photochemical smog in urban areas. PAN is largely responsible for the eye irritation experienced in smog episodes and, because of its phytotoxic nature, can damage agricultural crops. Other members of the peroxy acyl nitrate family include peroxy propyl nitrate, $C_2H_5C(O)O_2NO_2$, and peroxy benzoyl nitrate, $C_6H_5C(O)O_2NO_2$, which are found at levels ranging from a few percent to 30% of that of PAN [36]. The levels of PAN found in ambient air vary widely; highest levels are found in polluted urban air in the afternoon and can range up to 30 ppb, while levels as low as a few ppt are observed in cleaner air masses [37].

The reaction of alkyl peroxy radicals with NO is very important in the atmospheric degradation mechanism of alkanes. The reaction proceeds via two pathways.

$$RO_2 + NO \rightarrow RO + NO_2 \tag{26a}$$

$$RO_2 + NO + M \rightarrow RONO_2 + M \tag{26b}$$

Pathway (26a) is dominant and produces an alkoxy radical (RO) and an NO_2 molecule. Photolysis of NO_2 is an important source of tropospheric ozone and is responsible for the formation of photochemical smog in polluted air masses. Pathway (26b) gives alkyl nitrates which are much less reactive than NO or NO_2 and so sequester NO_x (NO and NO_2). Reaction channel (26b) represents a loss of both radicals and NO_x from the atmosphere and hence slows down the photochemical chain reactions that form ozone. In general, the relative importance of the nitrate producing channel increases with the size of "R" [38] (for example, with $R=CH_3$, $k_{26b}/(k_{26a}+k_{26b}) < 0.005$ [39], while for $R=tert$-butyl, $k_{26b}/(k_{26a}+k_{26b}) = 0.18$ [40]). For C_1-C_3 hydrocarbons the formation of nitrates is of minor importance while for C_4 and above, nitrate formation is significant.

The alkoxy radicals formed in pathway (26a) have very interesting atmospheric chemistry. The atmospheric fate of alkoxy radicals differs with the nature of the "R" group. Some alkoxy radicals (e.g. CH_3O) are lost solely via reaction with O_2, others undergo rapid decomposition via C-C bond scission. Long chain alkoxy radicals can undergo isomerization via intramolecular H-atom abstraction:

$$CH_3CH_2CH_2CH_2CHO(\cdot)CH_3 \rightarrow CH_3C(\cdot)HCH_2CH_2CH(OH)CH_3 \tag{27}$$

The resulting hydroxy alkyl radical then adds O_2 to become a hydroxy alkyl peroxy radical which reacts with more NO to give more NO_2 and another alkoxy radical capable of undergoing further isomerization:

$$CH_3CHO(\cdot)CH_2CH_2CH(OH)CH_3 \rightarrow CH_3CH(OH)CH_2CH_2C(\cdot)(OH)CH_3 \tag{28}$$

In this fashion the oxidation of long chain alkanes quickly gives multifunctional oxygenated products. These products often have low volatility and form aerosols which can be removed by wet or dry deposition to the ground. The isomerization reactions given above occur via a six membered transition state. Isomerization can also occur via five and seven membered states. The complexity associated with unraveling the precise degradation mechanism of any given alkane can be appreciated by considering the case of n-hexane. Initial OH radical attack leads to the formation of three different alkoxy radicals, each of which can either react with O_2, decompose, isomerize (via several possible pathways), or undergo a combination of these possible loss processes. Our understanding of the atmospheric chemistry of alkoxy radicals is rather crude at present and this is an area of active research. For a review of the current status of our knowledge in this area the reader should consult Ref. [8].

In polluted air masses typical of continental and urban areas the NO levels are high enough to dominate the loss of peroxy radicals. Alkanes and other volatile organic compounds (VOC) are present at high concentrations in polluted air. Ozone levels in such environments are controlled by the rate at which peroxy radicals are formed, the rate at which they convert NO to NO_2, and the extent to which they form alkyl nitrates and peroxyacylnitrates which sequester NO_x. The latter three factors are determined by the quantity and identity of the organic compounds which are present. There is great practical interest in establishing the relationship between emissions of alkanes and other volatile organic compounds (VOC) and NO_x into urban air masses and the formation of secondary air pollutants such as ozone. Over the past 3 decades great strides have been made in understanding the fundamental chemistry which is in operation in the formation of oxidants in urban air. It is now recognized that complex non-linear feedback processes relate VOC and NO_x emissions to ozone levels [41]. While the complexity of the situation defies easy solution, recognition of the need to reduce VOC and/or NO_x emissions has driven the search for new technology and new approaches which reduce the emission of these compounds. The automobile industry provides a good example of the progress made on this front. In the 1960s prior to control, the tailpipe emissions of a typical automobile in California were 8.8 g of hydrocarbons and 3.6 g of NO_x per mile. With the help of catalyst technology, tailpipe emissions from new vehicles in California in 1997 have been reduced to approximately 0.2 g of hydrocarbons and 0.3 g of NO_x per mile [42].

Regulations in the U.S. and Europe pertaining to the emission of reactive organic compounds into the atmosphere are moving away from standards in which only the total organic mass is considered, to standards which take into account the different reactivity of different compounds. This trend in regulations has generated a considerable interest in the relative ranking of organic compounds in terms of their ability to contribute to ozone formation in urban areas. Calculation of scales of "reactivity factors" [43] or "photochemical ozone creation potentials" (POCPs) [44–46] for organic compounds have received considerable prominence. These scales reflect the rate at which the organic compound gives peroxy radicals together with the efficiency with which these peroxy radicals contribute to, or hinder, ozone formation.

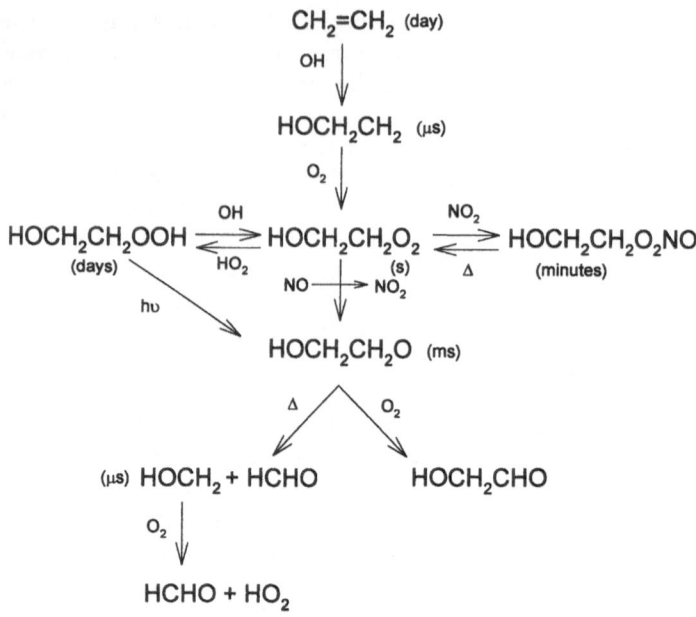

Fig. 4. Mechanism of the OH initiated atmospheric degradation of ethene

5
Degradation of Alkenes

More than 140 different alkenes have been identified in the atmosphere [25]. The sources of alkenes are similar to those for the alkanes with combustion of fossil fuel being a major source. The presence of unsaturated bonds makes these compounds much more reactive than the alkanes. The most persistent member of this class of compounds (ethene) has an atmospheric lifetime of the order of a day, while more typically the lifetimes for alkenes are measured in hours. As a result of their short lifetimes the atmospheric concentrations of alkenes are highly variable and decrease dramatically away from their source locations. As with the alkanes the reaction of OH radicals is an important loss mechanism. This reaction proceeds mainly via addition to the unsaturated bond as illustrated for ethene in Fig. 4. In one atmosphere of air at 298 K the dominant atmospheric fate of the alkoxy radical $HOCH_2CH_2O$ is decomposition via C-C bond scission, while reaction with O_2 makes a 20% contribution [47]. The fate of alkoxy radicals resulting from addition of OH to alkenes is generally decomposition via C-C bond scission [8]. Thus, the OH radical initiated oxidation of propene gives acetaldehyde and HCHO, oxidation of 1-butene gives propionaldehyde and HCHO, and oxidation of 2-butene gives acetaldehyde [48, 49].

In addition to OH radicals, unsaturated bonds are reactive towards O_3 and NO_3 radicals and reaction with these species is an important atmospheric degradation mechanism for unsaturated compounds. Table 4 lists rate constants for the reactions of O_3 and NO_3 radicals with selected alkenes and acetylene. To place such rate constants into perspective we need to consider the typical

Table 4. Rate constants and corresponding lifetimes for reaction of selected unsaturated compounds with O_3 and NO_3 [8]

Compound	$k(O_3)$[a]	$\tau(O_3)$[b]	$k(NO_3)$[a]	$\tau(NO_3)$[b]
ethene	1.6×10^{-18}	2 days	2.1×10^{-16}	220 days
propene	1.0×10^{-17}	11 h	9.5×10^{-15}	4.9 days
1-butene	9.6×10^{-18}	12 h	1.4×10^{-14}	3.3 days
1,3-butadiene	6.3×10^{-18}	18 h	1.0×10^{-13}	11 h
cyclohexene	8.1×10^{-17}	1.4 h	5.9×10^{-13}	1.9 h
limonene	2.0×10^{-16}	34 min	1.2×10^{-11}	6 min
2-methyl-1,3-butadiene (isoprene)	1.3×10^{-17}	8.7 h	6.8×10^{-13}	1.6 days
α-pinene	8.7×10^{-17}	1.3 h	6.2×10^{-12}	1.2 h
acetylene	1×10^{-20}	470 days	$<1 \times 10^{-16}$	>463 days

[a] Units of cm^3 molecule^{-1} s^{-1}.
[b] Calculated using $[O_3] = 100$ ppb and $[NO_3] = 10$ ppt. NO_3 radicals are only present at night, a calculated lifetime in excess of 12 h indicates that multiple nights exposure are needed for substantial loss.

ambient atmospheric concentrations of O_3 and NO_3 radicals. Typical ozone concentrations in pristine environments are 20–40 ppb while concentrations in the range 100–200 ppb are experienced in polluted air. The ambient concentration of NO_3 is limited by the availability of NO_x sources. In remote marine environments the NO_x levels are extremely low (a few ppt) and NO_3 radicals do not play an important role in atmospheric chemistry. In continental and urban areas the NO_x levels are much higher (up to several hundred ppb in polluted urban areas) and NO_3 radicals can build up to 5–100 ppt at night (NO_3 radicals are photolyzed rapidly and are not present during the day). For the purposes of the present discussion we have calculated the atmospheric lifetimes of selected unsaturated compounds in Table 4 in the presence of 100 ppb (2.5×10^{12} cm^{-3}) of O_3 and 10 ppt (2.5×10^8 cm^{-3}) of NO_3. Lifetimes in other environments can be evaluated by appropriate scaling of the data in Table 4. As seen from Table 4, the more reactive unsaturated compounds have lifetimes with respect to reaction with O_3 and NO_3 radicals of only a few minutes!

The reaction of ozone with alkenes proceeds via addition to the double bond giving a short lived "energy-rich" ozonide which decomposes via C–C bond scission to give carbonyl compounds and biradical species known as "Criegee" biradicals:

For symmetrical alkenes pathways "a" and "b" give the same products while unsymmetric alkenes give different products via pathways "a" and "b". The relative importance of decomposition pathways "a" and "b" are comparable for 1-alkenes [8]. For alkenes of the type $R_1R_2=CH_2$ decomposition via channel "b" to form the disubstituted biradical, $R_1R_2 C(\cdot)OO(\cdot)$, is slightly favored (by a ratio of $\approx 65:35$) over decomposition via channel "a" to give the unsubstituted biradical, $(\cdot)CH_2O_2(\cdot)$. Likewise, ozonides formed in reaction with alkenes of the general type $R_1CH=CR_3R_4$ show a slight preference ($\approx 65:35$) to decompose to give the di-substituted biradical [8].

The reaction of O_3 with alkenes is very exothermic. Assuming that the C-H bond dissociation energy in CH_3O_2 radicals is the same as that in CH_3OH (94.1 kcal mol^{-1}) it can be estimated that reaction (29) is exothermic by 27 kcal mol^{-1}.

$$O_3 + CH_2=CH_2 \rightarrow HCHO + (\cdot)CH_2O_2(\cdot) \tag{29}$$

Criegee biradicals formed in the reaction of ozone with alkenes carry significant internal excitation and the majority of them decompose before they can be collisionally stabilized. The biradicals are sufficiently excited that many different rearrangements and decomposition pathways are energetically allowed. For example, in the case of the $[CH_3C(\cdot)HOO(\cdot)]^*$ species formed during propene oxidation the IUPAC panel recommends the following channels (and yields) [23]:

$$[CH_3C(\cdot)HOO(\cdot)]^* + M \rightarrow CH_3C(\cdot)HOO(\cdot) + M^* \quad 15\% \tag{30a}$$

$$[CH_3C(\cdot)HOO(\cdot)]^* \rightarrow CH_3 + CO + OH \qquad 54\% \tag{30b}$$

$$[CH_3C(\cdot)HOO(\cdot)]^* \rightarrow CH_3 + CO_2 + H \qquad 8.5\% \tag{30c}$$

$$[CH_3C(\cdot)HOO(\cdot)]^* \rightarrow HCO + CH_3O \qquad 8.5\% \tag{30d}$$

$$[CH_3C(\cdot)HOO(\cdot)]^* \rightarrow CH_4 + CO_2 \qquad 14\% \tag{30e}$$

Channel (30b) is of considerable importance as it produces OH radicals. The OH radical yield varies between 10 and 100% depending on the particular alkene [8]. Although it has been known for many years that OH radicals are produced in the reaction of ozone with alkenes [50] it has only recently been recognized that this could be an important nighttime source of OH radicals in the atmosphere. Channel (30a) gives a stabilized biradical. The atmospheric fate of stabilized biradicals is dominated by reaction with water vapor which proceeds predominately to give carboxylic acids, e.g.

$$CH_3C(\cdot)HOO(\cdot) + H_2O \rightarrow CH_3C(O)OH + H_2O \tag{31}$$

In summary, the reaction of ozone with alkenes is important in the atmospheric degradation of alkenes. In all cases the reaction leads to rupture of the $\rangle C=C\langle$ double bond. The double bond is replaced by a carbonyl group on one side and a Criegee biradical on the other. The Criegee biradical is formed energetically excited and decomposes by a variety of different routes to give a complex mixture of oxygenated products (mainly carbonyls).

Table 4 lists kinetic data for reactions of NO_3 radicals with selected alkenes. As seen in Table 4 reaction with NO_3 radicals can play an important role in the atmospheric degradation of alkenes. Reaction proceeds via addition of NO_3 to the double bond giving a nitrooxy alkyl radical. The resulting energy-rich nitrooxy radical either decomposes to an oxirane or is stabilized by transferring its excitation during collisions with a third body (M). In the presence of air the thermalized nitrooxy alkyl radical adds O_2 rapidly (within a µs) to give a peroxy radical which can then react with NO, NO_2, HO_2, or other peroxy radicals ($R'O_2$) to give a variety of products.

$$NO_3 + {>}C{=}C{<} \longrightarrow \{{>}C(ONO_2){-}C(.){<}\}^* \xrightarrow{\;k_a\;} \text{oxirane} + NO_2$$

$$M \downarrow k_b \qquad \nearrow k_c$$

$$>C(ONO_2){-}C(.){<} + M^{\cdot}$$

$$O_2 \downarrow k_d$$

$$>C(ONO_2){-}COO(.){<}$$

Berndt and Böge [51, 52] have studied the relative kinetics in the mechanism given above. For reaction of NO_3 radicals with tetramethyl ethylene they report $k_b/k_a = (1.75 \pm 0.19) \times 10^{-19}$ and $k_d/k_c = (1.30 \pm 0.34) \times 10^{-16}$ cm^3 molecule^{-1}. Assuming k_d is of the order of 10^{-12} and that k_b approaches the gas kinetic rate limit of 10^{-10} cm^3 molecule^{-1} s^{-1} (rates typical for such processes [33, 53]) then k_c is of the order of 10^4 s^{-1} while k_a is of the order of 10^9 s^{-1}. Rate constant ratios k_b/k_a have been measured for the reaction of NO_3 with the following alkenes (in units of 10^{-19} cm^3 molecule): propene (1.53), 1-butene (1.9), *trans*-2-butene (3.3), 2-methyl propene (7.4), 2-methyl 2 butene (4.4), and 2,3-dimethyl 2-butene (1.8).

In the atmosphere the nitrooxy alkyl peroxy radical, $\langle C(ONO_2){-}COO({\cdot})\rangle$, behaves like other alkyl peroxy radicals and will react with NO_2, HO_2, and other peroxy radicals. Reaction of nitrooxy alkyl peroxy radical with NO is unlikely because the conditions necessary for the formation of NO_3 radicals (high O_3) are incompatible with the presence of significant amounts of NO. For unsymmetrical alkenes the addition of NO_3 radicals leads to the formation of two different peroxy radicals, e.g. for propene:

$$NO_3 + CH_3CH{=}CH_2 \longrightarrow CH_3CH(ONO_2){-}CH_2(.) \;+\; CH_3CH(.){-}CH_2ONO_2$$

$$\downarrow O_2 \qquad\qquad\qquad \downarrow O_2$$

$$CH_3CH(ONO_2){-}CH_2OO(.) \qquad CH_3CHOO(.){-}CH_2ONO_2$$

As discussed in Sect. 4, reaction of the peroxy radicals with NO_2 gives thermally unstable peroxy nitrates. Reaction with HO_2 gives hydroperoxides and possibly carbonyl compounds. Reaction with other peroxy radicals ($R'O_2$) gives alkoxy radicals, carbonyls, and alcohols. The alkoxy radicals will then either isomerize, react with O_2, or decompose (see Sect. 3). Thus, the NO_3 radical initiated atmospheric degradation of alkenes leads to oxiranes (generally in small yield), nitrooxy hydroperoxides, nitrooxy carbonyls, and nitrooxy-alcohols. For a detailed listing of products from individual alkenes the reader should consult Ref. [8].

In summary, alkenes are reactive compounds and are removed rapidly from the atmosphere by a variety of processes. Reaction with OH radicals, ozone, and NO_3 radicals all play important roles. These reactions proceed via addition to the unsaturated bond giving an adduct which decomposes and/or reacts with O_2 leading to the generation of a variety of transient radical species which react to form the first generation closed shell products (principally carbonyl compounds).

6
Degradation of Aromatics

As illustrated by Table 3 aromatic compounds such as toluene, benzene, o,m,p-xylenes and ethyl benzene are important pollutants in urban air. These compounds comprise a substantial fraction of automotive fuel and its use results in substantial emissions of aromatic compounds into urban atmospheres. In addition to the monocyclic aromatic compounds there is considerable interest in the chemistry of polycyclic aromatic hydrocarbons (PAH) such as naphthalene, anthracene and pyrene. The discussion here will be confined to monocyclic aromatics, but the degradation of PAHs has many similarities to that of the simpler one ring systems. Details can be found elsewhere [54].

Reaction of monocyclic aromatics with O_3 and NO_3 radicals is generally very slow and unimportant. Atmospheric degradation of aromatics is initiated by OH radical attack. The kinetics of the reaction of OH radicals with aromatic compounds are well established [15]. As seen from Table 2 the lifetime of aromatics with respect to reaction with OH is typically a few days or less. Reaction proceeds by addition to the ring and H-atom abstraction from either the substituent groups or possibly from the ring C-H sites. In all cases the addition channel is dominant. For benzene, toluene, o,m,p-xylene, and ethyl benzene the H-atom abstraction pathway accounts for 5–10% of the overall reaction [15, 55], the remaining 90–95% proceeds via addition. For toluene $k_a/(k_a + k_b) = 0.07$ [15, 55].

The benzyl radical formed in channel (a) behaves like an alkyl radical and in air adds O_2 rapidly (within a μs) to give a benzyl peroxy radical which either reacts with HO_2 radicals to give a hydroperoxide, with NO_2 to give a thermally unstable peroxy nitrate, or with NO to give a 10–12% yield of benzyl nitrate and 88–90% yield of an alkoxy radical which then reacts with O_2 to produce benzaldehyde [15, 56].

The majority of the reaction proceeds via the addition channel (b) to give a methylcyclohexadienyl radical. Cyclohexadienyl radicals are resonance stabilized and are relatively unreactive. Typical alkyl radicals add O_2 rapidly with rate constants of the order of 10^{-12}, in contrast the reaction of cyclohexadienyl and methyl-cyclohexadienyl radicals with O_2 proceed with rate constants of $(2-5) \times 10^{-16}$ cm^3 molecule^{-1} s^{-1} [57, 58]. It can be calculated that in one atmosphere of air the cyclohexadienyl radicals have a lifetime of approximately 1 ms with respect to reaction with O_2. It is uncertain as to whether other reactions compete with O_2 for cyclohexadienyl radicals. A simple order of magnitude calculation can be performed to provide insight into which species could compete. The collision frequency provides an upper limit for the rate constant of a gas phase neutral-neutral reaction of 10^{-10} cm^3 molecule^{-1} s^{-1}. To be competitive with the O_2 reaction the reaction partner must be present at a level $> 10^{13}$ molecule cm^{-3} (> 400 ppb). While this would appear to exclude the possibility that NO_2 plays any significant role in the atmospheric chemistry of cyclohexadienyl radicals there is substantial body of data from product studies which appears to show that reaction of NO_2 *is* important [15]. It is possible that the measured rate of reaction with O_2 is not relevant to atmospheric chemistry; for example, the reaction could give an adduct which decomposes back to reactants. Alternatively, perhaps there is some unforeseen problem in the interpretation of the product studies. Reaction with O_3 is a possible fate of cyclohexadienyl radicals which needs investigating. The atmospheric fate of cyclohexadienyl radicals is poorly understood and more work is needed in this area.

There have been a large number of smog chamber experiments which have focused on identifying the products formed during the atmospheric degradation of aromatics. This is challenging work as the simultaneous identification and quantification of the large number of chemically similar products is a difficult task. Table 5 lists the observed major products for benzene, toluene, and o-xylene. The atmospheric degradation of aromatic compounds proceeds via complex pathways which generate a large number of products. Of all the aromatic compounds, the atmospheric degradation of toluene has been studied most extensively and is best understood. Yet even for this molecule the observed identified products only account for approximately 60% of its loss. Despite their recognized importance and a great deal of scientific study the atmospheric degradation mechanisms of aromatic compounds remains unclear at the present.

Table 5. Products of the OH initiated atmospheric degradation of selected aromatic compounds, adapted from Atkinson [15]

Aromatic	Product	Molar Yield
Benzene	Glyoxal (HC(O)C(O)H)	0.21
	Phenol (C_6H_5OH)	0.24
	Nitrobenzene ($C_6H_5NO_2$)	0.03[a]
Toluene	Glyoxal (HC(O)C(O)H)	0.08–0.15
	Methyl glyoxal ($CH_3C(O)C(O)H$)	0.08–0.15
	Benzaldehyde (C_6H_5CHO)	0.05–0.12
	Benzyl nitrate ($C_6H_5CH_2ONO_2$)	0.008
	o-Cresol ($C_6H_4(CH_3)OH$)	0.13–0.22
	m,p-Cresol ($C_6H_4(CH_3)OH$)	0.05
	m-Nitrotoluene ($C_6H_4(CH_3)NO_2$)	0.01[a]
o-Xylene	Glyoxal (HC(O)C(O)H)	0.03–0.09
	Methyl glyoxal ($CH_3C(O)C(O)H$)	0.12–0.25
	Biacetyl ($CH_3C(O)C(O)CH_3$)	0.09–0.26
	o-Tolualdehyde ($C_6H_4(CH_3)CHO$)	0.05–0.17
	2-Methylbenzyl nitrate ($C_6H_4(CH_3)CH_2ONO_2$)	0.01[a]
	2,3-Dimethylphenol ($C_6H_3(CH_3)_2OH$)	0.10
	3,4- Dimethylphenol ($C_6H_3(CH_3)_2OH$)	0.06
	3-Nitro-o-xylene ($C_6H_3(CH_3)_2NO_2$)	0.005
	4-Nitro-o-xylene ($C_6H_3(CH_3)_2NO_2$)	0.01[a]
	Nitro-o-xylenes ($C_6H_3(CH_3)_2NO_2$)	0.07

[a] In limit of zero [NO_2].

7
Degradation of NO_x

NO_x refers to the sum of two of the most reactive nitrogen species NO and NO_2. NO_x plays a critical role in determining levels of ozone. Over the last 100 years it is believed that tropospheric background levels of ozone in the Northern hemisphere have increased by approximately a factor of two and that this increase has been strongly influenced by increased levels of NO_x [9].

The major sources of NO_x in the troposphere are fossil fuel combustion, biomass burning, microbiological activity in soil, lightning, oxidation of ammonia, and tropopause folding. Each year approximately 53 Tg N of NO_x is emitted into the atmosphere. Fossil fuel combustion contributes one half of this total while the contribution from biomass burning is 2.1–5.5 Tg N/yr [59]. N_2O emitted from biological activity in soil is long-lived in the troposphere and is transported to the stratosphere where it reacts with O(^1D) to produce NO. Hence N_2O, which is normally not included in any of the NO_x or NO_y families, can be a secondary source of NO_x. NO_y refers to "odd nitrogen" and is the sum of NO_x and all oxidized nitrogen species not including N_2O.

The chemical role played by NO_x in the atmosphere is extremely complex and will not be dealt with in complete detail in this chapter. The sources of NO_x have been mentioned briefly above. How is NO_x then removed from the atmosphere?

A "first-order" simplified atmosphere containing only NO, NO_2 and O_3 is characterized by the equations:

$$NO + O_3 \rightarrow NO_2 + O_2 \tag{32}$$

$$NO_2 + h\nu \; (\lambda < 420 \text{ nm}) \rightarrow NO + O \tag{6}$$

$$O + O_2 + M \rightarrow O_3 + M \tag{7}$$

NO reacts rapidly with O_3 giving NO_2. Using $k_{32} = 1.8 \times 10^{-14}$ cm^3 molecule^{-1} s^{-1} [21], and assuming a constant O_3 concentration of 30 ppbv (7×10^{11} cm^{-3}), then NO has a lifetime of 1.3 min with respect to reaction (32). NO_2 is a brown-colored gas and absorbs at 400–450 nm. The NO_2 photolysis rate, J_{NO_2}, in the troposphere depends on the cloud cover and is typically in the range $(0.3 - 1) \times 10^{-2}$ s^{-1}, giving a lifetime of NO_2 with respect to photolysis of 2 – 6 min [4]. In air, at one atmosphere, reaction (7) has an effective bimolecular rate constant of $k_7 = 1.5 \times 10^{-14}$ cm^3 molecule^{-1} s^{-1} [21], $[O_2] = 5.2 \times 10^{18}$ cm^{-3}, and O atoms have a lifetime of 13 ms with respect to reaction (7). As a result of the rapidity of reactions (6), (7), and (32), a photostationary steady state condition is set up where NO and O atoms are created and destroyed continually but maintain a steady concentration, i.e. $d[O]/dt = d[NO]/dt = 0$. As a result of their rapid interconversion it is convenient to lump NO and NO_2 together and discuss the atmospheric degradation of NO_x.

Reactions (6), (7), and (32) form the basis for what is known as the Leighton relationship, after Philip Leighton who wrote one of the first textbooks on air pollution in 1961 [60]. Applying steady state conditions one can derive:

$$\frac{[NO_2]}{[NO]} = \frac{k_{32}[O_3]}{J_{NO_2}}$$

where J_{NO_2} is the first order rate constant for photolysis of NO_2. When measurements of J_{NO_2}, $[NO_2]$, $[NO]$, and $[O_3]$ are performed in the field, the observed concentration ratio $[NO_2]/[NO]$ is often more than that calculated from the right hand side of the above equation [61, 62]. In other words, to maintain the observed $[NO_2]/[NO]$ mixing ratio there must be some species, other than O_3, that is capable of oxidizing NO to NO_2 which is not accounted for (i.e. missing) in the simple mechanism of reactions (6, 7, 32). This "missing oxidant" is peroxy radicals. In the presence of peroxy radicals, additional conversion of NO to NO_2 occurs via reaction of peroxy radicals with NO.

$$RO_2 \text{ (or } HO_2) + NO \rightarrow RO \text{ (or } OH) + NO_2 \tag{26}$$

When reaction (26) is included in the steady state analysis the predicted ozone concentration becomes:

$$\frac{[NO_2]}{[NO]} = \frac{k_{32}[O_3]}{J_{NO_2}} + \frac{k_{26}[RO_2]}{J_{NO_2}}$$

While reactions (6) and (32) interconvert NO and NO_2, they do not remove NO_x from the atmosphere. NO_x is removed from the atmosphere by the conversion of NO_2 into HNO_3 and aerosol NO_3^- which are then subject to wet and dry deposition [36]. There are a variety of reactions that convert NO_x into HNO_3. In the daytime the association reaction of OH radicals with NO_2 is important.

$$OH + NO_2 + M \rightarrow HNO_3 + M \tag{33}$$

Nighttime loss of NO_x is tied to N_2O_5 which acts as a reservoir. NO_2 reacts with O_3 to form NO_3 which is equilibrium with N_2O_5:

$$NO_3 + NO_2 + M \leftrightarrow N_2O_5 + M \tag{34}$$

N_2O_5 undergoes a reaction with water to give HNO_3:

$$N_2O_5 + H_2O \text{ (aq)} \rightarrow 2\,HNO_3 \text{ (aq)} \tag{35}$$

As discussed in the previous sections reaction of NO_3 radicals with organic compounds results in the formation of HNO_3 and organic nitrates. In the stratosphere ClO radicals react with NO_2 to give $ClONO_2$ which undergoes hydrolysis.

$$ClONO_2 + H_2O \text{ (aq)} \rightarrow HNO_3 \text{ (aq)} + HCl \text{ (aq)} \tag{36}$$

NO_2 can be converted to PAN ($CH_3COO_2NO_2$) and PAN-like compounds (RO_2NO_2):

$$CH_3C(O)O_2 + NO_2 + M \leftrightarrow CH_3C(O)O_2NO_2 + M \tag{37}$$

$$RO_2 + NO_2 + M \leftrightarrow RO_2NO_2 + M \tag{25}$$

PAN and PAN-like compounds are important reservoirs for NO_x and can transport NO_x over great distances (see Sect. 4). In the past 10–20 years significant progress has been made in understanding the atmospheric chemistry of NO_x and other nitrogen species [36]. Measurement of the individual nitrogen species has proven to be a valuable tool for evaluating photochemical processes occurring in an airmass.

8
Degradation of SO_x

SO_2 emitted from human activities amounts to about 140 Mtonnes S yr^{-1} [63]. There is considerable scatter in the estimates of natural emissions of sulfur compounds, ranging from 34 Mtonnes S yr^{-1} [64] to 267 Mtonnes S yr^{-1} [65]. One recent estimate is 100 Mtonnes per year including SO_x from sulfides [63]. Anthropogenic emissions of SO_2 are increasing in the developing countries while decreasing in USA, Japan and Western Europe. These anthropogenic emissions are mainly due to fossil fuel combustion. Efforts to reduce SO_2 emissions have been mainly through flue-gas desulfurization, use of low-sulfur fuels, and the removal of sulfur from fuels.

SO_2 is oxidized to sulfuric acid both by homogeneous gas-phase reactions and by multiphase processes when a precursor gas is dissolved in water and then subsequently oxidized. The routes to atmospheric sulfuric acid production has been studied extensively for decades. The most important gas-phase mechanism is oxidation by OH radicals:

$$SO_2 + OH + M \rightarrow HSO_3 + M \tag{38}$$

$$HSO_3 + O_2 \rightarrow HO_2 + SO_3 \tag{39}$$

$$SO_3 + H_2O + M \rightarrow H_2SO_4 + M \tag{40}$$

At 298 K in air at 760 Torr, $k_{38} = 9.6 \times 10^{-13}$ cm^3 molecule^{-1} s^{-1} [66] which leads to a lifetime of SO_2 with respect to reaction with OH of approximately 12 days. HO_2 produced in reaction (2) can oxidize NO to NO_2 allowing ozone to be formed together with H_2SO_4. There are other gas-phase routes to formation of sulfuric acid from SO_2, but they are of very limited importance in the atmosphere.

SO_2 is a moderately soluble gas and its solubility is enhanced by hydrolysis:

$$SO_2 + H_2O \leftrightarrow SO_2H_2O \tag{41}$$

$$SO_2 + H_2O \leftrightarrow H^+ + HSO_3^- \tag{42}$$

$$HSO_3^- \leftrightarrow H^+ + SO_3^{2-} \tag{43}$$

pH influences the overall solubility in cloud droplets. This has an impact on the efficiency of the aqueous phase SO_2 oxidation. Since sulfuric acid is a product, the oxidation reaction can become self-limiting. The predominant aqueous phase oxidation mechanism is oxidation by hydrogen peroxide:

$$HSO_3^- + H_2O_2 \leftrightarrow HSO_4^- + H_2O \tag{44}$$

$$HSO_4^- + H^+ \rightarrow H_2SO_4 \tag{45}$$

The average pH of cloud and rain water is less than 5, particularly in polluted areas and in remote ocean areas far from large dust concentrations which may neutralize acidity. At a pH between 3 and 5 the sulfuric acid production rate shows a positive response to the presence of H^+ ions [67], which counteracts the tendency for SO_2 to be less soluble.

There is sufficient H_2O_2 in clouds to oxidize the SO_2 which reaches cloud height and calculations estimate that the multiphase oxidation of SO_2 by H_2O_2 accounts for most of the sulfuric acid present in rain, and over 70% of the sulfate aerosol observed in the atmosphere [68]. The aerosol is produced after evaporation of cloud droplets. It has been found in experiments carried out in the real atmosphere that the overall rate of oxidation of SO_2 in clouds is many times faster than would be predicted from using laboratory-based measurements [69].

9
Degradation of CFCs and Halons

Chlorofluorocarbons (CFCs) give rise to perhaps the most well known environmental problem, depletion of the stratospheric ozone layer. CFCs were developed in the late 1920s in the search for more "friendly" refrigerants as

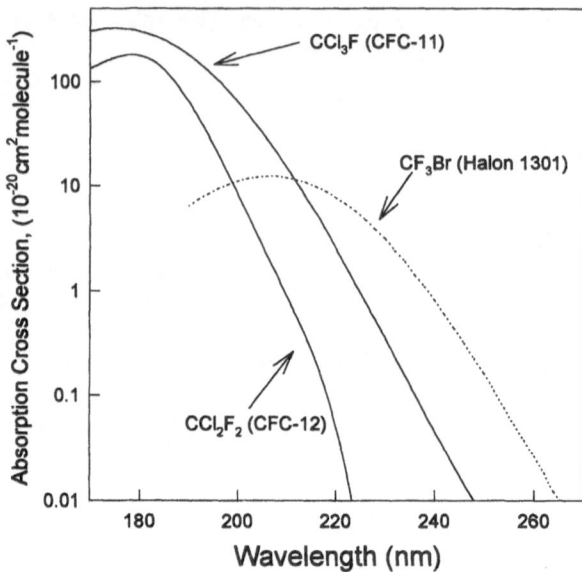

Fig. 5. UV absorption spectra (base e) for CFC-11, CFC-12 and Halon 1301 [21]

substitutes for ammonia and sulfur dioxide. The most widely used CFCs were CFC-11 ($CFCl_3$) and CFC-12 (CF_2Cl_2) which were employed as propellants in aerosol spray cans, foam blowing agents, solvents, and refrigerants. CFC production peaked in 1974 with a global annual production of CFC-11 and CFC-12 of 350,000 tonnes/yr and 400,000 tonnes/yr, respectively. The chemical inertness of CFCs which is one of their desired properties is also the property that causes them to be environmentally unacceptable. After their use CFCs are released to the atmosphere. Lovelock [70] was the first to detect CFCs in the troposphere. The amounts found in the atmosphere were identical, within experimental error, to the total amount manufactured and released. CFCs do not react with OH or NO_3 radicals or ozone and only absorb UV at short wavelengths which do not penetrate into the troposphere; there is no known loss mechanism of CFCs in the troposphere.

The UV absorption spectra of CFC-11 and – 12 and Halon 1301 (used as a fire extinguisher) are shown in Fig. 5. By comparison of Fig. 5 with the solar flux at different altitudes in Fig. 1 it can be seen that while photolysis of CFCs is not possible in the troposphere, it can occur slowly at 20 km and more rapidly at 40 km. The atmospheric degradation pathway of CFCs is transport to the stratosphere followed by UV photolysis:

$$CF_3Cl + h\nu \rightarrow CF_3 + Cl \tag{46}$$

with photodissociation coefficients at 40 km height of approximately $10^{-7}\,s^{-1}$ and $10^{-6}\,s^{-1}$ for CFC-12 and CFC-11, respectively. At 40 km the "local lifetimes" of CFC-11 is of the order of a month. The "local lifetime" should not be confused with the "atmospheric lifetime" of CFC-11 which is 50 years (see Table 6). The

Table 6. Atmospheric lifetimes, ozone depletion potential, and global warming potentials for selected HFCs and HCFCs

Compound	Lifetime [a] (years)	ODP [a]	GWP[a,b]
HFC-23 (CF_3H)	250	0	7.3
HFC-32 (CH_2F_2)	6.0	0	0.16
HFC-125 (CF_3CF_2H)	36	0	0.77
HFC-134a (CF_3CFH_2)	14	0	0.25
HFC-143a (CF_3CH_3)	55	0	1.1
HFC-227ea (CF_3CFHCF_3)	41	0	0.69
HFC-236fa ($CF_3CH_2CF_3$)	250	0	3.9
HCFC-22 (CHF_2Cl)	13.3	0.047	0.36
HCFC-123 (CF_3CCl_2H)	1.4	0.016	0.019
HCFC-124 (CF_3CFClH)	5.9	0.018	0.099
HCFC-141b ($CFCl_2CH_3$)	9.4	0.085	0.13
HCFC-142b (CF_2ClCH_3)	19.5	0.053	0.42
CFC-11 ($CFCl_3$)	50	1.0[c]	1.0[c]
CFC-12 (CF_2Cl_2)	105	0.95	3.1
CO_2			0.00076

[a] Taken from Pinnock et al. [77] and Wallington et al. [73].
[b] 500 year time horizon.
[c] By definition.

difference lies in the fact that only a small fraction of the total atmospheric mass resides at 40 km. The rate determining step in the degradation of CFCs is the time taken for air to pass through the stratosphere as part of the natural circulation. The measured altitude profiles of CFCs are consistent with their photolytic loss in the stratosphere. The radical fragments resulting from CFC photolysis are converted into HF, HCl, and CO_2. HF and HCl are soluble in water and are removed from the atmosphere by rain-out.

Rowland and Molina [71] were the first to recognize the threat of the released chlorine to the stratospheric ozone layer. The threat originates from the catalytic degradation cycle:

$$Cl + O_3 \rightarrow ClO + O_2 \tag{47}$$

$$ClO + O \rightarrow Cl + O_2 \tag{48}$$

$$net\ O + O_3 \rightarrow O_2 + O_2$$

A chlorine atom goes may go through this cycle many thousands of time before it is removed in the form of HCl following reaction with methane:

$$Cl + CH_4 \rightarrow HCl + CH_3 \tag{49}$$

Halons such as Halon-1211 (CF_2BrCl) and Halon-1301 (CF_3Br) are brominated CFCs which are used as a fire extinguishers. Like CFCs, Halons are chemically inert in the troposphere but photolyze in the stratosphere. Photolysis releases bromine atoms which can remove stratospheric ozone in a cycle that is analogous to the chlorine based cycle above. The bromine cycle can couple with the chlorine cycle in several ways, e.g.:

$$Br + O_3 \rightarrow BrO + O_2 \tag{50}$$

$$Cl + O_3 \rightarrow ClO + O_2 \tag{51}$$

$$BrO + ClO \rightarrow Br + Cl + O_2 \tag{52}$$

$$\text{net } O_3 + O_3 \rightarrow 3 O_2$$

On a per molecule basis bromine is more efficient than chlorine in destroying ozone.

Recognition of the adverse impact of CFCs on stratospheric ozone led to legislation in the US, Canada, and Scandinavia in the 1970s to limit the use of such compounds. Further legislation was forthcoming after the discovery of the Antarctic Ozone hole in 1985. In 1987 the so-called "Montreal Protocol" was signed enforcing international reduction in the production and use of CFCs. The Montreal protocol has been further tightened several times since 1987. The production and use of CFCs is now restricted on a global basis. The effect of these restrictions can already seen in the trends in atmospheric concentrations of CFCs. The prompt global action taken on this environmental problem has been extremely successful and should create optimism for solutions to future environmental problems. Within a relatively short time CFC replacement compounds have been identified, tested, and marketed. The results of a thorough investigation of their atmospheric chemistry is described in the next section.

10
Degradation of HFCs and HCFCs

Recognition of the adverse impact of chlorofluorocarbons (CFCs) on stratospheric ozone [71] has prompted an international effort to replace CFCs with environmentally acceptable alternatives [72]. Hydrofluorocarbons (HFCs) and hydrochlorofluorocarbons (HCFCs) are two classes of CFC replacements. In contrast to CFCs, HFCs and HCFCs contain one or more C-H bonds and so are susceptible to attack by OH radicals in the troposphere. As the result of a substantial research effort to define the environmental impact of HCFCs and HFCs we have a good understanding of the atmospheric degradation mechanism of such compounds [73,74].

10.1
Gas-Phase Chemistry

Reaction with OH radicals is the dominant loss process for all HFCs and HCFCs, accounting for >90% of the fate of these compounds. In the stratosphere photolysis and reaction with Cl and $O(^1D)$ atoms make minor contributions to the overall loss. A substantial kinetic database exists concerning the reaction of OH radicals with HFCs and HCFCs [21]. From this data atmo-

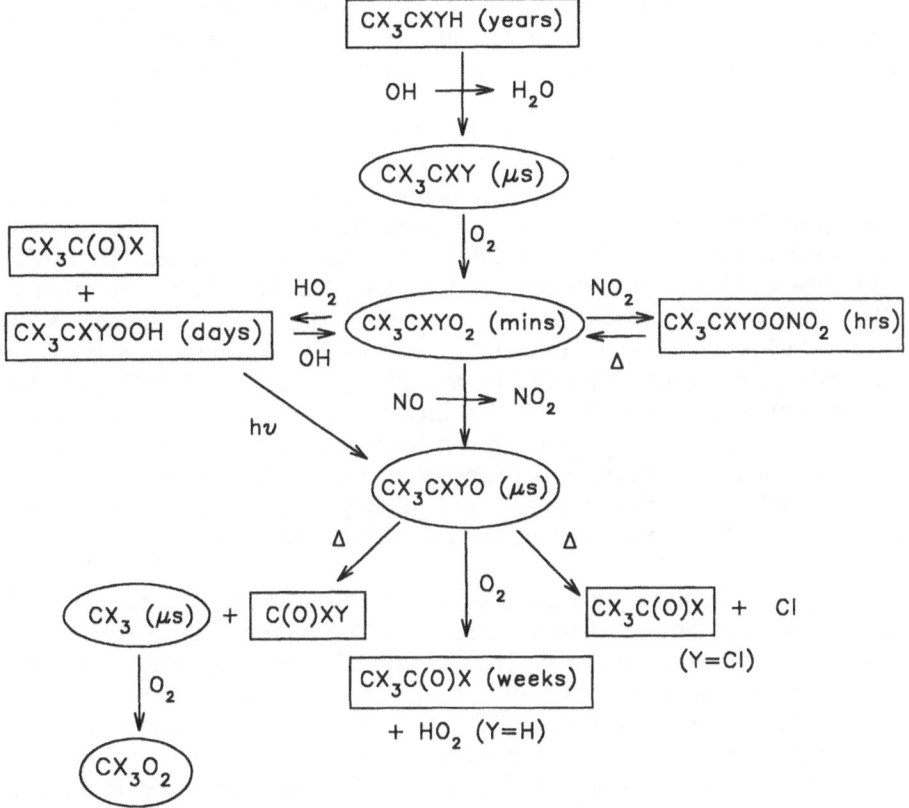

Fig. 6. Generic scheme for the oxidation of a C_2 halocarbon, X and Y represent Cl and, or, F. Values in *parentheses* are order of magnitude estimates for the lifetimes of the various species. Closed shell species are enclosed in *boxes*. Radical species are denoted by *ellipses*

spheric lifetimes can be calculated. Lifetimes range from 1 to 250 years and are listed in Table 6 along with those for CFC-11 and CFC-12 for comparison. Table 6 also lists the ozone depletion and global warming potentials for these compounds which reflect their relative ability to destroy stratospheric ozone [75, 76] or contribute to global warming [17, 73, 77].

A generic scheme for the atmospheric oxidation of a C_2 haloalkane is given in Fig. 6. Values in parentheses are order of magnitude lifetime estimates. Reaction with OH radicals gives a halogenated alkyl radical which reacts with O_2 to give the corresponding peroxy radical (RO_2). As discussed in previous sections, peroxy radicals can react with three important trace species in the atmosphere: NO, NO_2, and HO_2 radicals.

Peroxy radicals react rapidly with NO_2 to give alkyl peroxynitrates (RO_2NO_2). By analogy to the measured rate of reaction of CF_2ClO_2 and $CF_3CH_2O_2$ radicals with NO_2 [73] the lifetime of RO_2 radicals with respect to reaction with NO_2 is approximately 10 min. Alkyl peroxynitrates are thermally unstable and decompose to regenerate RO_2 radicals and NO_2. At room temperature in one atmosphere of air the peroxynitrates derived from HCFC-22 and HFC-134a have lifetimes of 24 s and <90 s, respectively. Thermal decomposition dominates the atmospheric chemistry of halogenated alkyl peroxynitrates.

The lifetime of CX_3CXYO_2 radicals with respect to reaction with HO_2 has been estimated to be 2–8 min [73]. The reaction of peroxy radicals with HO_2 radicals gives hydroperoxides and, in some cases, carbonyl products. Product data are available for two haloperoxy radicals: CH_2FO_2 and CF_3CFHO_2. Reaction of CH_2FO_2 radicals with HO_2 gives 30% yield of the hydroperoxide, CH_2FOOH, and 70% yield of the carbonyl product, $HC(O)F$ [29]. In the reaction of CF_3CFHO_2 with HO_2 radicals less than 5% of the products appear as the carbonyl $CF_3C(O)F$ and, by inference, >95% of the reaction proceeds to give the hydroperoxide $CF_3CFHOOH$ or the alkoxy radical CF_3CFHO [78]. The factors which determine the relative importance of the hydroperoxide and carbonyl forming channels are unknown. More work is needed in this area. The hydroperoxide $CX_3CXYOOH$ is expected to be returned to the CX_3CXYO_x radical pool via reaction with OH and photolysis. The fate of the carbonyl product $CX_3C(O)X$ produced in the $CX_3CXYO_2 + HO_2$ reaction is discussed later.

The peroxy radicals derived from HFCs and HCFCs react rapidly with NO to give NO_2 and an alkoxy radical RO. The lifetime of peroxy radicals with respect to reaction with NO is approximately 3–7 min. Numerous product studies of halocarbon oxidation have shown that the atmospheric fate of the alkoxy radical, CX_3CXYO, is either decomposition or reaction with O_2 [73]. Decomposition can occur either by C-C bond fission or Cl atom elimination. Reaction with O_2 is only possible when an α-H atom is available (e.g. in CF_3CFHO). In the case of the alkoxy radicals derived from HFC-32, HFC-125, and HCFC-22, only one reaction pathway is available. Hence, CHF_2O radicals react with O_2 to give $C(O)F_2$, CF_3CF_2O radicals decompose to give CF_3 radicals and $C(O)F_2$, and CF_2ClO radicals eliminate a Cl atom to give $C(O)F_2$. The alkoxy radicals derived from HFC-143a, HCFC-123, HCFC-124, HCFC-141b and HCFC-142b all have two or more possible fates, but one loss mechanism dominates in the atmosphere. For HCFCs 123 and 124 the dominant process is elimination of a Cl atom to give $CF_3C(O)Cl$ and $CF_3C(O)F$, respectively. For HFC-143a, HCFC-141b, and HCFC-142b reaction with O_2 dominates, giving CF_3CHO, $CFCl_2CHO$, and CF_2ClCHO respectively. The case of HFC-134a is the most complex. Under atmospheric conditions, the alkoxy radical derived from HFC-134a, CF_3CFHO, decomposes (to give CF_3 radicals and $HC(O)F$) and reacts with O_2 (to give $CF_3C(O)F$ and HO_2 radicals) at comparable rates. In the atmosphere 7–20% of the CF_3CFHO radicals formed in the $CF_3CFHO_2 + NO$ reaction react with O_2 to form $CF_3C(O)F$ while the remainder decompose to give CF_3 radicals and $HC(O)F$ [79].

Before moving on to consider the fate of the carbonyl products, it is appropriate to discuss the atmospheric fate of CF_3O radicals. The usual modes of alkoxy radical loss are not possible for CF_3O radical. Reaction with O_2 and decomposition via F atom elimination are both thermodynamically impossible under atmospheric conditions. Instead, CF_3O radicals react with NO and hydrocarbons.

$$CF_3O + NO \rightarrow C(O)F_2 + FNO \tag{53}$$

$$CF_3O + CH_4 \rightarrow CF_3OH + CH_3 \tag{54}$$

Reaction with NO yields $C(O)F_2$. $C(O)F_2$ does not react with any gas phase trace atmospheric species and its photolysis is slow [80]. $C(O)F_2$ is removed from the atmosphere by incorporation into water droplets and hydrolysis to give CO_2 and HF and by photolysis in the upper stratosphere to give FCO radicals and F atoms. FNO photolyzes to give NO and a F atom [81]. F atoms reversibly form FO_2 radicals by combining with O_2, and also react with CH_4 and H_2O to give HF which will be rained out of the atmosphere. The reaction of CF_3O radicals with hydrocarbons such as CH_4 produces CF_3OH. The CF_3O-H bond is unusually strong ($120\,kcal\,mol^{-1}$). CF_3OH is not attacked by any trace atmospheric radical [82] and is not photolyzed [83, 84]. CF_3OH undergoes heterogeneous decomposition to give $C(O)F_2$ and HF and reaction with atmospheric water droplets to give CO_2 and HF [85, 86]. Several years ago there was speculation that CF_3O radicals could participate in catalytic ozone destruction cycles [87]. Experimental studies have shown that this is not the case [88–92].

10.2
Reactions of Halogenated Carbonyl Intermediates

Thus far the oxidation of the halocarbons into halogenated carbonyl products has been discussed. While the gas phase oxidation mechanisms are complex, the carbonyl products are well established and are given in Table 7. The carbonyl products represent a convenient break point in our discussion. The sequence of gas phase reactions that follow from the initial attack of OH radicals on the parent halocarbon are sufficiently rapid that heterogeneous and aqueous processes play no role. In contrast, the lifetimes of the carbonyl products [e.g.

Table 7. Gas-phase atmospheric degradation products of HFCs and HCFCs [73]

Compound	Carbon Containing Degradation Products
HFC-23 (CF_3H)	$C(O)F_2$, CF_3OH
HFC-32 (CH_2F_2)	$C(O)F_2$
HFC-125 (CF_3CF_2H)	$C(O)F_2$, CF_3OH
HFC-134a (CF_3CFH_2)	$HC(O)F$, CF_3OH, $C(O)F_2$, $CF_3C(O)F$
HFC-143a (CF_3CH_3)	$CF_3C(O)H$, CF_3OH, $C(O)F_2$, CO_2
HFC-227ea (CF_3CFHCF_3)	$CF_3C(O)F$, CF_3OH, $C(O)F_2$
HFC-236fa ($CF_3CH_2CF_3$)	$CF_3C(O)CF_3$
HCFC-22 (CHF_2Cl)	$C(O)F_2$
HCFC-123 (CF_3CCl_2H)	$CF_3C(O)Cl$, CF_3OH, $C(O)F_2$, CO
HCFC-124 (CF_3CFClH)	$CF_3C(O)F$
HCFC-141b ($CFCl_2CH_3$)	$CFCl_2CHO$, $C(O)FCl$, CO, CO_2
HCFC-142b (CF_2ClCH_3)	CF_2ClCHO, $C(O)F_2$, CO, CO_2

HC(O)F, C(O)F$_2$, CF$_3$C(O)F] are relatively long. As discussed in the following section, incorporation into water droplets followed by hydrolysis plays an important role in the removal of halogenated carbonyl compounds [93]. In the case of HC(O)F, C(O)F$_2$, FC(O)Cl, and CF$_3$C(O)F reaction with OH radicals [94] and photolysis [80] are too slow to be of any significance. These compounds are removed entirely by incorporation into water droplets.

The gas phase oxidation mechanism for CX$_3$C(O)H and CF$_3$C(O)Cl is shown in Fig. 7. For CX$_3$C(O)H species reaction with OH radicals is important [95]. The lifetimes of CF$_3$C(O)H, CF$_2$ClC(O)H, and CFCl$_2$C(O)H with respect to OH attack have been estimated to be 24, 19, and 11 days, respectively [95]. Photolysis is probably also an important sink for CF$_3$C(O)H, CF$_2$ClC(O)H, and CFCl$_2$C(O)H [95]. Finally, scavenging by water droplets also probably plays a role in the atmospheric fate of these halogenated aldehydes. For CF$_3$C(O)Cl, reaction with OH is not feasible. Photolysis of CF$_3$C(O)Cl is important [96] and competes with incorporation of CF$_3$C(O)Cl into water droplets.

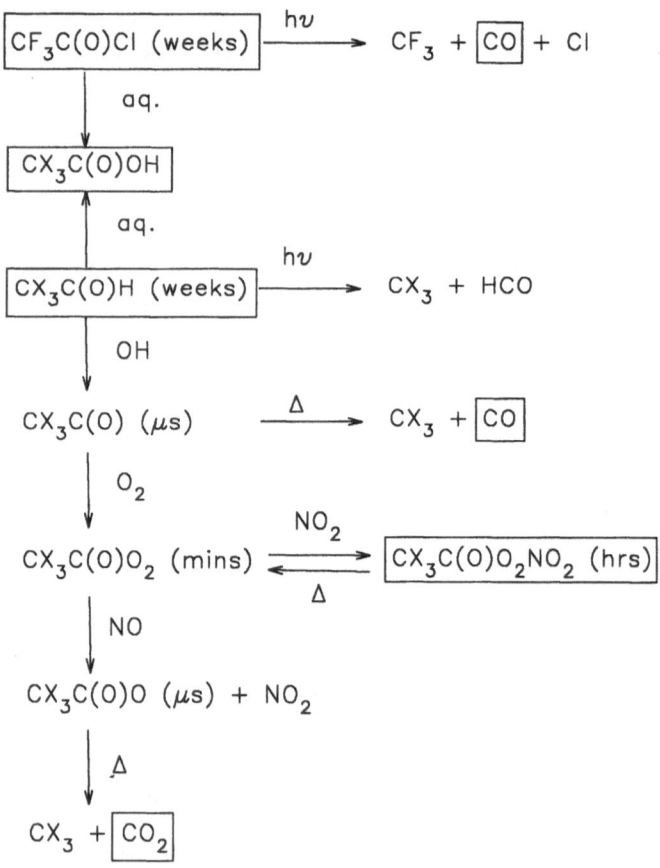

Fig. 7. Mechanism for the degradation of CX$_3$C(O)H species

As shown in Fig. 7, photolysis of $CF_3C(O)Cl$ gives CF_3, CO, and Cl. In addition, trace amounts (<1% yield) of CF_3Cl were reported. CF_3Cl is a long-lived compound that efficiently transports chlorine from the lower atmosphere to the stratosphere. However, the low yield of CF_3Cl from $CF_3C(O)Cl$ photolysis renders this pathway of negligible environmental significance. Following reaction with OH radicals, $CF_3C(O)$, $CF_2ClC(O)$, and $CFCl_2C(O)$ radicals can either react with O_2, or decompose to give CO and a halogenated methyl radical. Reaction with O_2 is essentially the sole atmospheric fate of $CF_3C(O)$ radicals [97,98] and possibly $CF_2ClC(O)$ and $CFCl_2C(O)$ radicals. The resulting $CX_3C(O)O_2$ radical can react with NO or NO_2. Reaction with NO_2 gives a halogenated acetyl peroxynitrate which undergoes thermal decomposition [97, 98] to regenerate $CX_3C(O)O_2$. Reaction with NO gives a $CX_3C(O)O$ radical which rapidly dissociates to give CX_3 radicals and CO_2 [98].

10.3
Heterogeneous and Aqueous Phase Chemistry

The final step in removal of any species from the atmosphere involves heterogeneous deposition to the earth's surface. Removal processes include wet deposition via rain-out (following uptake into tropospheric clouds) and dry deposition to the earth's surface, principally to the oceans. The rates of these processes are largely determined by the species' chemistries in aqueous solution. Heterogeneous lifetimes of the parent HFCs and HCFCs are of the order of hundreds of years because of their low aqueous solubility and reactivity.

The species listed in Table 8 are degradation products of the parent HFC and HCFC compounds that have removal rates in the gas phase (via reaction or photolysis) that are slow enough (days or longer) that heterogeneous processing might be significant. All the halogen containing species are thought to undergo aqueous interactions that are fast enough for efficient wet and dry deposition [99]. Estimates of tropospheric lifetimes for heterogeneous uptake into clouds and into the ocean are given in Table 8. For tropospheric cloud processing, the lower limit of 5 days is indicative of atmospheric transport limitations, i.e. the time taken to transport the species into the clouds. As seen from Table 8, heterogeneous removal of halocarbonyl species is rapid and is dominated by tropospheric cloud rain-out.

Table 8. Aqueous-phase atmospheric degradation products of HFCs and HCFCs [73]

Compound	Lifetime		Degradation Products
	Clouds (days)	Ocean (years)	
$C(O)F_2$	5–10	0.3–1.5	HF, CO_2
$C(O)ClF$	5–20	0.5–5.0	HF, HCl, CO_2
$CF_3C(O)F$	5–15	0.3–3.0	$CF_3C(O)OH$, HF
$CF_3C(O)Cl$	5–30	1.0–9.0	$CF_3C(O)OH$, HCl
$HC(O)F$	150–1500	80	HF, HCOOH

A substantial body of data concerning the atmospheric degradation of HFCs and HCFCs is available [73]. While some uncertainties exist, the current understanding of the atmospheric degradation of the commercially important HFCs and HCFCs is well established. HFCs have no impact on stratospheric ozone. HCFCs have small but non-negligible ozone depletion potentials. The direct global warming potentials of HFCs and HCFCs are approximately an order of magnitude less than those of the CFCs they replace. Finally, HFCs and HCFCs are sufficiently unreactive and are released in such small quantities that they do not contribute to urban smog formation [100].

11
Research Needs

There are several areas where further research is needed to better define the atmospheric degradation mechanisms and hence environmental impact of anthropogenic compounds. In general, there is a fairly complete data base concerning the kinetics of the reactions which initiate the oxidation of pollutants. Extensive data bases, structure activity relationships, and predictive techniques are available for the reaction of most anthropogenic molecules with OH and NO_3 radicals and O_3. However, our understanding of the subsequent reaction mechanisms and the identity of the oxidation products is much less well understood. Compared to kinetic studies there have been relatively few studies of the products of the atmospheric degradation reactions and there are large uncertainties associated with the atmospheric oxidation mechanism of common compounds, e.g. the aromatics. While there is a growing data base concerning peroxy radical atmospheric chemistry, significant uncertainties remain. In particular our understanding of the mechanism of reaction of peroxy with HO_2 radicals is very limited. Also there is surprisingly little available data concerning factors which influence the atmospheric fate of alkoxy radicals. The vast bulk of the available data concerning the atmospheric fate of alkoxy radicals is derived from experiments at room temperature and 700–760 Torr. Studies as a function of temperature and pressure are needed. Finally, it is becoming apparent that the formation of particulate matter in the atmosphere can have an adverse impact on human health [101]. Work is needed to better define the importance of gas-to-particle conversion in polluted urban air [102].

References

1. Francisco JS, Goldstein AN, Li Z, Zhao Y (1990) J Phys Chem 94:4791
2. Tromp TK, Ko MKW, Rodriguez JM, Sze ND (1995) Nature 376:327
3. Standard Atmosphere NOAA NASA USAF Washington DC 1976
4. Wallington TJ, Nielsen OJ (1997) Peroxy Radicals and the Atmosphere. In: Alfassi Z (ed) Peroxy Radicals. John Wiley
5. Ravishankara AR, Lovejoy ERJ (1994) Chem Soc Faraday Trans 90:2159
6. Dorn H-P, Brandenburger U, Brauers T, Ehhalt DH (1996) Geophys Res Lett 23:2537
7. Hofzumahaus A, Aschmutat U, Heßling M, Holland F, Ehhalt DH (1996) Geophys Res Lett 23:2541

8. Atkinson R (1997) J Phys Chem Ref Data 26:215
9. Crutzen PJ (1995) Ozone in the Troposphere chapter. In: Singh HB (ed) Composition Chemistry and Climate of the Atmosphere. Van Nostrand Reinhold, New York
10. Platt U, Perner D, Winer A M, Pitts JN Jr (1980) Geophys Res Lett 7:89
11. Cantrell CA, Stockwell WR, Anderson LG, Busarow KL, Perner D, Schmeltekopf A, Calvert JG, Johnston HS (1985) J Phys Chem 89:139
12. Platt U, LeBras G, Burrows JP, Moortgat G Nature (1990) 348:147
13. Wayne RP, Barnes I, Biggs P, Burrows JP, Canosa-Mas CE, Hjorth J, LeBras G, Moortgat GK, Perner D, Poulet G, Restelli G, Sidebottom H (1991) Atmos Environ 25 A:1
14. Becker E, Rahman MM, Schindler RN (1992) Ber Bunsenges Phys Chem 96:776
15. Atkinson R (1994) J Phys Chem Ref Data Monograph 2
16. Wayne RP (1991) Chemistry of Atmospheres, 2nd edn. Oxford University Press
17. Intergovernmental Panel on Climate Change (IPCC) (1995) The Science of Climate Change Cambridge. University Press, New York
18. Siegenthaler U, Joos F (1992) Tellus 44 B:186
19. Joos F, Bruno M, Fink R, Siegenthaler U, Stocker T, Le Quéré C, Sarmiento JL (1996) Tellus 48 B:397
20. Prinn RG, Weiss RF, Miller BR, Huang J, Alyea FN, Cunnold DM, Fraser PJ, Hartley DE, Simmonds PG (1995) Science 269:187
21. DeMore WB, Sander SP, Golden DM, Hampson RF, Kurylo MJ, Howard CJ, Ravishankara AR, Kolb CE, Molina MJ (1994) JPL Publication 94-26
22. Prather M, Spivakovsky CM (1990) J Geophys Res 95:18723
23. Atkinson R, Baulch DL, Cox RA, Hampson RF, Kerr JA, Troe J (1992) J Phys Chem Ref Data 21:1125
24. Atkinson R (1989) J Phys Chem Ref Data Monograph No 1
25. Graedel TE, Hawkins D, Claxton LD (1986) Atmospheric Chemical Compounds Academic Press New York
26. Madronich S, Calvert JG (1990) J Geophys Res 95:5697
27. Wallington TJ, Japar SM (1990) Chem Phys Lett 167:513
28. Wallington TJ, Japar SM (1990) Chem Phys Lett 166:495
29. Wallington TJ, Hurley MD, Schneider WF, Sehested J, Nielsen OJ (1994) Chem Phys Lett 218:34
30. Wallington TJ, Hurley MD, Schneider WF (1996) Chem Phys Lett 251:164
31. Burrows JP, Moortgat GK, Tyndall GS, Cox RA, Jenkin ME, Hayman GD, Veyret B (1989) J Phys Chem 93:2375
32. Wallington TJ, Hurley MD, Ball JC, Jenkin ME (1993) Chem Phys Lett 211:41
33. Lightfoot PD, Cox RA, Crowley JN, Destriau M, Hayman GD, Jenkin KE, Moortgat GK, Zabel F (1992) Atmos Environ 26 A:1805
34. Crutzen PJ (1979) Annu Rev Earth Planet Sci 7:443
35. Singh HB, Hanst PL (1981) J Geophys Res 8:941
36. Roberts JM (1995) Reactive odd-Nitrogen (NOy) in the Atmosphere.In: Singh HB (ed) Composition Chemistry and Climate of the Atmosphere. Nostrand Reinhold, New York
37. Altshuller AP (1993) J Air Waste Manag Assoc 43:1221
38. Atkinson R, Aschmann SM, Carter WPL, Winer AM, Pitts JN Jr (1982) J Phys Chem 86:4563
39. Pate CT, Finlayson BJ, Pitts JN Jr (1974) J Am Chem Soc 96:6554
40. Becker KH, Geiger H, Wiesen P (1991) Chem Phys Lett 184:256
41. Jeffries HE (1995) Photochemical Air Pollution. In: Singh HB (ed) Composition Chemistry and Climate of the Atmosphere. Van Nostrand Reinhold, New York
42. Chang TY, Chock DP, Hammerle RH, Japar SM, Salmeen IT (1992) Critical Rev Environ Cont 22:27
43. Carter WPL, Atkinson R (1989) Environ Sci Tech 23:864
44. Derwent RG, Jenkin ME (1991) Atmos Environ 25 A:1661
45. Simpson DJ (1995) Atmos Chem 20:163
46. Derwent RG, Jenkin ME, Saunders SM (1996) Atmos Environ 30:181

47. Niki H, Maker PD, Savage CM, Breitenbach LP (1981) Chem Phys Lett 80:499
48. Niki H, Savage CM, Breitenbach LPJ (1978) Phys Chem 82:135
49. Atkinson R, Tuazon EC, Carter WPL (1985) Int J Chem Kinet 17:725
50. Niki H, Maker PD, Savage CM, Breitenbach LP, Hurley MD (1987) J Phys Chem 91:941
51. Berndt T, Böge O (1994) Ber Bunsenges Phys Chem 98:869
52. Berndt T, Böge O (1995) J Atmos Chem 21:275
53. Wallington TJ, Dagaut P, Kurylo MJ (1992) Chem Rev 92:667
54. Atkinson R, Arey J (1994) Environ Health Perspect 102 Supplement 4117
55. Devolder P, Sawerysyn JP, Fittschein C, Goumri A, Elmaimouni L, Bigan B, Bourbon C
    (1997) In: Le Bras G (ed) Chemical Processes in Atmospheric Oxidation, LACTOZ
    report. Springer, Berlin Heidelberg New York, p 100
56. Nozière B, Lesclaux R, Hurley MD, Dearth MA, Wallington TJ (1994) J Phys Chem
    98:2864
57. Zellner R, Fritz B, Preidel M (1985) Chem Phys Lett 121:412
58. Knispel R, Koch R, Siese M, Zetzsch C (1990) Ber Bunsenges Phys Chem 94:1375
59. Crutzen PJ, Andreae MO (1990) Science 250:1669
60. Leighton P (1961) Photochemistry of Air Pollution Academic Press New York
61. Cantrell CA, Lind JA, Shetter RE, Calvert JG, Goldan PD, Kuster W, Fehsenfeld FC,
    Montzka SA, Parrish DD, Williams EJ, Buhr MP, Westberg HH, Allwine G, Martin R
    (1992) J Geophys Res 97:20671
62. Parrish DD, Trainer M, Williams EJ, Fahey DW, Hübler G, Eubank CS, Liu SC, Murphy PC,
    Albritton DL, Fehsenfeld FC (1986) J Geophys Res 91:5361
63. Ando J (1994) In: Calvert JG (ed) The Chemistry of the Atmosphere: Its Impacts on
    Global Change. Blackwell Scientific Publications
64. Eriksson E J (1963) Geophys Res 68:4001
65. Granat l, Rohde H, Halberg RO (1976) Ecol Bull SCOPE Report 7, 22:89
66. Wine PH, Thompson RJ, Ravishankara AR, Semmes DH, Gump CA, Torabi A, Nicovich
    JM (1984) J Phys Chem 88:2095
67. Martin LR, Damschen DE (1981) Atmos Environ 15:1615
68. Charlson RJ, Langner J, Rodhe H, Leovy CB, Warren SG (1991) Tellus 43 B:152
69. Gallagher MW, Downer RM Chourlarton TW, Gay MJ, Stromber I, Mills CS, Radojevic M,
    Tyler BJ, Bandy BJ, Penkett SA, Davis TJ, Dollard GJ, Jones BMR (1991) Atmos Environ
    25 A:2029
70. Lovelock JE, Maggs RJ, Wade RJ (1973) Nature 241:194
71. Molina M, Rowland FS (1974) Nature 249:810
72. World Meteorological Organization (1989) Global Ozone Research and Monitoring
    Project Report No 20, Scientific Assessment of Stratospheric Ozone Vol 1
73. Wallington TJ, Schneider WF, Worsnop DR, Nielsen OJ, Sehested J, DeBruyn WJ, Shorter
    JA (1994) Environ Sci Tech 28:320 A
74. Francisco JS, Maricq MM (1995) Atmospheric Chemistry of alternative halocarbons. In:
    Neckers DC, Volman DH, Bünau C von (eds) Advances in Photochemistry, Vol 20. John
    Wiley, New York
75. Weubbles DJ (1983) J Geophys Res 88:1433
76. Weubbles DJ, Connell PS, Patten KO (1995) Evaluating the potential effects of Halon
    replacements on the global environment. In: Miziolek AW, Tsang W (eds) Halon
    Replacements: Technology and Science Americian Chemical Society
77. Pinnock S, Shine KP, Smyth TJ, Hurley MD, Wallington TJ (1995) J Geophys Res
    100:23227
78. Maricq MM, Szente JJ, Hurley MD, Wallington TJ (1994) J Phys Chem 98:8962
79. Wallington TJ, Hurley MD, Fracheboud JM, Orlando JJ, Tyndall GS, Sehested J,
    Møgelberg TE (1996) J Phys Chem 100:18116
80. Nölle A, Heydtmann H, Meller R, Schneider W, Moortgat GK (1992) Geophys Res Lett
    19:281
81. Wallington TJ, Schneider WF, Szente JJ, Maricq MM, Sehested J, Nielsen OJ (1995) J Phys
    Chem 99:984

82. Schneider WF, Wallington TJ (1993) J Phys Chem 97:12783
83. Schneider WF, Wallington TJ, Minschwaner K, Stahlberg EA (1995) Environ Sci Tech 29:247
84. Molina LT, Molina M (1996) J Geophys Res Lett 23:563
85. Wallington TJ, Schneider WF (1994) Environ Sci Tech 28:1198
86. Lovejoy ER, Huey LG, Hanson DR (1995) J Geophys Res 100:18775
87. Biggs P, Canosa-Mas CE, Shallcross DE, Wayne RP, Kelly C, Sidebottom HW (1993) Proceedings of the STEP-HALOCSIDE/AFEAS Workshop University College Dublin Ireland, March, page 177
88. Nielsen OJ, Sehested J (1993) Chem Phys Lett 213:433
89. Wallington TJ, Hurley MD, Schneider WF (1993) Chem Phys Lett 213:442
90. Maricq MM, Szente JJ (1993) Chem Phys Lett 213:449
91. Fockenberg C, Saathoff H, Zellner R (1994) Chem Phys Lett 218:21
92. Ravishankara AR, Turnipseed AA, Jensen NR, Barone S, Mills M, Howard CJ, Solomon S (1994) Science 263:71
93. DeBruyn W, Duan SX, Shi XQ, Davidovits P, Worsnop DR, Zahniser MS, Kolb CE (1992) Geophys Res Lett 19:1939
94. Wallington TJ, Hurley MD (1993) Environ Sci Tech 27:1448
95. Scollard DJ, Treacy JJ, Sidebottom HW, Balestra-Garcia C, Laverdet G, LeBras G, MacLeod H, Téton S (1993) J Phys Chem 97:4683
96. Rattigan OV, Wild O, JonesR L, Cox AR (1993) J Photochem Photobiol A: Chemistry 73:1
97. Zabel F, Kirchner F, Becker KH (1994) Int J Chem Kinet 26:827
98. Wallington TJ, Sehested J, Nielsen OJ (1994) Chem Phys Lett 226:563
99. Wine PH, Chameides WL (1989) World Meteorological Organization Global Ozone Research and Monitoring Project Report No 20, Scientific Assessment of Stratospheric Ozone Vol 1, p 271–295
100. Hayman GD, Derwent RG (1997) Environ Sci Tech 31:327
101. Schwartz J, Dockery DW, Neas LM (1996) J Air Waste Manage Assoc 46:927
102. Odum JR, Jungkamp TPW, Griffin RJ, Flagan RC, Seinfeld JH (1997) Science 276:96
103. Intergovernmental Panel on Climate Change (IPCC) (1994) Radiative Forcing of Climate Change. Cambridge Univ Press, New York
104. Solomon S, Burkholder JB, Ravishankara AR, Garcia RR (1994) J Geophys Res 99:20929
105. Finlayson-Pitts BJ, Pitts JN Jr (1986) Atmospheric Chemistry: Fundamentals and Experimental Techniques. John Wiley, New York

4 Aquatic Photochemical Reactions in Atmospheric, Surface, and Marine Waters: Influences on Oxidant Formation and Pollutant Degradation

B. C. Faust

University of California at Los Angeles, Department of Civil and Environmental Engineering,
School of Engineering and Applied Science, Environmental Chemistry Laboratory
5732H Boelter Hall, Box 951593, Los Angeles, CA 90095-1593 USA. E-mail: bcfaust@ucla.edu

Abstract. The kinetic characterizations of direct photolysis and indirect photoreactions in natural waters are described, including the reactions of multiple species in rapid reversible equilibrium. The photochemical sources, fates, and environmental effects of various photo-reactants are discussed. It is concluded that in the oceans, hydroxyl radical ($^\bullet$OH) and the aquated electron (e_{aq}^-) will ultimately oxidize and reduce, respectively, nearly all organic pollutants that are otherwise not degraded before transport to the oceans. In atmospheric, and surface waters (fresh and saline), superoxide radical ion ($^\bullet O_2^-$) can be involved in the oxidation of reduced forms of transition metals (e.g. Cu(I), Fe(II)), and for certain transition metals (e.g. Cu(I)) can also be the dominant source of the reduced form of the transition metal. In atmospheric waters and probably also in surface (fresh and saline) waters, hydrogen peroxide (HOOH) is a significant source of hydroxyl radical, through the iron photo-Fenton's reaction. Hydrogen peroxide is the single most important oxidant for oxidizing sulfur dioxide to sulfuric acid during periods of cloudiness. Excited state triplets and organic peroxyl radicals derived from natural organic chromophores play significant roles for the oxidation of phenols in natural waters.

Keywords: Photoreactants (sources and sinks), cloud waters, surface waters (Fresh and marine)

Contents

The Handbook of Environmental Chemistry Vol. 2 Part L
Environmental Photochemistry (ed. by P. Boule)
© Springer-Verlag Berlin Heidelberg 1999

List of Symbols and Abbreviations

c	A chromophore, a molecule or functional group that absorbs sunlight.
c^+	A radical cation formed from a chromophore.
$^1(c)$	Singlet ground state of a chromophore.
$^1(c)^*$	Singlet excited state of a chromophore.
$^3(c)$	Triplet ground state of a chromophore.
$^3(c)^*$	Triplet excited state of a chromophore.
Cu(r)	A pool of reactive labile Cu(II) and Cu(I) species.
DOC	Dissolved organic compounds.
$\varepsilon_{i,\lambda}$	Base-10 molar absorptivity of the compound of interest ($M^{-1}\,cm^{-1}$).
Fe(r)	A pool of reactive labile Fe(III) and Fe(II) species.
f_i	Equilibrium molar fraction of the compound of interest present as the ith species.
f_L	Equilibrium molar fraction of the compound of interest that is present in the aqueous phase in a cloud or fog.
i	The compound (pollutant) of interest.
I_λ'	Spherically integrated solar irradiance (photons $cm^{-2}\,s^{-1}\,nm^{-1}$).
j_i	Apparent first-order rate constant (s^{-1}) for direct photolysis of the compound of interest in sunlight.
K_H	Henry's Law solubility equilibrium constant ($M\,atm^{-1}$).
$k_{x,i}$	Bimolecular rate constant for reaction of the photoreactant "x" with a the compound (pollutant) of interest "i" ($M^{-1}\,s^{-1}$):
$K_{x,Sn}$	Bimolecular rate constant for reaction of the photoreactant "x" with a natural scavenger "S_n" ($M^{-1}\,s^{-1}$).
LWC	Liquid water content of a cloud or fog, i.e. the volume fraction of a cloudy airmass that is liquid water.
$\Phi_{i,\lambda}$	Quantum efficiency for destruction of the compound of interest (molecules $photon^{-1}$).
R	Gas constant.
S_n	Natural scavenger of the photoreactant.
T	Temperature (Kelvin).
x	Photoreactant (e.g. $^\bullet OH$, $^\bullet O_2^-/HOO^\bullet$, $ROO^\bullet$, HOOH, ROOH, excited state triplets of organic chromophores, e_{aq}^-, $O_2(^1\Delta_g)$, etc.
$[x]_s$	Steady-state (photo-stationary state) concentration of the photoreactant.

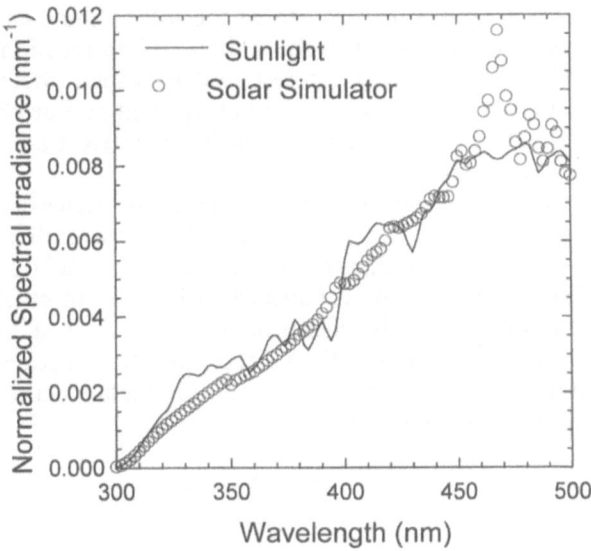

Fig. 1. Normalized spectral irradiance of authentic terrestrial sunlight, and of simulated sunlight from a solar simulator used in the author's research [1]. The area under each curve is unity. Adapted from Faust [1, 2]

1
Introduction

1.1
Spectral Composition of Terrestrial Sunlight

The energy present in sunlight can initiate photochemical reactions in aqueous atmospheric condensed phases and in surface and marine waters. Figure 1 illustrates the spectral composition of terrestrial sunlight from 300–500 nm, the principal wavelength range that has so far been found responsible for most abiotic photoreactions in sunlight [1–3]. In particular, the wavelength range of 300–400 nm corresponds to energies sufficiently large to break many types of chemical bonds: 300 nm corresponds to an energy of ≈400 kJ/mol and, and 400 nm is equivalent to ≈300 kJ/mol. Photoreactions of chemical species in sunlit atmospheric and surface waters can be rapid, because: (1) there is a large flux of photons in the solar actinic region of 300–400 mm, approximately 1 mole-of-photons m^{-2} h^{-1}; and (2) a significant percentage of inorganic and organic compounds absorb light in this spectral region.

1.2
Direct Photolysis vs Indirect Photoreaction

Photochemical reactions have been broadly classified into two categories: direct photolysis, and indirect photoreaction. During direct photolysis, the compound/pollutant of interest directly absorbs the solar photon and undergoes a "photo-lysis" (bond breakage). During indirect photoreaction, a substance other than the chemical species of interest absorbs a solar photon, and undergoes a photochemical reaction to form a photooxidant and/or photoreductant, which in turn reacts with the chemical species of interest and other species present in the natural water.

From the perspective of transforming the chemical species of interest, direct photolytic reactions should usually exhibit a higher overall efficiency than indirect photoreactions, since direct photoreactions do not "waste" energy on reactions not involving the chemical species of interest. In other words, from the perspective of destroying the chemical species of interest, indirect photoreactions are normally less efficient than direct photolyses, because usually only a small percentage of the photooxidant or photoreductant actually reacts with the chemical species of interest.

1.3
Direct Photolysis

The rate of direct photolysis of a single compound "i" in sunlight ($d[i]_T/dt$) can be characterized as follows:

$$d[i]_T/dt = - j_i [i]_T \tag{1}$$

where $[i]_T$ represents the total concentration (M) of the compound "i", and j_i is an apparent first-order photolysis rate constant (s^{-1}), that is described by:

$$j_i = \ln(10) \int_{290}^{800} (I_\lambda') (\Phi_{i,\lambda}) (\varepsilon_{i,\lambda}) \, d\lambda \tag{2}$$

where I_λ' is the spherically integrated solar irradiance (photons cm^{-2} s^{-1} nm^{-1}, or moles-of-photons cm^{-2} s^{-1} nm^{-1}), $\Phi_{i,\lambda}$ is the quantum efficiency for destruction of the parent compound "i" (molecules photon^{-1}, or moles mole-of-photons^{-1}), and $\varepsilon_{i,\lambda}$ is the base-10 molar absorptivity of the compound "i" (e.g. absorption cross section, M^{-1} cm^{-1}), and all of these quantities are dependent on the wavelength of light (λ, nm).

In some cases, the parent "compound" may actually exist as a family of species that are in a rapid reversible equilibrium with each other. Such cases include: the various forms of a weak organic/inorganic acid in different states of protonation; various metal complexes of a given metal in a given oxidation state; dissolved, surficial, and solid-phase species of a given compound; and even different species present in gas-drop equilibrium. In all of these cases, the "compound" exists as more than one species. Assuming that these species undergo chemical interchange on a time scale faster than direct photolysis, the total rate of photolysis of all forms of the compound "i" ($d[i]_t/dt$) can be expressed as [3, 4]:

$$d[i]_T/dt = - \left\{ \sum_i (j_i) (f_i) \right\} [i]_T \tag{3}$$

where $[i]_T$ now represents the total concentration of all forms (species) of "i", j_i now represents the apparent first-order photolysis rate constant for the "i^{th} species" (s^{-1}), and f_i represents the equilibrium molar fraction of the "ith" species ([concentration of the ith species]/$[i]_T$). The quantitiy $\sum_i (j_i) (f_i)$ represents an average (mean) photolysis rate constant for all species of "i", where the weighting is in direct proportion to the relative molar abundance of each species of "i" (f_i) λ. In a sense, Eq. (3) is the most general form of the direct photolysis rate expression, and includes the special case of a compound which exists only as one species ($i = 1, f_i = 1$, Eq. 1). From this analysis it is possible to define an average (mean) half-life solely for direct photolysis of all forms of "i" as:

$$\text{Half-life} = \ln(2)/\left\{\sum_i (j_i) (f_i)\right\} \tag{4}$$

It is a common mistake to judge the environmental significance of a photochemical reaction solely by the quantum efficiency/yield of the reaction. As seen from Eqs. 2 and 4, the direct-photolytic half-life of a given compound is a function of three factors: (1) the speciation of the compound (f_i), (2) the rate of sunlight absorption by the chemical (which is proportional to the quantity $I_\lambda \varepsilon_{i,\lambda}$), and (3) the transformation (loss) quantum efficiency ($\Phi_{i,\lambda}$).

1.4
Indirect Photoreactions

As noted above, during indirect photoreaction, a chromophore other than the chemical species of interest absorbs a solar photon, and undergoes a photochemical reaction to form a photooxidant or photoreductant, which in turn reacts with the chemical species of interest and other species present in the natural water. Thus, indirect photoreaction is somewhat similar, in a conceptual sense, to cometabolism in the context of biodegradation; in both cases the chemical species of interest is incidentally destroyed as a consequence of photochemical/metabolic reaction cycles involving other substances. A simplified representation of indirect photoreactions is as follows:

$$\text{chromophore} + \text{light} \rightarrow \rightarrow x \tag{5}$$

$$x + S_n \rightarrow \text{products} \tag{6}$$

$$x + i \rightarrow \text{products} \tag{7}$$

where "x" represents the photoreactant (a photooxidant or photoreductant), S_n represents the dominant natural scavenger(s) of the photoreactant, and "i" represents the compound of interest. It is important to note that the formation of the photoreactant often involves one or more thermal reactions (involving e.g. O_2, H_2O, H^+, other photoreactants, etc.) (reaction 5). For simplicity these thermal reactions are not shown here, but this does not diminish the utility or applicability of the kinetic model. Most of the photoreactant molecules react with natural scavengers (reaction 6), but a small percentage of the photoreactants react with the compound of interest "i" (reaction 7) (excluding the special case where the compound of interest "i" is a dominant natural scavenger).

For this type of model, the steady-state approximation is commonly employed to solve for the steady-state concentration of "x": $d[x]/dt \ll$ rate of formation of "x", rate of consumption of "x" (note $d[x]/dt \neq 0$). Utilizing the steady-state approximation and the assumption that the photoreactant "x" is primarily scavenged by natural scavengers (and not normally by the compound of interest), the following expression can be derived for the steady-state (photo-stationary state) concentration of "x", $[x]_s$ [5]:

$$[x]_s = \left(\frac{(\text{Photoformation rate of "x"})}{\sum_n (k_{x,Sn}[S_n]} \right) \tag{8}$$

where $[S_n]$ is the concentration of the nth natural scavenger (M), $k_{x,Sn}$ represents the bimolecular rate constant for reaction of photoreactant "x" with a given natural scavenger S_n ($M^{-1}s^{-1}$), and the summation sign ($\sum$) indicates that there may be more than one important natural scavenger of "x". In nearly all cases almost all of the photoreactant will be primarily scavenged by only a few (often only 1 or 2) natural scavengers.

The rate of destruction of the compound of interest "i" by the photoreactant "x" is given by:

$$d[i]_T/dt = - k_{x,i}[x]_s[i]_T \tag{9}$$

where $k_{x,i}$ is the bimolecular rate constant for reaction of "x" with "i" ($M^{-1}s^{-1}$), and $[x]_s$ is the steady-state concentration of "x" (M). From this expression it can be seen that the half-life (s) for destruction of the compound of interest "i" by reaction with "x" is given by:

$$\text{Half-life} = \ln(2)/(k_{x,i}[x]_s) \tag{10}$$

For the case where the compound of interest "i" exists as a family of rapidly reversible species, the rate of reaction of the photoreactant "x" with all forms of the compound of interest can be expressed as:

$$d[i]_T/dt = -\sum_i \{k_{x,i} f_i\} [x]_s [i]_T \tag{11}$$

In this case the average (mean) half-life solely for photoreaction of all forms of the compound of interest is:

$$\textit{Half-life} = \ln(2)/(\sum_i \{k_{x,i} f_i\} [x]_s) \tag{12}$$

As seen from Eq. 12, the half-life for indirect photoreaction of the compound of interest is related to three factors: (1) the speciation of the compound of interest (f_i), (2) the steady-state concentration of the photoreactant x ($[x]_s$), which is itself a function of the rate of sunlight absorption by the precursor chromophore(s), and the concentration(s) of natural scavenger(s) of x, and (3) reactivity of the compound with the photoreactant ($k_{x,i}$).

1.5
The Photochemical Zone

The rate of elementary photoreactions are proportional to the solar irradiance (Eq. 2). As for direct photolysis, it is seen that the speciation of the compound affects its overall indirect photoreaction rate, and hence is affected by the attenuation of sunlight. Photochemical reactions involving wavelengths of light in the 300–400 nm range occur throughout the entire 100-kilometer depth of the atmosphere. However, the attenuation of sunlight is drastically greater in surface and marine waters than it is in the atmosphere.

The interaction (absorption, scattering) of sunlight with compounds present in a natural water, gives rise to a depth-dependent profile in the spectral composition and irradiance of sunlight, which, in turn, causes depth-dependent differences in the chemical composition, reactions, and specification of substances in the water column. The "photochemical zone" for a given photochemical reaction in a surface or marine water can be defined as the maximum depth for the light level corresponding to 0.1% of the incident (to the surface of the water) light for the wavelength range of interest. For absorption coefficients of 0.1–0.5 m^{-1}, which are typical values for surface and marine waters at 400 nm, this corresponds to a photochemical zone depth of 6–30 meters. Since the attenuation of sunlight by natural waters increases from 400 to 300 nm, the photochemical zone for photoreactions of 300-nm light will be considerably shallower than that for 400-nm light. In any case, it can be seen that abiotic photoreactions in surface and marine waters will occur in the top meters of these water bodies, the "photochemical zone."

The "photochemical zone" for abiotic photochemical reactions in natural waters is directly analogous to: (1) the "photic zone" for photosynthetic reactions in natural waters, and (2) the thin "oxic layer" in natural aquatic sediments (in lakes, rivers, and oceans). It is a well-defined spatial region of the water column within which highly dynamic reactions occur – in this case sunlight mediated abiotic photochemical reactions. These abiotic photochemical reactions are often oxidation-reduction (redox) reactions, the so-called "photoredox" reactions. However, other types of abiotic photochemical reactions are also possible (e.g. photohydrolysis, etc.).

1.6
Effects of Particulate Matter and Cloud Drops

Particulate matter exhibits two main effects on sunlight-mediated reactions in surface waters. One, particles alter the irradiance (I_λ') through absorption and scattering (a physical effect). And two, particles can effect the speciation (f_i) of the pollutant (a chemical effect). Both of these factors influence the overall reactivity. Depending on the environment and the specific reactions, the physical and chemical effects can either reinforce or oppose each other [6–10].

Most of the particle-induced attention of irradiance is usually due to absorption by dissolved substances, although light scattering by particulates in water bodies with high suspended solids concentrations can affect (decrease,

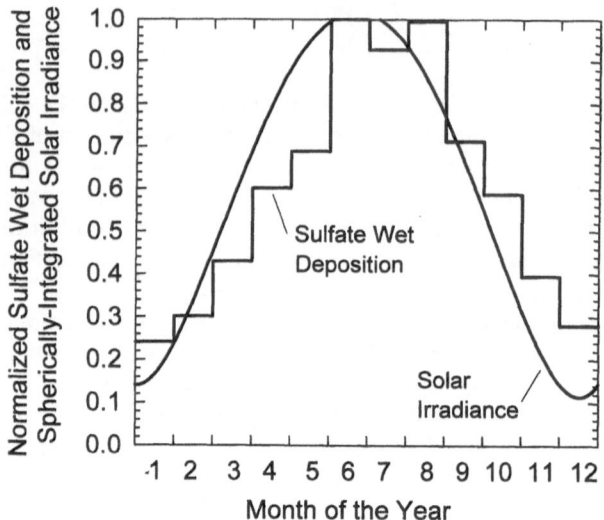

Fig. 2. Normalized monthly regional wet deposition (1976–1979) of sulfate over the northeastern United States (38–44° N, sites in New York, Pennsylvania, Virginia) [11], and normalized spherically-integrated solar irradiance (40° N, 310–315 nm) [12]. Other variables can not explain this seasonal trend: (1) there is no discernable trend in sulfate wet deposition and precipitation amount [11], (2) sulfur emissions vary by < 25 % from January to July [13], and (3) the height of the mixing layer increases by approximately 2-fold from January to July (i.e. giving slower deposition rates for a given deposition velocity) [14]. Adapted from Faust [2, 4]

and apparently even increase) photolysis rates [6]. The specific (photo)chemical effects of natural particles on photoreaction rates have not been systematically elucidated.

1.7
Photoreactants as Limiting Reagents

Figure 2 demonstrates that the trends in regional wet deposition of sulfate and solar ultraviolet irradiance in the northeastern Unites States exhibit similar seasonal trends [11–14]. Since all photo-oxidants responsible for oxidizing SO_2 to sulfuric acid are ultimately derived from a photochemical reaction, the similarity in the trends shown in Fig. 2 suggests that the availability of one or more photooxidants is limiting the wet deposition of sulfuric acid in the northeastern United States (as noted in the caption to Fig. 2, other variables can not explain this trend). So the interplay of the chemical timescales for oxidation of SO_2 to sulfuric acid with the typical physical timescales for transport and deposition affects the amount of sulfuric acid deposited on the northeastern United States and eastern Canada, and photochemical reactions that produce photooxidants appear to be controlling the chemical (SO_2 oxidation) timescales.

    The effect of cloud drops on the solar irradiance is complicated to treat theoretically. Nevertheless, theoretical calculations suggest that irradiance values inside clouds can sometimes exceed clear-sky values due to internal reflections within a cloud drop and to multiple scattering between cloud drops [15, 16]. Measurements of ultraviolet irradiance values in clear and cloudy air are needed to quantify this effect.

2
Sources, Sinks Daytime Steady-State Concentrations and Environmental Significance of Photoreactants in Atmospheric, Surface (Fresh), and Marine Waters

There are numerous environmental aquatic oxidants and reductants that are formed from photochemical or thermal redox processes – far too many to discuss in this review. Nevertheless, it is useful to provide some indication of the likely reactivity of many oxidants, while focussing only on a few.

One-electron reduction potentials of oxidants and radicals are reasonable predictors of their reactivities in electron transfer reactions, although they are not reliable predictors for reactivity in H-atom abstraction reactions. Table 1 lists the one-electron reduction potentials of various environmental oxidants and free radicals [17–31]. Although Table 1 contains information on many more oxidants than are discussed here, such a list should be useful for future research and reference, and, hence is included here.

Table 1. Standard one-electron reduction potentials ($E_h°$) of some environmental oxidants and free radicals in aqueous solution (25 °C)

Oxidant or Free Radical	Reaction (in Water)[a]	$E_h°$ (volts vs NHE)[b]
$^•OH$[c]	$^•OH + e^- \leftrightarrow OH^-$	1.77–1.91
	$^•OH + e^- + H^+ \leftrightarrow H_2O$	2.59–2.85
$^•O^-$	$^•O^- + e^- + H_2O \leftrightarrow 2OH^-$	1.64–1.87
O_2	$O_2 + e^- \leftrightarrow {}^•O_2^-$	−0.16
$^•O_2^-$	$^•O_2^- + e^- + H_2O \leftrightarrow HOO^- + OH^-$	0.25
$HOO^•$[c]	$HOO^• + e^- + H^+ \leftrightarrow HOOH$	1.48–1.50
O_3	$O_3 + e^- \leftrightarrow O_3^-$	1.01
NO_2	$NO_2 + e^- \leftrightarrow NO_2^-$	1.03
$NO_3^•$	$NO_3^• + e^- \leftrightarrow NO_3^-$	2.3–2.7
$SO_3^{•-}$	$SO_3^{•-} + e^- \leftrightarrow SO_3^{2-}$	0.63
$SO_4^{•-}$	$SO_4^{•-} + e^- \leftrightarrow SO_2^{2-}$	2.5–3.1
$C(O)O^{•-}$	$C(O)O^{•-} + e^- + H^+ \leftrightarrow HC(O)O^-$	1.49
$HOCl$	$HOCl + e^- \leftrightarrow Cl^- + {}^•OH$	−0.04–(+0.26)
$HOBr$	$HOBr + e^- \leftrightarrow Br^- + {}^•OH$	−0.36–(−0.045)
Cl_2	$Cl_2 + e^- \leftrightarrow Cl_2^{•-}$	0.70
Br_2	$Br_2 + e^- \leftrightarrow Br_2^{•-}$	0.58
I_2	$I_2 + e^- \leftrightarrow I_2^{•-}$	0.2
$Cl_2^{•-}$	$Cl_2^{•-} + e^- \leftrightarrow 2Cl^-$	2.09

Table 1 (Continued)

Oxidant or Free Radical	Reaction (in Water)[a]	$E_h°$ (volts vs NHE)[b]
$Br_2^{•-}$	$Br_2^{•-} + e^- \leftrightarrow 2Br^-$	1.50–1.62
$I_2^{•-}$	$I_2^{•-} + e^- \leftrightarrow 2I^-$	1.06–1.13
$ClO^•$	$ClO^• + e^- \leftrightarrow ClO^-$	1.5–1.8
$BrO^•$	$BrO^• + e^- \leftrightarrow BrO^-$	–0.24
ClO_2	$ClO_2 + e^- \leftrightarrow ClO_2^-$	0.934
BrO_2	$BrO_2 + e^- \leftrightarrow BrO_2^-$	1.2
Fe^{3+}	$Fe^{3+} + e^- \leftrightarrow Fe^{2+}$	0.77
Cu^{2+}	$Cu^{2+} + e^- \leftrightarrow Cu^+$	0.16
Phenoxyl radical ($PhO^•$)	$4\text{-}PhO^• + e^- \leftrightarrow 4\text{-}PhO^-$	0.79
4-X-phenoxyl radical ($4\text{-X-}PhO^•$)[d]	$4\text{-X-}PhO^• + e^- \leftrightarrow 4\text{-X-}PhO^-$	0.174–0.68
4-Y-phenoxyl radical ($4\text{-Y-}PhO^•$)[e]	$4\text{-Y-}PhO^• + e^- \leftrightarrow 4\text{-Y-}PhO^-$	0.76–1.22
$RO^•$[g]	$RO^• + e^- \leftrightarrow RO^-$	1.23[f]
	$RO^• + e^- + H^+ \leftrightarrow ROH$	2.18[f]
$RC(O)O^•$[g]	$RC(O)O^• + e^- \leftrightarrow RC(O)O^-$	1.96[f]
	$RC(O)O^• + e^- + H^+ \leftrightarrow RC(O)OH$	2.24[f]
$ROO^•$[g]	$ROO^• + e^- \leftrightarrow ROO^-$	0.77[f]
	$ROO^• + e^- + H^+ \leftrightarrow ROOH$	1.47[f]
$RC(O)OO^•$[g]	$RC(O)OO^• + e^- \leftrightarrow RC(O)OO^-$	1.12[f]
	$RC(O)OO^• + e^- + H^+ \leftrightarrow RC(O)OOH$	1.60[f]

$e^-(aq) + H_2O(l) \leftrightarrow H^• + OH^-$ $K = 3.9 \times 10^{-5}$ M [31]

[a] All species except e^- are fully hydrated. One must distinguish between $e^-(aq)$, the hydrated electron, and e^- which designates the normal (standard) hydrogen electrode (NHE).
[b] NHE = normal hydrogen electrode. Data are from [17–31].
[c] pK_a values are: 11.9 ($^•OH$), 4.8 ($HOO^•$)
[d] X = OH, NH$_2$, CH$_3$, OCH$_3$, N(CH$_3$)$_2$
[e] Y = F, Cl, Br, I, C(O)O$^-$, C(O)CH$_3$, CN, NO$_2$
[f] estimated value.
[g] R = alkyl group [30].

2.1
Hydroxyl Radical ($^•OH$)

Hydroxyl radical is one of the strongest known aqueous oxidants. It has a comparatively large standard one-electron reduction potential (Table 1). Moreover, unlike $^•OH$, none of the other species listed in Table 1 with reduction potentials higher than $^•OH$ readily undergoes H-atom abstraction reactions. Hydroxyl radical reacts with nearly all organic compounds via two main mechanisms: (1) H-atom abstraction, and (2) addition to a double bond.

Hydrogen-atom abstraction can be represented as:

$$^•OH + R\text{-}H \rightarrow H_2O + R^• \tag{13}$$

where $R^•$ would most often represent a carbon-centered radical, whose main fate would be to react with aqueous O_2 at near diffusion-controlled rates [32] to propagate a free-radical chain oxidation process of the organic compound. Addition of $^\bullet$OH to a double bond is another oxidation pathway, and can be represented by:

$$^\bullet OH + {>}C{=}C{<} \rightarrow {>}C(OH){-}C(^\bullet){<} \tag{14}$$

where, again, the carbon-centered radical will react with O_2 to propagate the free-radical oxidation of the organic compound.

The sources, sinks (fates), and steady-state concentrations of $^\bullet$OH in natural waters are listed in Table 2 [33–40]. Hydroxyl radical reacts at near diffusion-controlled rates with most organic compounds [41]. Hence, nearly all organic compounds (even biologically refractory chemicals) will ultimately be oxidized by $^\bullet$OH in surface waters and the ocean. For a compound with a rate constant of $1.0 \times 10^9 M^{-1}s^{-1}$ (a factor of ~10 below the diffusion-controlled limit) for reaction with $^\bullet$OH, and using the steady-state $^\bullet$OH concentrations reported in Table 2, the half-lives of the compound from $^\bullet$OH-mediated oxidation in the different environmental compartments are:

(1) *Inside* cloud drops: ~1–12 days. This is an important sink if a significant fraction of the compound's mass is present in the cloud drop. Ignoring solid-water partitioning within the cloud drop, the equilibrium fraction of a compound's mass that is present in the aqueous phase in a cloud (f_L) is given by:

$$f_L = \{1 + [K_H(LWC)RT]^{-1}\}^{-1} \tag{15}$$

where K_H is the Henry's Law solubility constant ($A(g) \leftrightarrow A(aq), M\,atm^{-1}$), and LWC is the cloud liquid water content (typically $2 \times 10^{-7} - \times 10^{-6}$ liter/liter). Only compounds that exhibit higher Henry's Law constants or that otherwise partition significantly into cloud drops will be significantly affected by aqueous-phase oxidation by $^\bullet$OH. Of course, compounds reacting at the diffusion-controlled rate (approximately 10 times faster) would exhibit even shorter half-lives than the values given above (~0.1 times the values given). The half-life of $^\bullet$OH in cloud waters has been measured as 2–11 µs, and is controlled by reactions with dissolved organic compounds, HSO_3^-, and Cl^- [40]. The half-life of $^\bullet$OH in cloud waters is too short to allow significant drop-to-gas transfer (which occurs on a millisecond time scale).

(2) The surface layer of fresh waters: 0.8–800 years. Depth-averaged half-lives will be even longer. Thus, except for the most extreme conditions (e.g. low pH with high iron concentrations), in fresh waters $^\bullet$OH-mediated oxidation of compounds is likely to be a negligible process (unless a unique product is formed by this oxidation), except for water bodies with multi-year water residence times (e.g. larger lakes). Of course, half-lives are shorter (by 0.1-fold) for compounds that react with $^\bullet$OH at the diffusion-controlled limit.

(3) The surface layer of the ocean and marine waters: 80–8000 years. Again, depth-averaged half-lives will be even longer. However, the water residence time in the sunlit portion of the ocean is large. So, $^\bullet$OH-mediated oxidation

Table 2. Significant known sources and sinks, and typical daytime steady-state concentrations of hydroxyl radical ($^\bullet$OH) in atmospheric and surface (fresh and marine) waters

Natural Water Type	Significant $^\bullet$OH Sources	Significant $^\bullet$OH Sinks (Fates)	Typical Daytime (Surface) Steady-State Aqueous Concentrations
Continental Cloud[a]	$^\bullet$OH(g) $\leftrightarrow$ $^\bullet$OH(aq) Fe(II) + HOOH $\rightarrow$ Fe(III) + $^\bullet$OH(aq) + OH$^-$ O_3(aq) + $^\bullet O_2^-$ (+ H$^+$) $\rightarrow \rightarrow$ $^\bullet$(OH(aq) + 2O_2 Fe(OH)$^{2+}$ + hν $\rightarrow$ Fe^{2+} + $^\bullet$OH(aq) NO$_3^-$ + hν (+ H$^+$) $\rightarrow$ NO$_2$ + $^\bullet$OH(aq)	DOC[b] HSO$_3^-$/HSO$_2^-$ Cl$^-$/Br$^-$[c]	$1 \times 10^{-14} - 1 \times 10^{-12}$ M
Fresh Surface Waters	NO$_3^-$ + hν (+ H$^+$) $\rightarrow$ NO$_2$ + $^\bullet$OH NO$_2^-$ + hν (+ H$^+$) $\rightarrow$ NO + $^\bullet$OH Fe(II) + HOOH $\rightarrow$ Fe(III)+ $^\bullet$OH + OH$^-$	DOC[b] HCO$_3^-$/CO$_3^{2-}$	$1 \times 10^{-18} - 1 \times 10^{-15}$ M
Marine Waters	NO$_3^-$ + hν (+ H$^+$) $\rightarrow$ NO$_2$ + $^\bullet$OH NO$_2^-$ + hν (+ H$^+$) $\rightarrow$ NO + $^\bullet$OH DOC + hν $\rightarrow \rightarrow$ $^\bullet$OH? Fe(II) + HOOH $\rightarrow$ Fe(III) + $^\bullet$OH + OH$^-$	Br$^-$/Cl$^-$ DOC[b] HCO$_3^-$/CO$_3^{2-}$	$1 \times 10^{-19} - 1 \times 10^{-17}$ M

[a] The liquid water content of a tropospheric cloud is typically $(0.2–1.0) \times 10^{-6}$ vol/vol.
[b] DOC $\equiv$ dissolved organic compounds.
[c] Only in marine coastal atmospheres and/or atmospheres impacted by incinerator emissions.
Information is from [33–40].

of compounds will represent an eventual fate for nearly all unreactive compounds that are formed, washed into, or deposited (from the atmosphere) into the ocean.

As noted above, the overall half-life of a compound due solely to indirect photoreaction is a function of its speciation (Eq. 12). Thus particle-water speciation (partitioning) of a compound might be expected to affect the overall reactivity of a given compound. It appears that dissolved species can account for the reactivity of a compound with $^{\bullet}OH$ that is formed in the aqueous phase, and that particulate-bound species contribute little to the overall oxidation of a compound by $^{\bullet}OH$ that is formed in the aqueous phase [10]. However, $^{\bullet}OH$ radical is also formed photochemically at solid-water interfaces, and in such cases, interfacial-formed $^{\bullet}OH$ is known to oxidize surface-bound compounds [42].

2.2
Superoxide Radical Ion (Hydroperoxyl Radical) ($^{\bullet}O_2^-$/HOO$^{\bullet}$) and Organic Peroxyl Radicals (ROO$^{\bullet}$)

From Table 1 it is seen that superoxide radical anion ($^{\bullet}O_2^-$) is a comparatively weak oxidant, and, in fact, can often act as a reductant (vida infra). By comparison, the hydroperoxyl radical HOO$^{\bullet}$ (pK$_a$ = 4.8) [43] as well as organic peroxyl radicals (EOO$^{\bullet}$) are normally stronger oxidants than $^{\bullet}O_2^-$, based on their higher one-electron reduction potentials (Table 1). However, both HOO$^{\bullet}$ and ROO$^{\bullet}$ are less reactive and more selective oxidants than $^{\bullet}OH$. Table 3 presents information on the sources, fates, and steady-state concentrations of $^{\bullet}O_2^-$/HOO$^{\bullet}$ in natural waters [34, 36, 37, 44–46].

The primary importance of $^{\bullet}O_2^-$/HOO$^{\bullet}$ is as a source of HOOH, which is a precursor to $^{\bullet}OH$:

$$^{\bullet}O_2^-/HOO^{\bullet} + {}^{\bullet}O_2^-/HOO^{\bullet} \rightarrow HOOH + O_2 \tag{16}$$

where, more frequently, this dismutation reaction is catalyzed by redox-active transition metals, such as Cu(II)/Cu(I) and in lower pH systems (e.g. cloud waters) Fe(III)/Fe(II) [3, 43, 47, 48]:

$$^{\bullet}O_2^-/HOO^{\bullet} + Cu(II)/Fe(III) \rightarrow O_2 (+ H^+) + Cu(I)/Fe(II) \tag{17}$$

$$^{\bullet}O_2^-/HOO^{\bullet} + Cu(I)/Fe(II)(+ H^+) \rightarrow HOOH + Cu(II)/Fe(III) \tag{18}$$

In principal, reactions 17 and 18 represent potential sources and sinks (respectively) of Cu(I) and Fe(II) [3, 47, 48], in competition with direct photolysis (ligand-to-metal electron transfer reactions) of Cu(II) and Fe(III) complexes [3, 49]. Based on the direct photoreactivity of Cu(II)/amino-acid complexes, half-lives of Cu(II) due to direct photoreduction are estimated to be 1–100 days [49]. Based on the reactivity of $^{\bullet}O_2^-$ with Cu(II)/arginine complexes [48], and on the steady-state concentrations of $^{\bullet}O_2^-$ reported in seawater (Table 3), half-lives of Cu(II) due to reduction by $^{\bullet}O_2^-$ are estimated as

Table 3. Significant known sources and sinks, and typical daytime steady-state concentrations of hydroperoxyl radical (HOO$^{\bullet}$) and superoxide radical anion ($^{\bullet}O_2^-$) and in cloud, fresh, and marine waters

Natural Water Type	Significant HOO$^{\bullet}$/$^{\bullet}O_2^-$ Sources	Significant HOO$^{\bullet}$(aq)/$^{\bullet}O_2^-$ Sinks (Fates)	Typical Daytime Steady-State Aqueous Concentrations (Surface)
Continental Cloud[a]	HOO$^{\bullet}$(g) $\leftrightarrow$ HOO$^{\bullet}$(aq) RC(O)R' + PhOH + hv (+ O$_2$) $\rightarrow$ $\rightarrow$ $^{\bullet}O_2^-$ + products Fe(dicarboxylate)$_n^{(3-2n)+}$ + hv $\rightarrow$ $\rightarrow$ $^{\bullet}O_2^-$ + Fe(II) + products	HOO$^{\bullet}$(aq)/$^{\bullet}O_2^-$ Cu(II)/Cu(I), Fe(III)/Fe(II) DOC[b]	$1 \times 10^{-9} - 4 \times 10^{-8}$ M
Fresh and Marine Surface Waters	DOC + hv (+ O$_2$) $\rightarrow$ $\rightarrow$ $^{\bullet}O_2^-$ + oxidized DOC	HOO$^{\bullet}$(aq)/$^{\bullet}O_2^-$ Cu(II)/Cu(I), DOC	$10^{-9} - 10^{-8}$ M?

[a] The liquid water content of a tropospheric cloud is typically $(0.2 - 1.0) \times 10^{-6}$ vol/vol.
[b] DOC ≡ dissolved organic compounds.
Information is from [33, 36, 37, 44–46].

0.04–0.4 min. So indirect photoreduction could dominate over direct photo-reduction of Cu(II), unless the speciation of Cu(II) in natural waters is predominantly by species that are not reactive with $^{\bullet}O_2^-$ but that photolyze in sunlight. In this regard it is interesting to note that Cu complexes of phytochelatin [50] and HS^- [51, 52] are thought to exist in natural waters.

In contrast, Fe(III) species absorb sunlight at a much greater rate than do Cu(II) complexes [3] and the reaction of $^{\bullet}O_2^-$ with Fe(III)/organic-complexes is much slower than with inorganic Fe(III) [43]. So in natural waters where Fe(III) complexation is dominated by organic complexes [53–55], it was concluded that direct photoreduction of Fe(III) species may be a more significant source of Fe(II) than its indirect photoreduction by $^{\bullet}O_2^-$ [48]. In atmospheric waters direct photoreduction of Fe(III) to Fe(II) is also likely to dominate, with indirect photoreduction by $^{\bullet}O_2^-$ as a significant additional source [3, 56].

Oxidation of Cu(I)/Fe(II) to Cu(II)/Fe(III), respectively, is likely to be dominated by reactions with $^{\bullet}O_2^-/HOO^{\bullet}$ [3, 48, 56], HOOH, and O_2 (vide infra). Thus, a rapid photochemical cycle is established, in which Fe(III) and Cu(II) are rapidly photoreduced (by direct and indirect photoreactions), on time scales of seconds to minutes, and the photoformed Fe(II) and Cu(I) are rapidly reoxidized by oxidants (primarily $^{\bullet}O_2^-/HOO^{\bullet}$ and HOOH) [2, 3], also on time scales of seconds to minutes [3, 40, 48]. Hence it is useful to define a pool of reactive labile metal species (e.g. [Fe(r)], r = II, III; [Cu(r)], r = I, II; and so forth for other redox-active metals) that participate in the rapid photo-redox reactions that give rise to the steady-state concentrations of labile metals (vide infra).

Certain organic peroxyl radicals (ROO$^{\bullet}$), such as the acetylperoxyl radical (CH$_3$C(O)OO$^{\bullet}$), can be strong oxidants. In fact, the acetyl radical has been quantified in seawater [57] and the acetyl and the acetylperoxyl radicals are formed during the direct photolysis of biacetyl [CH$_3$C(O)-C(O)CH$_3$] [58]. Different pieces of information indicate that the acetylperoxyl radical is a highly oxidizing species [58]: (1) its one-electron reduction potential is similar to that of the highly-oxidizing CCl$_3$OO$^{\bullet}$ radical, (2) it reacts more rapidly than HOO$^{\bullet}$ with $^{\bullet}O_2^-$, (3) it reacts more rapidly than any other studied ROO$^{\bullet}$ radicals (including CCl$_3$OO$^{\bullet}$) with N,N,N',N'-tetramethyl-p-phenylenediamine, and (4) it is likely to be responsible for the oxidation of added formate in biacetyl photoreactions. Organic peroxyl radicals are capable of oxidizing easily oxidized compounds such as phenols [58], but as will be seen later, triplet excited states of aromatic carbonyl compounds probably play a larger role in the oxidation of phenols.

2.3
Hydrogen Peroxide (HOOH) and Organic Hydroperoxides (ROOH)

Hydrogen peroxide itself is not very reactive with organic compounds. However, it reacts with Fe(II) and is a significant source of $^{\bullet}OH$ in natural waters (Table 4), and especially in cloud waters [40, 60–61] (Table 4). Moreover, it undergoes an acid-catalyzed reaction with bisulfite (HSO$_3^-$) that represents

Table 4. Significant known sources and sinks, and typical daytime steady-state concentrations of hydrogen peroxide (HOOH) in cloud, fresh, and marine waters

Natural Water Type	Significant HOOH Sources	Significant HOOH Sinks (Fates)	Typical Daytime Steady-State Aqueous Concentrations (Surface)
Continental Cloud[a]	$HOOH(g) \leftrightarrow HOOH(aq)$ $RC(O)R' + PhOH + h\nu\,(+\,O_2) \rightarrow \rightarrow HOOH + products$ $Fe(dicarboxylate)_n^{(3-2n)+} + h\nu \rightarrow \rightarrow HOOH + Fe(II) + products$	$HOO^{\bullet}(aq)/{}^{\bullet}O_2^{-}$ Cu(II)/Cu(I), Fe(III)/Fe(II)	$1\times10^{-6} - 1\times10^{-4}$ M
Fresh and Marine Surface Waters	$DOC^{b} + h\nu\,(+\,O_2) \rightarrow \rightarrow HOOH + oxidized\ DOC$	$HOO^{\bullet}(aq)/{}^{\bullet}O_2^{-}$ Cu(II)/Cu(I), DOC	$10^{-8} - 10^{-6}$ M ?

[a] The liquid water content of a tropospheric cloud is typically $(0.2-1.0) \times 10^{-6}$ vol/vol.
[b] DOC $\equiv$ dissolved organic compounds.
Information is from: References cited in Table 1 of [60], and references cited in Table 1 of [61].

the single most important mechanism for the atmospheric oxidation of sulfur dioxide to sulfuric acid in periods of cloudiness [62]:

$$SO_2(aq) \leftrightarrow H^+ + HSO_3^-$$ (19)

$$HSO_3^- + HOOH \leftrightarrow HOOS(O)O^-$$ (20)

$$HOOS(O)O^- + H^+ \leftrightarrow H_2SO_4$$ (21)

The sources of HOOH to surface waters are considered to be photochemical reactions of natural dissolved organic matter [61, 63–66]. Rates of HOOH photoformation in fresh water lakes were found to be inversely correlated with the total iron concentration, but to increase with increasing dissolved organic carbon concentration and fluorescence, indicating that dissolved organic chromophores are the precursors of HOOH [66].

Sinks for HOOH in fresh and marine waters include: (1) destruction by other photoreactants, such as Fe(II) [39], and (2) destruction by microorganisms [65]. Different water chemistries and active cell populations of the natural waters could give rise to different mechanisms of HOOH decay.

In cloud waters, HOOH photoproduction rates are weakly correlated with dissolved (< 0.5 µm) iron concentrations, but are more strongly correlated with dissolved (< 0.5µm) organic carbon concentrations [60]. Two principle mechanisms of HOOH photoformation have been postulated: (1) photolysis of aromatic aldehydes and ketones in the presence and absence of phenolic compounds [67], and (2) photolysis of Fe(III)-oxalate complexes [49]. Both mechanisms undoubtedly contribute to HOOH photoformation in atmospheric waters. Wavelength studies of authentic cloud waters indicated that photolysis of Fe(III)-oxalate complexes was not the predominant source of photoformed HOOH in the samples, but experiments did not verify if oxalate (possibly a labile species) was present or had decayed in the samples by the time of the photochemical experiments [68].

All of these studies were carried out with bulk cloud waters, and the extent to which the chemical composition of individual cloud drops varies is not known. Since most (but not all) iron is present in aerosol of aerodynamic diameter larger than 2 µm [69], and since most dicarboxylates are present in aerosol of aerodynamic diameter less than 2 µm [70], it is not clear how much oxalate is present together with iron in an individual cloud drop. The average life time of a cloud drop is 20–30 min, yet macroscopic clouds persist for hours to days (even weeks in special cases), so this probably helps to promote interaction of oxalate and iron within a given drop. Conversely, since aromatic carbonyls and phenols are derived largely from the same sources (e.g. wind-blown terrestrial organic matter, wood/lignin combustion) it is likely that they will both be present in the same aerosol particle and hence cloud drop. Much more work is needed on the characterization of redox and photochemically active components of size-classified cloud, fog, and aerosols drops.

2.4
Excited State Triplets of Organic Chromophores

Few studies have been carried out on this type of photoreactant as it relates to natural waters. One study reported steady-state concentrations of 10^{-15} to 10^{-13} M in natural waters [71]. Studies have found that triplet excited states of aromatic ketones (benzophenone, 2-acetophenone, 3'-methoxy-aceto-phenone) oxidize phenolic compounds, by electron transfer and H-atom abstraction at pH 8 [72] for surface waters, and at pH 3.7 for cloud waters forming HOOH [73]. However, in particular for the oxidation of methoxyphenols, it was suggested that a longer-lived photooxidant (perhaps an organic peroxyl radical) could be responsible for the phenol oxidation at lower phenol concentrations [72].

For the acidic conditions of clouds, fogs, and aerosols, an additional factor in triple-mediated oxidations must be considered. It is noteworthy that the triplet excited states of aromatic carbonyls have drastically different pK_a values than the corresponding ground state triplet, with shifts of up to 8 log units. Thus, even though pK_a values of ground state triplets of aromatic aldehydes and ketones are $\ll 0$, the pK_a values of their excited triplet states range up to ≈ 4 [67]. This is significant because the protonated form of the triplet exhibits a different net reactivity towards phenol (as affected by reactivity with phenol, and by the life time of the triplet), often higher than the deprotonated form of the triplet [67]. The difference in the net reactivities of the protonated and deprotonated triplet states gives rise to a highly pH-dependent quantum yield of HOOH photoformation and phenol oxidation [67].

2.5
Hydrated Electron (e_{aq}^-)

The main source of hydrated electron (e_{aq}^-) in natural waters is considered to be photoionization of dissolved organic chromophores (c) [74, 75, and references cited therein]:

$$c + light \rightarrow (c^+ \bullet\bullet\bullet e_{aq}^-)_{cage} \rightarrow c^+ + e_{aq}^- \qquad (22)$$

where c and c^+ represent a chromophore and its radical cation, respectively. Nearly all investigations of e_{aq}^- have utilized laser flash photolysis to characterize its quantum yields. In one comparative study of the same dissolved organic matter (Suwanee River), it was found that quantum yields of "the electron" were nearly 100-fold higher for laser flash photolysis studies than for steady-state illuminations using an e_{aq}^- scavenger [74]. This indicates that the spectroscopic techniques used in laser flash photolyses may have responded to partially separated electrons that were still associated with the organic matter within the solvent water cage ($c^+ \bullet\bullet\bullet e_{aq}^-)_{cage}$, or that biphotonic processes active in high-irradiance laser light greatly increased the yield of e_{aq}^- [74, 75]. For all discussions here, only results of the steady-state illuminations are considered.

The main sink of e_{aq}^- is considered to be reaction with O_2:

$$e_{aq}^- + O_2 \rightarrow {}^\bullet O_2^-$$ (23)

and secondarily with nitrate. Thus, the half-life of e_{aq}^- is $\approx 0.1\,\mu s$ in natural waters. There is disagreement as to whether reaction 23 is the major source of ${}^\bullet O_2^-$ in natural waters.

Measurements of the steady-state concentrations of the hydrated electron (e_{aq}^-) have not been made in natural waters. But it is possible to estimate steady-state concentrations for the surface of a lake in summer, using the measured steady-state photoformation rates together with a knowledge of the O_2 concentration and reactivity with e_{aq}^-. By this approach the steady-state e_{aq}^- concentration in surface waters was estimated to be of the order $1 \times 10^{-17}\,M$ [74]. Even for a pollutant with a diffusion-controlled rate constant of $\approx 1 \times 10^{10}\,M^{-1}\,s^{-1}$, this gives a pollutant half-life for indirect photoreaction with of e_{aq}^- of ≈ 80 years.

This half-life is long compared to the residence of water in rivers and nearly all lakes. Hence, reactions of the hydrated electron, like those of the hydroxyl radical, do not have significant impact on pollutant degradation in smaller water bodies and rivers. However, in larger lakes, and certainly in the oceans, reactions of e_{aq}^- (like those of ${}^\bullet OH$) represent an eventual sink for many compounds that are otherwise not degraded in the environment. In particular reactions of e_{aq}^- with fully halogenated organic compounds (e.g. ozone-depleting chlorofluorocarbons) represent a potentially more benign pathway for their degradation in the environment. It is noteworthy that taken together, e_{aq}^- and ${}^\bullet OH$ represent eventual reductive and oxidative sinks, respectively, for pollutants in natural waters.

2.6
Singlet Molecular Oxygen [$O_2(^1\Delta_g)$]

Singlet molecular oxygen is formed from energy transfer reactions of excited triplet states of natural organic chromophores (c):

$$^1(c) + light \rightarrow {}^1(c)^* \xrightarrow{ISC} {}^3(c)^* \xrightarrow{O_2} {}^1(c) + O_2\,(^1\Delta_g)$$ (24)

where superscripts 1 and 3 represent singlet and triplet states of the chromophore "c", * denotes an excited state, and ISC represents intersystem crossing. The principal fate of $O_2(^1\Delta_g)$ in natural waters is physical quenching by water solvent itself, giving heat and ground state O_2. The half-life of $O_2(^1\Delta_g)$ in natural waters is $\approx 3.0\,\mu s$.

Steady-state concentrations of $O_2(^1\Delta_g)$ measured in surface waters range from $(0.4 - 30) \times 10^{-14}\,M$ [76]. Singlet molecular oxygen is a selective oxidant, reacting rapidly with deprotonated phenolic compounds [77], but it has been independently concluded that $O_2(^1\Delta_g)$ is not the principal pathway for oxidizing phenolic compounds in natural waters [33, 59, 72].

3
Conclusion

In a general sense, on average, direct photolyses are more efficient at destroying compounds in natural waters, because energy is not wasted on creating a secondary photoreactant (a portion of which only reacts with the pollutant of intererst). But indirect photoreactions, when considered collectively (involving all the photoreactants) exhibit the widest range of effects on chemical substances in the aquatic environment, because the indirect photoreactants themselves exhibit a very wide range in chemical behavior (e.g. from the aquated electron to the hydroxyl radical).

References

1. Faust BC (1993) Rev Sci Instrum 64:577
2. Faust BC (1994) Environ Sci Technol 28:217A
3. Faust BC (1994) A review of the photochemical redox reactions of iron (III) species in atmospheric, oceanic, and surface waters.: influences on geochemical cycles and oxidant formation. In: Helz GR, Zepp RG, Crosby DG (eds) Aquatic and Surface Photochemistry. Lewis, Ann Arbor, chap 1, p 3
4. Faust BC (1996) Environ Sci and Technol 30:1919
5. Hoigné J, Faust BC, Haag WR, Scully FE Jr, Zepp RG (1989) Aquatic humic substances as sources and sinks of photochemically produced transient reactants. In: Suffet IH, MacCarthy P (eds) Aquatic humic substances. Am Chem Soc, Washington, Advances in Chemistry Series 219, p 363
6. Miller GC, Zepp RG (1979) Water Res 13:453
7. Oliver BO, Cosgrove EG, Carey JH (1979) Environ Sci Technol 13:1075
8. Winterle JS, Tse D, Mabey WR (1987) Environ Toxicol Chem 6:663
9. Mudambi AR, Hassett JP (1988) Chemosphere 17:1133
10. Sedlak DL, Andren AW (1994) Water Res 28:1207
11. The MAP3S/RAINES Research Community (1982) Atmos Env 16:1603
12. Peterson JT (1976) Calculated actinic fluxes (290–700 nm) for air pollution photo-chemistry applications. Report EPA-600/4-76-025. U.S. Environmental Protections Agency, Washington, DC, 1976
13. Husar RB, Wilson WE (1993) Environ Sci Technol 27:12
14. Samson PJ, Small MJ (1984) Atmospheric trajectory models for diagnosing the sources of acid precipitation. In: Schnoor JL (ed) Modeling of total acid precipitation impacts. Butterworth Publ, Boston, Acid Precipitation Series, vol. 9
15. Madronich S (1987) J Geophys Res 92:9740 and present volume
16. Bott A, Zdunkowski W (1987) J Opt Soc. Am A 4:1361
17. Steenken S, O'Neill P, Schulte-Frohlinde D (1977) J Phys Chem 81:26
18. Milazzo G, Caroli S, Sharma VK (1978) Tables of Standard Electrode Potentials. Wiley. New York
19. Henglein A (1980) Radiat Phys Chem 15:151
20. Eberson L (1982) Adv Phys Org Chem 18:79
21. Schwarz HA, Dodson RW (1984) J Phys Chem 88:3643
22. Koppenol WH, Liebman JF (1984) J Phys Chem 88:99
23. Koppenol WH, Butler J (1985) Adv Free Radical Biol & Medicine 1:91
24. Kläning UK, Sehested K, Holcman J (1985) J Phys Chem 89:760
25. Alfassi ZB, Huie RE, Neta P (1986) J Phys Chem 90:4156
26. Alfassi ZB, Huie RE, Mosseri S, Neta P (1988) Radiat Phys Chem 32:85
27. Surdhar PS, Mezyk SP, Armstron DA (1989) J Phys Chem 93:3360

28. Lind J, Shen X, Eriksen TE, Merényi G (1990) J Am Chem Soc 112:479
29. Jiang PY, Katsumura Y, Ishigure K, Yoshida Y (1992) Inorg Chem 31:5135
30. Merényi G, Lind J, Engman L (1994) J Chem Soc Perkin Trans 2:2551
31. Stanbury DM (1989) Adv Inorg Chem 33:69
32. Ross AB, Neta P (1982) Rate constants for reactions of aliphatic carbon-centered radicals in aqueous solution, NSRDS-NBS-70. U.S. National Bureau auf Standards
33. Mill T, Hendry DG, Richardson H (1980) Science 207:886
34. Chameides WL, Davis DD (1982) J Geophys Res 87:4863
35. Haag WR, Hoigné J (1985) Chemosphere 14:1659
36. Jacob DJ (1986) J Geophys Res 91:9807
37. Jacob DJ, Gottlieb EW, Prather MJ (1989) J Geophys Res 94:12,975
38. Mopper K, Zhou X (1990) Science 250:661
39. Moffett JW, Zafiriou OC (1993) J Geophys Res 98:2307
40. Arakaki T, Faust BC (1998) J Geophys Res 103:3487
41. Buxton GV, Greenstock CL, Helman CL, Ross AB (1988) J Phys Chem Ref Data 17:513
42. Sun L, Schindler KM, Hoy AR, Bolton JR (1994) Spin-trap EPR studies of intermediates involved in photodegradation reactions on TiO_2: is the process heterogeneous or homogeneous? In: Helz GR, Zepp RG, Crosby DG (eds) Aquatic and Surface Photochemistry. Lewis, Ann Arbor, chap 27, p 409
43. Bielski BHJ, Cabelli DE, Arudi RL, Ross AB (1985) J Phys Chem Ref Data 14:1041
44. Faust BC, Allen JM (1992) J Geophys Res 97:12,913
45. Micinski E, Ball LA, Zafiriou O (1993) J Geophys Res 98:2299
46. Zuo Y, Hoigné J (1992) Environ Sci Technol 26:1014
47. Sedlak DL, Hoigné J (1994) Environ Sci Technol 28:1898
48. Voelker BM, Sedlak DL (1995) Marine Chem 50:93
49. Hayase K, Zepp RG (1991) Environ Sci Technol 25:1273
50. Kawaguchi S, Maita Y (1990) Bull Environ Contam Toxicol 45:893
51. Luther III GW, Rickard DT, Theberge S, Olroyd A (1996a) Environ Sci Technol 30:671
52. Luther III GW, Rickard DT, Theberge S, Olroyd A (1996b) Environ Sci Technol 30:3640
53. Van den Berg CMG (1995) Marine Chem 50:139
54. Rue EL, Bruland KW (1995) Marine Chem 50:117
55. Wu J, Luther III GW (1995) Marine Chem 50:159
56. Arakaki T (1996) PhD Dissertation, Duke University
57. Blough N, Zepp RG (1995) Reactive oxygen species in natural waters. In: Foote CS, Valentine JS (eds) Active oxygen in chemistry, Chapman and Hall, chap 8, p 280
58. Faust BC, Powell K, Rao CJ, Anastasio C (1997) Atmos Environ 31:497
59. Faust BC, Hoigné J (1987) Environ Sci Technol 21:957
60. Anastasio C, Faust BC, Allen JM (1994) J Geophys Res 99:8231
61. Cooper WJ, Zika RG, Petasne RG, Plane JMC (1988) Environ Sci Technol 22:1156
62. Gunz DW, Hoffmann MR (1990) Atmos Environ A 24:1601
63. Cooper WJ, Zika RG (1983) Science 220:711
64. Cooper WJ, Lean D (1992) Encyclopedia of Earth Systems Science 2:527
65. Cooper WJ, Shao C, Lean DRS, Gordon AS, Scully FE Jr (1994) Factors affecting the distribution of HOOH in surface waters. In: Baker LA (ed) Environmental chemistry of lakes and reservoirs. American Chemical Society, Washington, Advances in Chemistry Series 237, p 391
66. Scully NM, McQueen DJ, Lean (1996) Limnol Oceanog 41:540
67. Anastasio C, Faust BC, Rao CJ (1997) Environ Sci Technol 31:218
68. Arakaki AT, Anastasio C, Shu PG, Faust BC (1995) Atmos Environ 14:1697
69. Davidson CI, Osborn JF (1986) The sizes of airborne trace metal containing particles. In: Nriagu JO, Davidson CI (eds) Toxic metals in the atmosphere, Wiley, Chap 12, p 355

70. Ludwig J, Klemm O (1988) Tellus 40 B: 340
71. Zepp RG, Schlotzhauer P, Sink MR (1985) Environ Sci Technol 19:74
72. Canonica S, Jans U, Stemmler K, Hoigné J (1995) Environ Sci Technol 29:1822
73. Anastasio C (1994) PhD dissertation, Duke University
74. Zepp RG, Braun AM, Hoigné J, Leenheer JA (1987) Environ Sci Technol 21:485
75. Kumamoto Y, Wang J, Fujiwara K (1994) Bull Chem Soc Jpn 67:720
76. Haag W, Hoigné J (1986) Environ Sci Technol 20:341
77. Tratnyek PG, Hoigné J (1991 Environ Sci Technol 25:1596

5 Singlet Oxygen in the Environment

Richard A. Larson and Karen A. Marley

Department of Natural Resources and Environmental Sciences, University of Illinois at Urbana-Champaign, 1101 W. Peabody Drive, Urbana, IL 61801, USA. *E-mail: ralarson@uiuc.edu*

Singlet oxygen ($^1O_2$) is produced in the environment almost exclusively by photochemical pathways. Although it can be detected in many environmental compartments, its potential roles in chemical reactions under environmental conditions are not completely understood. For example, in the atmosphere, it is formed at lower elevations (< 40 km) by photolysis of ozone and by energy transfer to ground-state oxygen from photoexcited donor species. In the ionosphere, $^1O_2$ participates in emission processes that contribute to the airglow, but in the stratosphere and troposphere its importance appears to be negligible. In natural surface waters, it is formed with about 1% efficiency from dissolved humic materials, but its deactivation by water molecules is so rapid that very few chemical reactions are able to compete. Its steady-state concentrations in most natural waters appear to center around 10^{-13} M. In water bodies polluted by light-absorbing chemicals such as dyestuffs, it is possible that it might be somewhat more important. In hydrophobic environments such as aquatic surface layers, especially those containing aromatic petroleum derivatives, $^1O_2$ might play a quite significant role due to its increased concentration and lifetime. Some solid phases, such as plant, soil, and mineral surfaces, could be sites where $^1O_2$ is formed, but little is known about its importance; and similar statements could be made about $^1O_2$ in living cells, despite many decades of investigation of the mechanisms of "photodynamic toxicity."

Keywords: Sensitized photooxidation, Humic materials, Airglow, Surface layers, Photodynamic toxicity.

Contents

The Handbook of Environmental Chemistry Vol. 2 Part L
Environmental Photochemistry (ed. by P. Boule)
© Springer-Verlag Berlin Heidelberg 1999

1
Introduction

The formation, properties, and reactions of singlet oxygen ($^1O_2$) are covered in detail elsewhere in this series. Singlet oxygen is an energetically excited state of ground-state molecular oxygen. The ground-state form is a triplet ($^3O_2$) in quantum-mechanical terms, having unpaired valence electrons, but it is only marginally more stable than two singlet (spin-paired) states, namely the $^1\Delta_g$ species (22 kcal/mol above the ground state) and the $^1\Sigma_g^+$ (37 kcal/mol).

The usual method by which $^1O_2$ is produced is photochemical; light absorbed by an intermediate compound, or photosensitizer, converts it to an excited triplet state, which in favorable instances is capable of efficient energy transfer to ground-state oxygen, producing the singlet form. Either of the excited singlet states may be produced by the energy transfer event, but there is little evidence that the $^1\Sigma_g^+$ form persists long enough in solution for it to take part in any important processes other than its virtually instantaneous conversion to the $^1\Delta_g$ form. In the remainder of this chapter we will generally employ the shortened term, $^1O_2$, to refer specifically to the $^1\Delta_g$ species, which to the best of our knowledge is the only environmentally important form, except for the ionosphere where the $^1\Sigma_g^+$ state persists long enough to emit some photons.

Singlet oxygen has attracted much attention over the past 35 years because of its unusual and heightened chemical reactivity toward many compounds of biochemical interest However, its importance (or lack of importance) in environmental chemistry has been subject to much debate, and many questions about its role remain unanswered. Photochemical reactions in the environment may proceed by any of several mechanistic pathways, some of which are referred to as "direct" (proceeding via the absorption of light by a substance). Others, however, may be "sensitized" or "indirect," involving energy or electron transfer reactions to a substance other than that which has absorbed the light. The latter class of reactions includes those in which $^1O_2$ is involved.

2
Singlet Oxygen in the Atmosphere

2.1
Tropospheric Singlet Oxygen

Direct production of singlet molecular oxygen by absorption of energy by the triplet (ground-state) form is of little or no importance in the lower atmosphere. Only the very weak absorptions ("atmospheric oxygen bands") in the red (760 nm) and infrared (1070 and 1270 nm) regions permit the forbidden transition to occur, and they are so insignificant that formation of $^1O_2$ is negligible. Ozone, however, which is present as a minor constituent of even unpolluted tropospheric air, and does absorb some near-surface solar UV, gives rise to some $^1O_2$ (together with atomic O) at wavelengths < 320 nm [1]:

$$O_3 + h\nu \rightarrow {}^1O_2 + O$$

At wavelengths from 290–310 nm, the quantum yield for the reaction is quite high (~ 0.9), but it drops sharply at longer wavelengths [2].

The principal sources of tropospheric singlet oxygen, however, are energy transfers from other electronically excited molecules to ground-state oxygen. For example, excited NO_2 is capable of bringing about this conversion:

$$*NO_2 + {}^3O_2 \rightarrow NO_2 + {}^1O_2$$

The quantum efficiency for this conversion is wavelength-dependent, but can be as high as 0.07 [3].

The relatively small number of singlet oxygen molecules in the troposphere (probably $<10^8$ molecules/cm^3: [4]) normally do not induce chemical transformations involving other molecules. The vast majority of them are deactivated to the ground state by collision. Therefore, early suggestions that the species might have importance in photochemical air pollution appear invalid. Even for the most reactive olefins, only about 1/1000 would be available for reaction with $^1O_2$ in the presence of normal concentrations of $HO \cdot$ and ozone [1].

It is conceivable that $^1O_2$ might be formed at the surface of aerosols or other atmospheric particles, particularly those containing adsorbed sensitizers such as polycyclic aromatic hydrocarbons (PAHs; [5]). In an environment of this nature, photooxidation reactions could conceivably occur at significant rates, particular if the sensitizer itself was reactive with $^1O_2$. (Examples of this reaction are discussed in Sect. 5.1). This sort of process could have public health consequences, since some oxidized PAHs are known to be potent mutagens.

There is also a possibility that $^1O_2$ could be formed in tropospheric water droplets [6]. This suggestion is discussed more fully in section 3 of this chapter.

2.2
Singlet Oxygen in the Upper Atmosphere

Since ozone occurs at much higher concentrations above the troposphere, and its photolysis is a highly important reaction in the stratosphere, one would expect a significant flux of $^1O_2$. In addition, because of the much reduced density of molecules above the troposphere, $^1O_2$ has a greatly increased lifetime at these elevations. In the stratosphere the collisional lifetime is on the order of 1–4 min, and in the ionosphere the population may persist with an average collisional duration of nearly an hour. These factors combine to afford peak concentrations of $^1O_2$ in excess of 10^{10} molecules cm^{-3} at elevations of 50–60 km. This is roughly two orders of magnitude higher than typical tropospheric levels and represents a concentration in the parts-per-million range.

Nevertheless, few chemical reactions of much importance have been identified in these regions. In the D (lower) region of the ionosphere, roughly 70 km above the earth's surface, $^1O_2$ chiefly undergoes reactions involving ionic species. It contributes in part to the formation of positive ions by undergoing photoionization,

$$^1O_2 + h\nu \rightarrow O_2^+ + e^-$$

and also provides free electrons by detaching them from and deactivating superoxide [2]:

$$^1O_2 + O_2^- \rightarrow 2O_2 + e^-$$

The decay of $^1O_2$ to ground-state $^3O_2$, a moderately important process at >70 km, contributes to an effect that has been called the "airglow" (sometimes broken down further into nightglow and dayglow). This phenomenon, which includes faint illuminance surrounding the earth, is too feeble to observe from the surface of the planet, but its visible portion can be seen by astronauts. It is due to photon-emitting transitions of vibrationally or electronically excited species including helium, hydrogen, and oxygen atoms and molecules. The forbidden transition

$$O_2(^1\Delta_g) \rightarrow O_2(^3\Sigma_g^-)$$

occurs with emission of photons at 1270 and 1580 nm (in the infra-red region). The latter emission is readily observable even by surface instruments, since it occurs in an isolated spectroscopic region where little absorption by atmospheric constituents occurs. The higher-lying excited state of $^1O_2(^1\Sigma_g^+)$ similarly emits photons at 760 nm as it decays; this wavelength is just at the visible\ infrared boundary. The emission can be detected using rocket-borne instruments.

3
Singlet Oxygen in Natural Waters

More than 20 years ago, it was first demonstrated that singlet oxygen could be produced in natural surface waters [7]. The method used was a chemical trapping experiment, in which fresh waters from the southeastern United States were exposed to sunlight. The trap was 2,5-dimethylfuran (1), which had previously been shown to react rapidly with $^1O_2$ and to form well-characterized products. Further evidence for the intermediacy of $^1O_2$ were the kinetic effects of DABCO (2, a $^1O_2$ physical quencher) and D_2O ($^1O_2$ lifetime extender) on the rates of the observed reaction. The higher rates of $^1O_2$ formation were found in the more highly colored water samples, indicating that the natural aquatic humic material was the active sensitizing species. An estimated steady-state $^1O_2$ concentration of $10^{-12}-10^{-13}$ M was calculated for these samples.

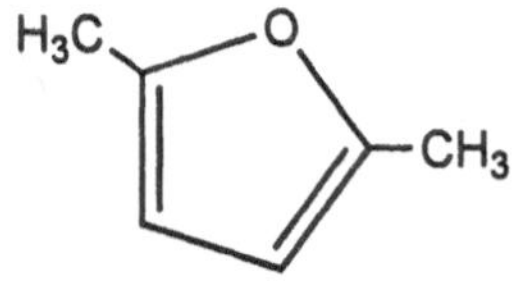

(1)

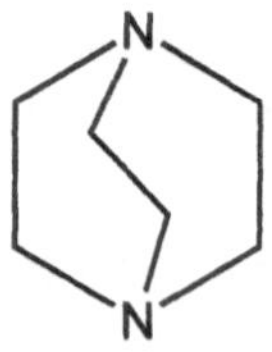

(2)

The role of humic material in the formation of $^1O_2$ in water was confirmed in a series of investigations in which humic materials of various origins were added to distilled water [8–10]. When the humic solutions were optically matched to an equivalent absorbance at 366 nm, they showed similar rates of $^1O_2$ formation and sunlight-photosensitized substrate reaction (within a factor of ca. 2).

Using similar techniques, several other research groups have confirmed that natural dissolved organic matter (DOM) can act as a sensitizer for $^1O_2$ production [11–15]. These researchers generally agree that roughly 1% of the sunlight absorbed by humic materials produces $^1O_2$. The Swiss group [13], using somewhat different kinetic techniques, found about the same steady-state concentration for $^1O_2$ as did the original discoverers; about $1-3 \times 10^{-14}$ M mg^{-1} dissolved organic C. The Dutch group, however, reported that in their experiments the disappearance of the DMF acceptor was faster by about an order of magnitude than could be accounted for by pure $^1O_2$ reactions [12]. They postulated that some of the DMF was being removed by photochemically induced free-radical reactions, and accordingly proposed a somewhat lower steady-state $^1O_2$ concentration, on the order of $10^{-13}-10^{-14}$ M. In any event, there is consensus that one should not normally expect $^1O_2$ to be present in high concentrations in a natural water body.

From several lines of evidence, it also appears unlikely that $^1O_2$ can be a very important species in photooxygenation reactions of organic compounds in natural waters. First of all, the lifetime of $^1O_2$ in water is extremely short, about 4 μsec, corresponding to a rate constant of 2.3×10^5 s^{-1} [16]. Collisions with water molecules result in reconversion of $^1O_2$ to $^3O_2$. This would mean that for a chemical reaction to compete efficiently with solvent quenching, its rate constant would have to be at least two orders of magnitude higher, and very few substrates have been identified with such high rate constants [17]. Secondly, little light penetrates into colored natural waters, which are those most likely to generate high fluxes of $^1O_2$. Therefore, such reactions would be important only in the upper few millimeters or, at most, centimeters of the water body. In addition, the interactions of $^1O_2$ with organic compounds include not only chemical reactions, but also physical quenching, in which $^1O_2$ is deactivated but no new compounds are produced. Finally, the excited states of humic materials may be involved in other types of photochemical reactions with dissolved substrates, which could be more important than $^1O_2$ processes in the disappearance of these compounds.

The measured rate constants for reactions of $^1O_2$ in water tend to be somewhat greater than those reported in the literature for organic solvent reactions [17]. However, since its steady-state concentration in natural waters is so low, its role as an active oxidizing agent in aqueous solution appears to be limited to only the most reactive acceptor molecules (such as tryptophan, histidine, phenolates, and perhaps a few sulfur compounds and reduced transition metal ions; [17–18]). For phenolic compounds, reaction rates increase with pH and the fraction of the phenol present as its anionic form [17]; the dissociated form is from one to two orders of magnitude more reactive. However, for most phenols (whose pKa's are around 10), only a minuscule fraction is ionized at environmental pHs. In a few cases, reactions of phenols with $^1O_2$ appear to be reversible, due to the formation of stable endoperoxides that thermally revert to the starting material upon standing after illumination [19].

Photochemical pathways that do not involve $^1O_2$ may also be expected to play important roles in the disappearance of many organic compounds from water, further reducing the apparent role of $^1O_2$ in their loss. For example, several phenolic compounds were shown to be photodegraded by direct electron transfer from the phenols to the excited triplet states of aquatic DOM rather than by intermediate mechanisms involving energy transfer from the DOM triplets to oxygen and $^1O_2$ formation [20].

Dissolved humic material is also well-known to occur in atmospheric water droplets; it accounts for a major fraction of the solar visible and ultraviolet light absorption by these particles. Like other humic material, it is probable that $^1O_2$ is generated by illumination of these chromophores. Other light-absorbing (and $^1O_2$-generating) substances such as soots and polycyclic aromatic hydrocarbons might also be present.

Reported experiments using several cloudwaters showed evidence for light-induced formation of both peroxyl radicals and $^1O_2$. Kinetic data using traps for peroxyl radicals and $^1O_2$ permitted an estimated steady-state $^1O_2$ concentration of $10^{-12}-10^{-13}$ M (under midsummer, midday conditions), or roughly similar to that in surface waters [6]. Therefore, it would seem likely that in this environment, quenching by water would also be much more important than chemical reactions for all except the most reactive acceptors.

It is possible that $^1O_2$ could be more important in wastewaters, especially dye wastewaters that contain sensitizing pigments. For example, an effluent from a wastewater treatment plant in northern Georgia, which treated a high volume of textile wastes, appeared visibly red and contained pigments that absorbed a significant fraction of visible light (absorbance per cm ≈ 0.2). Measurements in this wastewater indicated that the steady-state $[^1O_2]$ was approximately 10^{-12} M, or roughly one to two orders of magnitude higher than in unpolluted freshwaters. Furthermore, phenols added to the wastewater were photodegraded considerably faster than in waters containing only natural DOM [21].

A pathway for $^1O_2$ production that does not require light is the reaction of chlorine (or hypochlorite) with hydrogen peroxide [22]:

$$Cl_2 + H_2O_2 \rightarrow 2\,HCl + {}^1O_2$$

This reaction might have some importance in water treatment where aqueous chlorine could contact dissolved H_2O_2, but given the low concentration of the latter species in almost all natural waters and even wastewaters [23], it is unlikely to be of much consequence. Khan and Kasha [24] report that $^1O_2$ is formed even on mere acidification of HOCl, and suggest the possibility that deleterious health effects such as DNA lesions could ensue due to the consumption of chlorinated drinking water. Other workers, however [25], have identified many other reactions of HOCl in the digestive tract, and it remains to be seen whether the formation of $^1O_2$ in a complex environment such as the human stomach could compete significantly.

4
Singlet Oxygen in Aquatic Surface Layers and Micelles

In organic solvents, the rate of quenching of $^1O_2$ by solvent decreases greatly relative to water; in CCl_4, for example, its lifetime is 700 microseconds [26]. It is conceivable that $^1O_2$ might be of more importance in hydrophobic environments, such as surface microlayers or petroleum films, where its lifetime would be longer and ground-state oxygen and sensitizers would be present in higher concentration [27]. Numerous classes of oxidized products including hydroperoxides, phenols, and carbonyl compounds were detected when petroleum products, especially refined fuel oils, were exposed to sunlight either by themselves or in films over water, and the rate of oxidation was diminished in the presence of $^1O_2$ quenchers such as β-carotene [28]. Lichtenthaler et al. [29] calculated that $[^1O_2]$ could be as high as 6×10^{-9} M at the surface of a crude oil film, roughly five orders of magnitude higher than in water. This is due not only to the longer lifetime of $^1O_2$, but also to its efficient production; using selective quenchers, they estimated the quantum yield of formation to be about 0.6.

Other means of producing hydrophobic regions in an aquatic medium include micelle or microemulsion formation. Many photolysis reactions in micelles have long been known to proceed by different mechanisms compared to solution processes. Not only solvent differences but also surface phenomena may be important in these environments; sensitizer and acceptor molecules may occupy discrete locations within the micelle which may facilitate or inhibit particular photochemical pathways. In a typical example, pyrene incorporated into either dodecyltrimethylammonium chloride or sodium dodecyl sulfate micelles produced $^1O_2$ with a much higher quantum yield than in methanol solution, and also sensitized the decomposition of 1,3-diphenylisobenzofuran (3) with a much higher rate constant, due to the selective incorporation of both sensitizer and substrate within the micelles [30].

(3)

Sensitized reactions of nitrophenols in water-in-oil emulsions, similar to those that occur in oil spills mixed by wave action, were found to be kinetically enhanced relative to aqueous solution reactions [31]. In either cationic or anionic emulsions, the nitro compounds were photolyzed 10–100 times more efficiently relative to the rates in water.

5
Singlet Oxygen in the Solid Phase

5.1
Mineral Surfaces

Many reactive oxidants such as ozone and hydroxyl radical have been shown to form on illuminated minerals, especially oxides of zinc, titanium, and magnesium [32]. Production of $^1O_2$ on metal oxide surfaces has been demonstrated [33] by the use of specifically deuterated tetramethylethylene (TME; 4). The oxygenated products derived from this compound on illuminated Al_2O_3, MgO, and SiO_2 surfaces were formed at close to the ratio expected for a $^1O_2$ pathway having an isotopic discrimination ratio of H/D = 1.45.

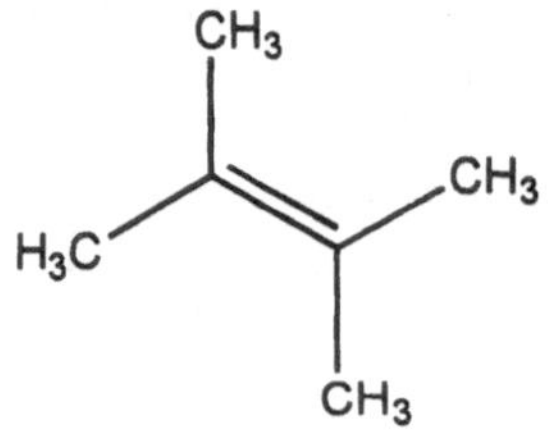

(4)

Another mechanism of formation of $^1O_2$ on surfaces is via sorbed sensitizers. In this mechanism, the solid acts only as an "inert" support and not as a source of reactive oxygen species. The products of illumination in some of these cases, however, may be different at the air-solid interface than in solution. Illustrations of the process have been given by an Oak Ridge group. Phenanthrene (5), for

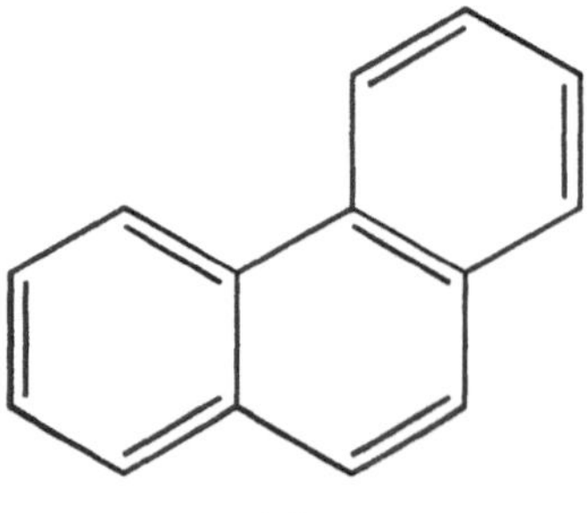

(5)

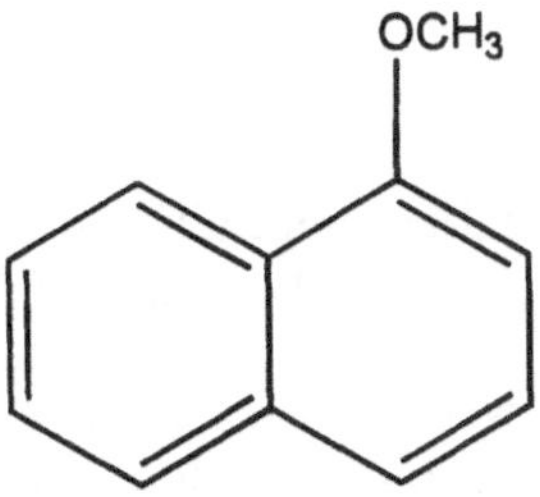

OCH₃

(6)

example, underwent photoalteration when sorbed to silica gel, affording a mixture of products characteristic of $^1O_2$. The authors proposed that the excited state of the hydrocarbon transferred energy to ground-state O_2 in its vicinity, which then reacted with it [34]. In contrast, 1-methoxynaphthalene (6) gave photoproducts typical of both $^1O_2$ and electron transfer mechanisms on silica [35].

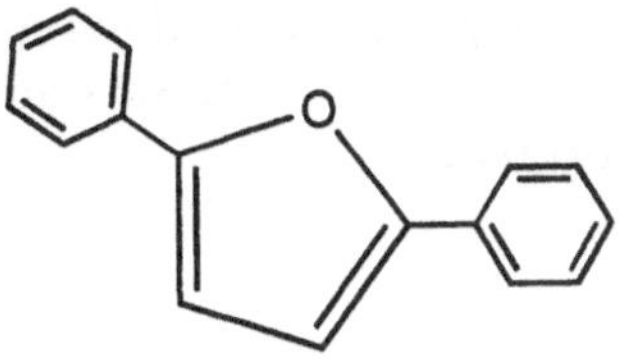

(7)

Although TiO$_2$ is a well-known semiconductor, engaging in electron-transfer and hole-transfer redox reactions when illuminated with photons more energetic than its band-gap energy, it may also produce $^1O_2$ at its surface by a process that does not require light. When 2,5-diphenylfuran (7) was deposited on the surface of TiO$_2$ and heated to ca. 60 °C, *cis*-dibenzoylethylene (8: a product characteristic of $^1O_2$ reactions) was formed. The reaction was shown to be inhibited by a nickel complex reported to be a $^1O_2$ quencher, but not by the radical quencher BHT [36].

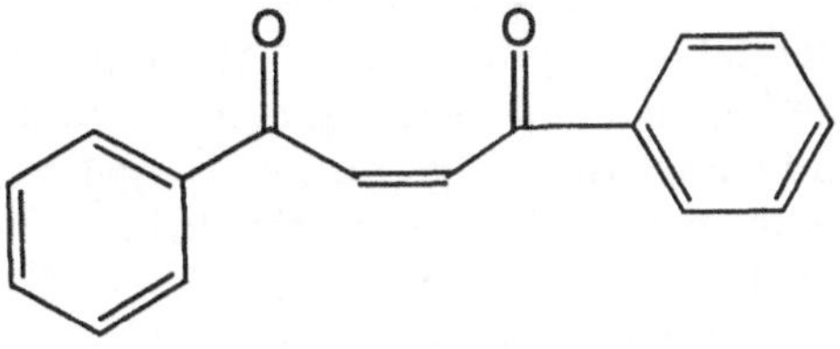

(8)

5.2
Soil Surfaces

Many soils incorporate high levels of organic matter, especially in their surface horizons. These colored and UV-absorbing materials presumably would include chromophores corresponding to those found in aquatic humic substances. Chlorophylls and porphyrins, also well-known sensitizers, are also likely to be present in the soil surface environment. Under illumination, therefore, it might be expected that these chromophores would undergo energy transfer to adjacent oxygen molecules to produce $^1O_2$.

The formation of singlet oxygen at soil surfaces was proposed by Gohre and Miller in 1983 [37]. These workers illuminated a Montana soil containing surface-sorbed TME (4) or dimethylfuran (1) with sunlight and monitored the rates and products of disappearance of the compound. Under midday sunlight intensities, the half-lives of the two compounds were surprisingly short (85 and 120 min respectively). The light and oxygen requirements for their loss strongly suggested that $^1O_2$ was an intermediate; furthermore, experiments in which the soil was heated at 50 °C gave negative results, suggesting that radical pathways of oxygenation were not important. Further evidence for the $^1O_2$ pathway was obtained by using more specific $^1O_2$ acceptors; deuterated TME (4) and 1,2-dimethylcyclohexane (9) [38]. The products formed when these compounds were sorbed to soils were typical of those formed by $^1O_2$ pathways and distinct from those characteristic of free-radical processes.

(9)

There are indications that $^1O_2$, when formed on soil surfaces, can diffuse to greater depths, even as far as 2 mm [39]. Azo and quinone dyes sorbed to soils were shown to be photodegraded in the upper 0.5 mm by an oxygen-independent, and thus presumably direct, photolytic process; however, at slightly greater depths, the kinetics and products (including demethylated amines) observed were consistent with a $^1O_2$ pathway [40].

Further evidence for the potential for migration of $^1O_2$ from the soil surface was obtained using a technique in which a sensitizing material (a dye or soil sample) is physically separated by a narrow air gap from a target molecule. It was demonstrated that the pyrethrin insecticide bioresmethrin (10) could be degraded by an oxidant (presumably $^1O_2$, but see the discussion of this issue in Sect. 6) that was released from illuminated soil [41]. The loss of bioresmethrin was slowed if β-carotene was present, consistent with a $^1O_2$-requiring pathway.

(10)

5.3
Plant Surfaces

The formation of singlet oxygen on plant material surfaces has been reported occasionally. The photodegradation of wood surfaces is complex and probably includes free-radical mechanisms, but some of the peroxyl radical intermediates could be formed by $^1O_2$ pathways, such as decomposition of sensitizer-formed peroxides [42]. The addition of tertiary amine-type quenchers to the surface of illuminated wood inhibited its weathering.

The separated-sensitizer technique was employed to show that the leaves of some phototoxic plants released $^1O_2$. Plant leaves held above a solution of furfuryl alcohol, a highly reactive acceptor, were illuminated from below and the loss rate of the alcohol was measured. It was suggested that $^1O_2$ could act as a defense against leaf-feeding insects [43]. Details of this experiment are discussed in section 6.

6
Singlet Oxygen in Cells

Certain compounds taken up by living cells become "activated" upon absorption of energy from sunlight. These light-absorbing molecules may either dissipate the absorbed energy without chemical alteration, or take part in reactions such as electron transfer and H atom abstractions, or energy transfer.

The early history of phototoxicity focused on the action of so-called photodynamic dyes, for example rose bengal, methylene blue and acridines. When irradiated with visible light, these dyes were active in killing bacteria. In addition, it was determined that for maximal effect, oxygen was also required [44]. Experiments with rose bengal by Kautsky [45] at the same time that Mulliken [46] was describing the molecular orbital calculations of the low-lying energy states of oxygen, led to conjectures that the active molecule of the dye-sensitized action was singlet oxygen, although alternative mechanisms were also proposed.

In the 1960s work was described giving parallel results between dye-sensitized oxidations and chemiluminescent peroxide decompositions, thus leading to the conclusion that the same intermediate was involved in both systems, singlet oxygen [47, 48]. (For background information on this concept, see the book by Schaap [49]).

Again, since phototoxicity is often enhanced by the presence of oxygen, its importance has received fundamental attention. Oxygen is an efficient quencher of fluorescence and the excited triplet state of a wide variety of molecules (polycyclic aromatic hydrocarbons, porphyrins, chlorophylls, etc.) and in the latter process $^1O_2$ is formed. Researchers at the Radiation Chemistry Data Center of the University of Notre Dame have compiled excellent reviews of the methods and quantum yields of singlet oxygen formation from at least 900 compounds [50] and rate constants for reactions with over 5100 substrates [51]. With the detection of $^1O_2$ in so many simple systems such as solution reactions, the assumption has been that it must be a primary or key intermediate in phototoxic events inside the cell. However, the measured yields of $^1O_2$ have not been correlated with biological effects. In response to this, arguments are made regarding the site of $^1O_2$ generation coupled with its partitioning into the complexity of the biological microenvironment [52].

Many investigators have also rationalized the role of oxygen by alternative proposals, such as 1) electron transfer to oxygen to form superoxide, H_2O_2 and subsequent •OH or 2) near diffusion-controlled reactions of O_2 with carbon-centered radicals, formed after electron transfer or H atom abstraction to form peroxyl radicals.

A few examples will illustrate the complexities of establishing the role of $^1O_2$ in the mechanisms of phototoxicity.

Case 1. Rose bengal is a sensitizing dye that absorbs visible light and has become the primary benchmark of singlet-oxygen producing systems. The ϕ (quantum yield) of the triplet excited state has been quoted as one, although this has been readjusted to be "high, but not unity"; and the ϕ of $^1O_2 = 0.75$ in water [53]. In the presence of oxygen, its reaction with alkenes to give the expected "ene" reaction as well as 1,4-cycloaddition and 1,2-cycloaddition reaction products has been offered as good evidence of $^1O_2$ formation [54].

When rose bengal is coated on an inert surface, separated by a small distance from the substrate, and irradiated, thus avoiding direct electron/energy transfer reactions, the conclusion has been drawn that only an air-diffusible species could be responsible for any observed effects [45; 55–56]. Since ground state oxygen is not reactive enough, singlet oxygen was inferred. Consequently, rose bengal has been used to demonstrate "pure singlet oxygen cytotoxicity", in such systems as DNA strand breakage [57–58], enzyme inactivation [59–60] and lipid oxidation [61]. However, it has now been well established that at least some •O_2^- is formed upon irradiation of the compound [62–63], although there are still questions regarding the exact mechanism and quantum yield [64–65]. Under conditions in which oxygen was excluded, it was shown that in the presence of electron donors, such as ascorbate and NADH, that the radical anion $^-$RB• was formed [66]. Using β-amino alcohols Epling and Jackson [67] re-evaluated experiments in which rose bengal was purported to be cleaving these compounds by a $^1O_2$ mechanism, finding however that electron transfer was the key step in the light-induced reactions. These authors then proposed that electron transfer from the reduced species of the exciplex could then transfer an electron to oxygen.

Following these observations, Dahl [68] then allowed that volatile species, such as perhaps H_2O_2 may be involved in the separated-sensitizer method, although Camoirano et al. [69] tried to distinguish between "dry" and aqueous rose bengal, suggesting that when it was bound in the dry state that only $^1O_2$ was formed. At least the authors did note "Rose Bengal bleached during drying." As many others have observed, when dyes are bleached there is no more photoactivity since its light absorption goes to zero. Soumillion [70] has reviewed the electron transfer reactions of light-induced reactions of rose bengal and other dyes.

Case 2. Porphyrins are molecules based on the tetrapyrrole system. These compounds have been extensively investigated for use in photodynamic therapy culminating with the recent approval of Photofrin II as a therapeutic agent in the treatment of esophageal cancer [71]. Although porphyrins may exist in the metal-free state such as "hematoporphyrins" or "protoporphyrins" (depending on the nature of the side chains), many metal ions can form stable complexes with the four pyrrolic nitrogen atoms. Examples of these are common in biological tissues, such as ferroprotoporphyrin in blood and chlorophyll in plants.

Singlet oxygen quantum yields are reported to be high: $\phi = 0.02 - 0.8$ for hematoporphyrin in aqueous systems [50]. Complications due to aggregate formation and competing photochemical pathways however, have led to the disparate values [72] thus demonstrating the role of electron transfer. Although early work had stressed the role of $^1O_2$ in phototherapy treatment of cancers [73] there is recent evidence suggesting the role of metals such as iron or copper. Whether H_2O_2 is produced directly from the porphyrin or by electron transfer with substrates, such as tryptophan or ascorbate, reduced metal ions can then catalyze the well-known Haber-Weiss reaction, yielding •OH and initiate subsequent oxidative reactions [74–75].

Case 3. Furocoumarins. Although the major emphasis of furocoumarin research has focused on [2+2] cycloadditions to DNA, the role of singlet oxygen has still been mentioned as significant. Numerous researchers have questioned, for example, the importance of DNA adduct formation in many phototoxic reactions of furocoumarins, such as "accelerated" sunburn. One suggested mode of action was that after covalent addition to DNA the intact furocoumarin molecule could still be photochemically active and thus sensitize $^1O_2$ formation [76]. Not only has •O_2^- as well as $^1O_2$ been determined [77] but additional work has shown that furocoumarins can cleave DNA, possibly by free-radical mechanisms [78–80]. (See also reviews [81–83].)

Even casual contact with a furocoumarin-containing plant in the presence of light can initiate a severe sunburn reaction. These compounds have also been implicated in the role of plant defense against insects and plant pathogens [84]. Using a method similar to the separated-sensitizer substrate method, Berenbaum and Larson [43] showed that two furocoumarin-containing plants induced loss of a well-characterized singlet oxygen acceptor, furfuryl alcohol (FFA) and concluded that singlet oxygen flux from the leaf surface could be an important toxicant in plant defense. Two caveats: 1) it has now been shown that FFA is not selective for $^1O_2$ but also reacts with peroxyl radicals [85] and 2) these plants contain an array of other photoactive molecules, for example, prickly ash also contains β-carboline alkaloids for which photoinduced electron transfer reactions may be important [86–87]. Additionally, even though plants such as wild parsnips and limes do contain small amounts of furocoumarins, the presence of synergistic compounds such as myristicin in the wild parsnip [88] or heretofore unknown, but newly described, photoactive compounds such as citral in *Citrus* plants [89–90] demonstrate that phototoxicity is complex and has no single, pervasive mode of action, such as $^1O_2$.

References

1. Finlayson-Pitts B and Pitts JN Jr (1986) Atmospheric chemistry. John Wiley, New York
2. Wayne RP (1991) Chemistry of atmospheres, 2nd ed. Clarendon, Oxford (UK)
3. Frankiewicz TC, Berry RS (1973) J Chem Phys 58:1787
4. Demerjian KL, Kerr JA, Calvert JG (1974) Adv Environ Sci Technol 4:1
5. Eisenberg WC, Taylor K, Murray RW (1984) Carcinogenesis 5:1095
6. Faust BC, Allen JM (1992) J Geophys Res 97:12913
7. Zepp RG, Wolfe NL, Baughman GL, Hollis RC (1977) Nature 267:421
8. Zepp RG, Baughman GL, Schlotzhauer PF (1981) Chemosphere 10:109
9. Zepp RG, Baughman GL, Schlotzhauer PF (1981) Chemosphere 10:119
10. Zepp RG, Schlotzhauer PF (1981) Chemosphere 10:479
11. Wolff CJM, Halmans MTH, van der Heide HB (1981) Chemosphere 10:59
12. Baxter RM, Carey JH (1982) Freshwat Biol 12:285
13. Haag WR, Hoigné J, Gassmann E, Braun AM (1984) Chemosphere 13:641
14. Momzikoff A, Santus R, Giraud M (1983) Marine Chem 11:1
15. Frimmel FH, Bauer J, Putzien J, Murasecco P, Braun AM (1987) Environ Sci Technol 21:541
16. Rodgers MA, Snowden PT (1982) J Amer Chem Soc 104:5541
17. Scully FE Jr, Hoigné J (1987) Chemosphere 16:681
18. Braun AM, Frimmel FH, Hoigné J (1986) Internat J Environ Anal Chem 27:137
19. Tratnyek PG, Hoigné J (1994) J Photochem Photobiol A 84:153
20. Canonica S, Jans U, Stemmler K, Hoigné J (1995) Environ Sci Technol 29:1822
21. Tratnyek PG, Elovitz MS, Colverson P (1994) Environ Toxicol Chem 13:27
22. Arnold SJ, Ogryzlo EA, Witzke H (1964) J Chem Phys 40:1769
23. Cooper WJ, Zika RG, Petasne RG, Plane JM (1988) Environ Sci Technol 22:1156
24. Khan AU, Kasha M (1994) Proc Natl Acad Sci USA 91:12362
25. Scully FE Jr, White WN (1991) Environ Sci Technol 25:820
26. Merkel PB, Kearns DR (1972) J Amer Chem Soc 94:7244
27. Larson RA, Bott TL, Hunt LL, Rogenmuser K (1979) Environ Sci Technol 13:965
28. Larson RA, Hunt LL (1978) Photochem Photobiol 28:553
29. Lichtenthaler RG, Haag WR, Mill T (1989) Environ Sci Technol 23:39
30. Miyoshi N, Tomita G (1978) Z Naturforsch B 33:622
31. Borsarelli CD, Durantini EN, Garcia NA (1996) J Chem Soc Perkin Trans 2 2009
32. Kormann CD, Bahnemann DW, Hoffmann MR (1988) Environ Sci Technol 22:798
33. Gohre K, Miller GC (1985) J Chem Soc Faraday Trans 1 81:793
34. Barbas JT, Sigman ME, Dabestani R (1996) Environ Sci Technol 30:1776
35. Sigman ME, Barbas JT, Chevis EA, Dabestani R (1996) New J Chem 20:243
36. Slawson V, Adamson AW, Stein RA (1978) Lipids 13:128
37. Gohre K, Miller GC (1983) J Agric Food Chem 31:1104
38. Gohre K, Scholl R, Miller GC (1986) Environ Sci Technol 20:934
39. Hebert VR, Miller GC (1990) J Agric Food Chem 38:913
40. Adams RL, Weber EJ, Baughman GL (1994) Environ Toxicol Chem 13:889
41. Clements P, Wells CHJ (1992) Pestic Sci 34:163
42. Hon DNS, Chang ST, Feist WC (1982) Wood Sci Technol 16:193

43. Berenbaum MR, Larson RA (1988) Experientia 44:1030
44. Gaffron H (1927) Ber Deut Chem Ges 60:2229
45. Kautsky H, de Bruijn H, Neuwirth R, Baumeister W (1933) English translation in Singlet molecular oxygen, Ed. Schaap AP, Dowden Hutchinson, and Ross, Inc, Stroudsburg, PA 1976
46. Mulliken RS (1932) Rev Mod Phys 4:51
47. Khan AU and Kasha M (1970) J Am Chem Soc 92:3293
48. Foote CS, Wexler S, Ando W, Higgins R (1968) J Am Chem Soc 90:975
49. Schaap AP, Ed (1976) Singlet molecular oxygen, Dowden Hutchinson, and Ross, Inc, Stroudsburg, PA
50. Wilkinson F, Helman WP, Ross AB (1993) J Phys Chem Ref Data 22:113
51. Wilkinson F, Helman WP, Ross AB (1995) J Phys Chem Ref Data 24:663
52. Lissi EA, Encinas MV, Lemp, E, Rubio MA (1993) Chem Rev 93:699
53. Neckers DC (1989) J Photochem Photobiol A Chem 47:1
54. Blossey EC, Neckers DC, Thayer AL, Schaap AP (1973) J Am Chem Soc 95:5820
55. Midden WR, Wang SY (1983) J Am Chem Soc 105:4129
56. Krishna CM, Lion Y, Riesz P (1987) Photochem Photobiol 45:1
57. Di Mascio P, Wefers H, Do-Thi HP, Lafleur MVM, Sies H (1989) Biochem Biophys Acta 1007:151
58. Ciulla TA, Van Camp JR, Rosenfeld E, Kochevar IE (1989) Photochem Photobiol 49:293
59. Tomlinson G, Cummings MD, Hryshko L (1986) Biochem Cell Biol 64:515
60. Escobar JA, Rubio MA, Lissi EA (1996) Free Rad Biol & Med 39:285
61. Yasaei PM, Yang GC, Warner CR, Daniels DH, Ku Y (1996) J Am Oil Chem Soc 73:1177
62. Srinivasan VS, Podolski D, Westric N, Neckers DC (1978) J Am Chem Soc 100:6513
63. Rodgers MAJ, Lee P (1984) J Phys Chem 88:3484
64. Linden SM, Neckers DC (1988) Photochem Photobiol 47:543
65. Lambert CR, Kochevar IE (1996) J Am Chem Soc 118:3297
66. Sarna T, Zajac J, Bowman MK, Truscott TG (1991) J Photochem Photobiol A Chem 60:295
67. Epling GA, Jackson ML (1991) Tetrahedron Lett 32:7507
68. Dahl TA (1992) in Environmental aspects of aquatic and surface photochemistry, Helz GR, Zepp RG, Crosby DG, eds, Lewis Publishers, Boca Raton FL p 241
69. Camoirano A, de Floria S, Dahl TA (1993) Environ Molec Mutagen 21:219
70. Soumillion JP (1993) Topics Curr Chem 168:93
71. Key S, Marble M (1996) Cancer Biotech Weekly, 15 Jan 1996, p 17
72. Tanielian C, Wolff C, Esch M (1996) J Phys Chem 100:6555
73. Kasha M (1985) In: Frimer AY (ed) Singlet O_2, vol 1, Physical-chemical aspects. CRC Press, Boca Raton FL, p 1
74. van Steveninck J, Tijssen K, Boegheim JPJ, van der Zee J, Dubbelman TMAR (1986) Photochem Photobiol 44:711
75. Bachowski GJ, Girotti AW (1988) Free Rad Biol & Med 5:3
76. De Mol NJ, Beijersbergen van Henegouwen GMJ, van Beele B (1981) Photochem Photobiol 34:661
77. Joshi PC, Pathak MA (1983) Biochem Biophys Res Commun 112:638
78. Kagan J, Chen X, Wang TP, Forlot P (1992) Photochem Photobiol 56:185
79. Shim SC, Yun MH (1992) Abstr 20th Annu Meet Am Soc Photobiol 51S
80. Oroskar AA, Gasparro FP, Peak MJ (1993) Photochem Photobiol 57:648
81. Potapenko AY (1991) J. Photochem. Photobiol B9:1
82 Midden WR (1988) In: Psoralen DNA photobiology vol 2, CRC Press Boca Raton FL, p 1
83. Larson RA, Marley KA (1994) in Environmental oxidants, John Wiley, New York, p 269
84. Berenbaum MR (1978) Science 201:532
85. Haag WR, Mill T (1987) Environ Toxicol Chem 6:359
86. Larson RA, Marley KA, Tuveson RT, Berenbaum MR (1988) Photochem Photobiol 48:665
87. Biondic MC, Erra-Balsells R (1994) J Photochem Photobiol A Chem 77:149
88. Neal J (1989) J Chem Ecol 15:309
89. Asthana A, Larson RA, Marley KA, Tuveson RT (1992) Photochem Photobiol 56:211
90. Green ES, Berenbaum MR (1994) Photochem Photobiol 60:459

6 The Photochemistry of PAHs and PCBs in Water and on Solids

Richard M. Pagni[1] and Michael E. Sigman[2]

[1] Department of Chemistry, The University of Tennessee, Knoxville, TN 37996–1600, U.S.A.
  E-mail: *pagni@novell.chem.utk.edu*
[2] Chemical and Analytical Sciences Division, Oak Ridge National Laboratory, Oak Ridge, TN
  37831 U.S.A. E-mail: *sigmanme@ornl.gov*

The photochemistry of numerous polycyclic aromatic compounds (PAHs) and polychlorobiphenyls (PCBs) has been studied in water and on solids in environmental and laboratory settings. Photoproducts have been identified in many instances. In the presence of O_2, the
photochemistry of PAHs is dominated by oxidation, with either singlet O_2 or superoxide
serving as oxidant. The factors which dictate whether energy transfer or electron transfer
occurs from the PAH excited state to O_2 are not well understood. The photochemistry of PCBs
in water is dominated by hydrolysis in which Cl is replaced by OH, even in the presence of O_2.
This is a consequence of the large oxygen-hydrogen bond dissociation energy and dielectric
constant of water. Photoisomerization of chlorobiphenyls also occurs in water. On solids the
photochemistry of PCBs is different. In the presence of O_2, hydrogen is replaced by OH. These
reactions appear to involve PCB radical cations. A variety of successful photochemical sche-
mes have been developed, involving solids and water, for the destruction of PCBs. Notable in
this regard are the numerous studies on the photocatalytic decomposition of PCBs on TiO_2.

Keywords: Photochemistry, Aqueous photochemistry, Interfacial photochemistry, Polychlorobiphenyls, Polycyclic aromatic hydrocarbons.

Contents

The Handbook of Environmental Chemistry Vol. 2 Part L
Environmental Photochemistry (ed. by P. Boule)
© Springer-Verlag Berlin Heidelberg 1999

1
Introduction

The United States Environmental Protection Agency (EPA) has included sixteen polycyclic aromatic hydrocarbons (PAHs) and seven polychlorinated biphenyl (PCB) Aroclors on their list of priority pollutants [1]. These PAHs and PCB Aroclors are listed in Table 1. The list of 129 EPA priority pollutants was established by the Clean Air and Water Act of 1977. Shortly after the Act was passed, a report was issued by Mabey et al. that gave equilibrium and kinetic constants necessary for evaluating the transformation and transport of 114 of the priority pollutants in aquatic systems [2]. The distribution and fate of pollutants in the environment are of special significance due to associated health concerns. Eight of the 16 priority PAHs are typically considered probable human carcinogens, as indicate in Table 1 [3, 4]. PAHs, as a class of compounds, are widespread in the environment. One interesting study of soil samples collected from Rothamsted Experiment Station in southeast England since the mid-1800s reported an approximate 4 fold increase in the total PAH load in top-soil, with some compounds showing considerably greater increases [5]. Another study estimates the total momentary PAH content of the open Baltic

Table 1. PAH and PCB Aroclors contained on the EPA list of 129 Priority Pollutants

PAH	PCB Aroclors
1. Acenaphthene	1. Aroclor 1016
2. Acenaphthylene	2. Aroclor 1221
3. Anthracene	3. Aroclor 1232
4. Benzo[a]anthracene[a]	4. Aroclor 1242
5. Benzo[b]fluoranthene[a]	5. Aroclor 1248
6. Benzo[k]fluoranthene[a]	6. Aroclor 1254
7. Benzo[g,h,i]perylene[a]	7. Aroclor 1260
8. Benzo[a]pyrene[a]	
9. Chrysene[a]	
10. Dibenzo[a,h]anthracene[a]	
11. Fluoranthene	
12. Fluorene	
13. Indeno[1,2,3-cd]pyrene[a]	
14. Naphthalene	
15. Phenanthrene	
16. Pyrene	

[a] Possible or probable human carcinogen.

Sea at 5.4 tons [6]. The ubiquitous nature of these materials, and potential health risks associated with their presence, makes an understanding of their photochemistry in the environment important.

A large volume of work, both preceding and following passage of the 1977 Clean Air and Water Act, has focused on the mechanisms and products derived from photochemical transformations of PAHs and PCBs in aqueous solutions, on solids and in environmental settings. This paper provides a review of some of those studies. We have chosen to focus on mechanistic aspects of the photochemistry of PAHs and PCBs. Some general background information is also given which we feel is important to the non-specialist reading this review. The PAH studies discussed in this review focus on mechanistic chemical studies, largely wherein photoproducts are identified. Photophysical and purely transient spectroscopic studies are not discussed.

2
Background

2.1
Artificial Light Sources and Solar Irradiance

Much of the photochemistry discussed in this review was conducted by irradiating samples with conventional low-intensity (non-laser) light sources. An overview of conventional light sources has been give by Gould [7]. Some of the studies discussed here were performed in sunlight and relied upon the best available solar irradiance data for kinetic analyses. A review and tabulation of solar irradiance data have been given by Leifer [8]. These two resources contain important fundamental information heavily utilized throughout the work described here.

2.2
Kinetics and Mechanisms

2.2.1
Photolysis in Environmental Solutions

The kinetics treatment of photochemical processes occurring in aquatic environmental systems has been thoroughly discussed elsewhere [8]. Much of the theory and practice of determining kinetic rates and quantum yields comes from the work of Zepp and Cline [9]. In optically dilute solutions, as is often the case for organic pollutants in environmental aquatic systems, direct photo-degradation generally proceeds by first-order kinetics. The photoreaction rate constant in optically dilute systems is typically interpreted with the aid of Equation 1, where k is the first-order rate constant (units of s^{-1}), ϕ is the wavelength-independent quantum yield, ε_λ and Z_λ are the molar extinction of the pollutant and the irradiance (units of photons cm^{-2} s^{-1}) at wavelength λ and j is a conversion factor ($j = 6.022 \times 10^{20}$).

$$k = 2.303\, j^{-1} \phi \Sigma \varepsilon_\lambda Z_\lambda \tag{1}$$

In a case where the pollutant is absorbing all of the incident light, the rate of photoreaction becomes zero-order and the rate of photoreaction is equal to the product of ϕ and the rate of light absorption by the system, typically the photon flux (Einstein s^{-1}). Determination of the irradiance can be done with standard chemical actinometers or with integrating (broad-band) or spectral radiometers. Several chemical actinometers systems have been developed specially for use in sunlight [8, 10]. A detailed procedure for determining wavelength averaged quantum yields and simple computer code to assist in the calculation have been reported [11, 12].

The effects of suspended sediments on photolysis rates of dissolved pollutants have been investigated by Zepp and Miller for naturally occurring sediments in six natural waters [13]. Specific diffuse attenuation coefficients for all samples tested were found to be very similar, and photolysis rates within the photolytic zone of turbid systems were generally faster than in pure water. The increased photolysis rates can be attributed to light scattering which effectively increases the pathlength of light. Models required to account for photochemical behavior of pollutants in natural waters are more complex than those models required for simple laboratory systems. Descriptions of these more complex models have been given by several researchers [14–16].

2.2.2
Oxidation by Singlet Molecular Oxygen

The lowest energy singlet state of molecular oxygen, $O_2(^1\Delta_g)$, often referred to simply as "singlet molecular oxygen," is a powerful oxidizing agent. The chemical and physical properties of singlet molecular oxygen have been discussed extensively in other books and reviews and only a general discussion is given here [17, 18]. Singlet molecular oxygen can be generated in laboratory or environmental samples by energy transfer from an excited electronic state of a sensitizing agent to molecular oxygen in its lowest energy electronic state, $O_2(^3\Sigma_g^-)$ [19]. Rapid generation of $O_2(^1\Delta_g)$ has been demonstrated to occur in natural waters where its formation can be sensitized by naturally occurring humic substances [20, 21]. The energy transferred to oxygen in solution can come from the lowest electronically excited singlet state, S_1, of the sensitizing agent, provided the energy gap between the sensitizer's lowest excited singlet and triplet states (S_1-T_1) exceeds 7882 cm^{-1} (the 0,0 excitation energy of $O_2(^1\Delta_g)$). However, some studies indicate that $O_2(^1\Delta_g)$ may not be formed by quenching a sensitizer S_1 state at inorganic oxide solid-air interfaces [22]. Encounters between a sensitizer's lowest electronically excited triplet state and $O_2(^3\Sigma_g^-)$ can also lead to $O_2(^1\Delta_g)$ formation for a statistical 1/9 of all encounters. Energy transfer to oxygen from a sensitizer T_1 state is also subject to the constraint that the sensitizer triplet, T_1, lies greater than 7882 cm^{-1} above the sensitizer ground electronic state (T_1-$S_0 \geqslant$ 7882 cm^{-1}). Additionally, singlet molecular oxygen is known to be generated upon irradiation of environmental constituent, non-transition-metal oxide, surfaces such as silica gel, aluminum oxide and magnesium oxide in the presence of oxygen [23]. The formation of singlet molecular oxygen in these systems is thought to arise from molecular oxygen quenching of excitons on the metal oxide surface.

Once generated, $O_2(^1\Delta_g)$ can decay back to its lowest electronic state, $O_2(^3\Sigma_g^-)$, or react with organics which are susceptible to attack by this oxidant. Singlet molecular oxygen reactions in solution have recently been reviewed [18]. The lifetime of $O_2(^1\Delta_g)$ in water is $3-5\,\mu s$ (decay rate constant of $2\times10^5-3.3\times10^5\,s^{-1}$) [19]. At very low substrate concentrations, as typified by PAHs in water, it is highly probable that $O_2(^1\Delta_g)$ in solution will relax back to the ground-state before encountering a PAH to oxidize. For example, the rate constant of $O_2(^1\Delta_g)$ addition to anthracene in benzene is $1.5\times10^5\,M^{-1}s^{-1}$. Assuming that the same rate constant holds in water, the pseudo first-order rate for $O_2(^1\Delta_g)$ quenching by addition to anthracene (at a typical anthracene aqueous concentration of 3×10^{-7} M) would be $4.5\times10^{-2}\,s^{-1}$, which is insignificant relative to the $O_2(^1\Delta_g)$ decay rate. However, this does not preclude anthracene (or other PAHs) oxidation by $O_2(^1\Delta_g)$ in water. An alternative mechanism involves a $O_2(^1\Delta_g)$ molecule encountering the exact same sensitizing PAH molecule that brought about its formation, with subsequent oxidation of that PAH. The efficiency of this process at low sensitizer concentrations is a function of the probability of re-encounter of the oxygen-sensitizer "geminate pair", and has been addressed by Stevens et al [24].

In the gas phase, two $O_2(^1\Delta_g)$ molecules can quench to give two ground-state oxygen molecule and a photon of 634 nm light, the dimol reaction; however, the probability for this reaction occurring is low due to the low steady-state concentration of $O_2(^1\Delta_g)$. Alternatively, $O_2(^1\Delta_g)$ can be quenched by collisions with a ground-state oxygen molecule or with a surface. Gas phase studies have determined the ground-state oxygen quenching rate constant to be $1\times10^3\,M^{-1}s^{-1}$ and the wall (surface) quenching rate constant to be $0.2\,s^{-1}$ [25].

Bimolecular rate constants for reaction with $O_2(^1\Delta_g)$ have been measured for only a limited number of PAHs from Table 1 [26]. However, rate constants have been estimated for the reaction of $O_2(^1\Delta_g)$ with each PAH in Table 1 [2]. Of these estimated rate constants, many exceed $1\times10^7\,M^{-1}\,h^{-1}$.

2.2.3
Single Electron Transfer and Free Radical Mechanisms

Transient spectroscopic studies of humic substances in water have revealed that three short-lived species are formed upon laser excitation [27, 28]. The three species are identified as the solvated electron and the associated radical cation and a triplet state of the humic substance. Solvated electrons have also been identified in transient studies of dissolved organic matter from natural water bodies and soils [29]. The radical cation can serve as a one electron oxidant and the solvated electron can reduce pollutants to initiate radical reactions. The solvated electron is also readily scavenged by oxygen, to produce superoxide, or by other electron deficient species.

Alkoxy radicals (RO), alkyl peroxyl radicals (RO$_2$) and hydroxyl radical (HO) are also oxidants, in addition to singlet molecular oxygen and radical cations, which can readily be formed in natural settings. Nanomolar alkyl peroxyl radical concentrations have been reported in natural waters [30]. The most important reaction of these radicals with unsubstituted PAHs and PCBs is addition to the aromatic system. If phenols, amines or aliphatic substituents are present, these radicals (RO•, RO$_2$•, or HO) can abstract a hydrogen atom to generate a new radical from the substrate. Rate constants for reactions of hydrated electrons, hydrogen atoms and hydroxyl radicals in aqueous solution have been tabulated by Buxton and coworkers [31].

3
Fundamental/Model Studies of PAHs and PCBs

3.1
Photochemistry of PAHs

3.1.1
Aqueous Photochemistry of PAHs

Many studies have been made of the photochemistry of PAHs in solution, although most studies have employed organic solvents, rather than water. The inherently low aqueous solubility of PAHs, and associated analytical problems, has significantly contributed to the lack of attention which these systems have received. Another discussion of PAH aqueous photochemistry by Kochany has recently been published [32]. A review of the photochemistry of petroleum in water by Payne and Phillips has also discussed some aspects of PAH photochemistry that are covered in this review [33]. Additionally, reviews and reports of environmental phototoxicity of PAHs have addressed selected mechanistic aspects of the aqueous photochemistry of this class of pollutant [34–36].

Quantum yields for photodepletion of a series of PAHs in water have been reported by several groups. Zepp and Schlotzhauer measured the photodepletion quantum yields for 11 PAHs (Table 2) irradiated at 313 and 366 nm. The resulting quantum yields were significantly larger than the values calculated assuming an oxidation mechanism involving $O_2(^1\Delta_g)$-PAH geminate pair re-encounter [37], as discussed above. The quantum yields were reported to be rather insensitive to the presence of oxygen and, furthermore, PAH triplet state quenchers did not alter the photodepletion rate. These researchers concluded that, excluding photoreactions in petroleum slicks, $O_2(^1\Delta_g)$ does not play a significant role in PAH phototransformations in water. The quantum yields measured in the laboratory were used by Zepp along with Eq. 1 and solar irradiance data to predict photodegradation lifetimes in natural waters. Picel et al. measured the sunlight-irradiated degradation rates of four PAHs (fluoranthene, pyrene, benzo[a]anthracene and benzo[a]pyrene) in pure water and in aqueous coal-oil systems [38], and found fair agreement with the lifetimes computed by Zepp [37]. In most cases Picel found the photodegradation rate significantly depressed in aqueous coal-oil system relative to that in pure water. The decreased rate in the aqueous coal-oil system was attributed partly to light screening effects and partly to unexplained inhibitory effects of phenolic components.

Mill et al. measured photodepletion quantum yields for benz[a]anthracene and benzo[a]pyrene and several nitrogen containing PAH analogs in natural waters from oligotrophic and eutropic sources and compared these results with those obtained in pure water [39]. In contrast to Zepp's results, oxygen was found to inhibit the photolysis of both non-heteroatom-containing PAHs (benz[a]anthracene and benzo[a]pyrene) by 40–50%. In pure water, a complete material balance was achieved for benzo[a]pyrene which yielded a 1:3 ratio of 3,6- and 1,6-benzo[a]pyrenequinones and a trace quantity of the 6,12-quinone. The only product identified for benz[a]anthracene was 7,12-benz[a]anthracenequinone, which accounted for 30% of the photolyzed starting material. Photolysis rates for both PAHs were observed to decrease in natural waters by 40–50% relative to the measured rates in pure water.

Table 2. Disappearance quantum yields for PAH in pure water

PAH	Disappearance Quantum Yield × 10^3 (Wavelength nm)	Initial Concentration (M)
Naphthalene	15 ± 1 (313)[a]	1.2×10^{-4}
	60 ± 6 (313)[c]	5.0×10^{-7}
1-Methylnaphthalene	18 ± 1 (313)[a]	1.2×10^{-4}
2-Methylnaphthalene	5.3 ± 0.2 (313)[a]	8.8×10^{-5}
Phenanthrene	10 ± 1.6 (313)[a]	8.4×10^{-7}
	31.2 ± 3.6 (313)[c]	5.0×10^{-7}
Anthracene	3.0 ± 0.2 (366)[a]	2.3×10^{-7}
	3.11 ± 0.6 (365)[c]	4.0×10^{-7}
9-Methylanthracene	7.5 ± 0.5 (366)[a]	3.1×10^{-7}
9,10-Dimethylanthracene	4.0 ± 0.4 (366)[a]	1.3×10^{-7}
Pyrene	2.0 ± 0.3 (313)[a]	3.9×10^{-7}
	2.2 ± 0.3 (366)[a]	3.9×10^{-7}
	0.398 ± 0.038 (313)[c]	5.0×10^{-7}
Fluoranthene	0.12 ± 0.01 (313)[a]	6.6×10^{-7}
	0.002 ± 0.002 (366)[a]	6.6×10^{-7}
Chrysene	2.8 ± 0.7 (313)[a]	8.2×10^{-9}
Naphthacene	13 ± 5 (436)[a]	2.2×10^{-9}
Benz[a]anthracene	3.2 ± 0.1 (313)[b,d]	7.9×10^{-8}
	3.4 ± 0.2 (366)[b,d]	8.6×10^{-8}
Benzo[a]pyrene	0.89 ± 0.08 (313)[b,d]	1.1×10^{-6}
	0.54 ± 0.1 (366)[b,d]	5×10^{-8}
7,12-Dimethylbenzo-[a]anthracene	184 ± 2 (313)[c]	5.0×10^{-7}
Benzo[e]pyrene	2.04 ± 0.38 (365)[c]	4.0×10^{-10}

[a] Zepp, Ref. 41.
[b] Mill, Ref. 39.
[c] Shevchuk, Ref. 43.
[d] Error estimated from reported uncertainty in first-order rate of PAH disappearance.

146   R. M. Pagni and M. E. Sigman

Table 2 (Continued)

Acenaphthene

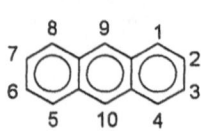

Acenaphthylene

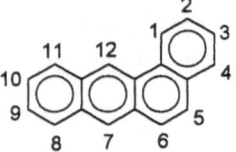

Fluorene

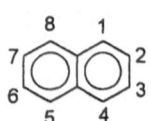

Naphthalene

Anthracene

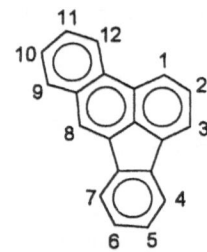

Benzo[a]anthracene

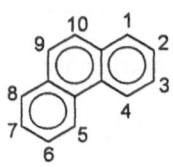

Phenanthrene

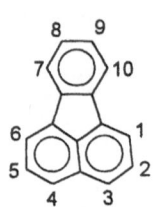

Fluoranthene

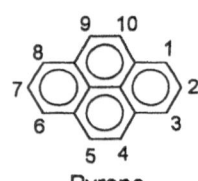

Benzo[b]fluoranthene

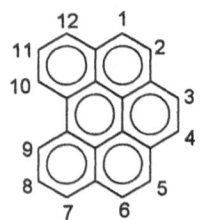

Benzo[k]fluoranthene

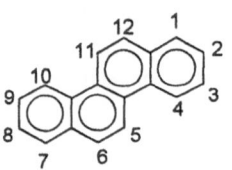

Chrysene

Pyrene

Benzo[ghi]perylene

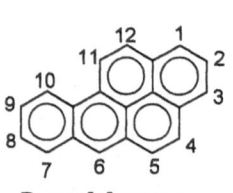

Benzo[a]pyrene

Dibenzo[ah]anthracene

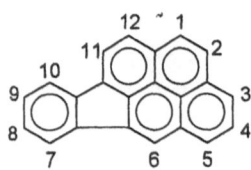

Indeno[1,2,-cd]pyrene

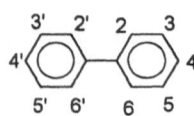

Biphenyl

Katz et al. have also studied the photolysis of benzo[a]pyrene in 50% methanol/water solutions, in pond water and in pond water with added natural brown humic matter [40]. These researchers found that air or oxygen was required for photolysis to proceed in any of the solutions. In pond water, in the presence of air, added fulvic acid brought about a four fold increase in the photolysis rate. Katz identified the 1,6-, 3,6- and 6,12-quinones of benzo[a]-pyrene as photoproducts. In related work, Zepp and Schlotzhauer have studied sunlight-induced algal transformations of pyrene, phenanthrene, fluoranthene and naphthalene in water [41]. Most of the PAHs reacted faster with algae present than in distilled water. Anilines and parathions were also studied and their rates of photolysis were found to be unaffected by heat-killing the algae, although the mechanism of rate enhancement was not determined. The authors determined that the overall effect of the algae would be low when extrapolated to algae concentrations typically found in natural water bodies.

Paalme and coworkers examined the relative rates of photolysis of 20 PAHs in water and correlated the rates with the differences between the ground-state oxidation potentials of the PAHs and the energies of their lowest triplet states [42]. The reported rates were normalized relative to the photodegradation rate for benzo[a]pyrene. Correlation of the relative rates for 15 PAHs with the quantity $(E_{ox} - E_T)$ gave a correlation coefficient of $r = 0.955$. Quantum yields were not reported and the measured rates were not corrected to account for the variable molar extinctions among the PAHs studied. The reported correlation with $(E_{ox} - E_T)$ did not correct for intersystem crossing efficiencies of the different PAHs. Corrections for variable intersystem crossing efficiencies are frequently not taken into account because these values are typically assumed to be nearly equal for most PAHs. The effects of oxygen on the photolysis rates and photoproducts were not reported.

Shevchuk has examined the photolysis quantum yields (Table 2) for six PAHs and benzene in water and has proposed two mechanisms of photooxidation [43]. Oxidation mediated by singlet molecular oxygen and oxidation by single electron transfer from the lowest energy triplet state of the PAHs to oxygen were proposed. The singlet molecular oxygen reaction was proposed to account for oxidation of the linear, anthracene-like, PAHs, while electron transfer oxidation was proposed for the PAHs which are poor acceptors of singlet molecular oxygen. The most probable electron acceptor in the system was proposed to be molecular oxygen. Correlation of the natural logarithm of the quantum yield with the difference between the PAH ground state oxidation potential and the triplet excitation energy was cited as evidence for electron transfer coming from the lowest triplet state of the PAH. The quantity $(E_{ox} - E_T)$, as also used by Paalme et al. [42], partly characterizes the electron-donor capability of the excited state. Rehm's and Weller's expression for the free energy change for excited-state electron transfer includes a similar term, as well as terms accounting for the reduction potential of the electron acceptor and the dielectric stabilization of the medium [44].

Ramamurthy and coworkers reported the facile dimerization of *trans*-stilbene in water at concentrations as low as 1×10^{-6} M [45]. The dimerization of *trans*-stilbene is rather inefficient in organic solvents, and the enhanced dimerization efficiency observed in water was attributed to stilbene ground-state association promoted by a "hydrophobic effect." Enhanced rates for the Diels-Alder reaction in water have also been attributed to a "hydrophobic effect" [46]. Products from *trans*-stilbene photochemical oxidation were not reported in Ramamurthy's study. The combined yield of the two dimers, formed in an approximate 1:1 ratio, accounted for 22% of the identified final photo-products. The other identified compounds included *cis*-stilbene (11%), phenan-threne (33%) and recovered *trans*-stilbene (33%).

In related work, Yates and coworkers have studied the photolysis of aryl alkenes and aryl alkynes in neutral and acidic water [47, 48]. The proposed photochemical oxidation mechanism involves protonation of the aromatic hydrocarbon's excited state in a Markovnikov fashion, followed by hydration and deprotonation. Strong electron withdrawing groups, i.e. nitro, attached to the aromatic ring can lead to anti-Markovnikov protonation of the excited state. Unsubstituted aryl alkenes, for example, gave the corresponding 1-arylethanol upon photolysis.

$$Ar-CH=CH_2 \xrightarrow[(+H^+)]{h\nu} Ar-\overset{+}{C}H-CH_3 \xrightarrow[(-H^+)]{H_2O} Ar-\overset{\overset{\displaystyle OH}{|}}{C}H-CH_3$$

Photolysis of naphthalene and alkylated naphthalenes in distilled water and simulated sea water has been reported by Fukuda et al [49]. The rate of photolysis in simulated sea water was found to be several times faster than the rate in distilled water, with naphthalene showing the greatest enhancement. The rate enhancement was found to be well correlated with the NaCl concentration, although NaI and NaBr did not give similar results. Photolysis products were determined for 2-isopropylnaphthalene; however, a material balance was not given. The observed products included 2-(2-naphthyl)-2-propanol, phthalic acid and benzoic acid. It was shown that humic substances did not sensitize the photodecomposition of these PAHs.

This chemistry differs from that of aryl alkenes in that the alkyl substituent can not be protonated to form the corresponding benzylic cation; however, the benzylic radical can be formed by deprotonation of the benzilic carbon in an aryl radical cation, as shown in the example below.

$$Ar-CH(CH_3)_2 \xrightarrow[(-e^-)]{h\nu} \left[Ar-CH(CH_3)_2\right]^{\ddagger} \xrightarrow[(-H^+)]{} Ar-\overset{\bullet}{C}H(CH_3)_2$$

Photolysis of naphthalene in pure water has been shown to proceed efficiently (Table 2) under irradiation by low intensity (conventional non-laser) light sources. Additionally, Steenken et al. have examined the transients formed from laser photolysis of naphthalene in water [50]. Naphthalene is one of the few PAHs with sufficient solubility in water (without added organic co-solvent) to allow this type of study. The radical cation of naphthalene was observed to be formed upon laser excitation at 248 nm, and laser power dependence studies showed radical cation formation to occur by a biphotonic process. However, monophotonic ionization has been reported for naphthalene, anthracene and related PAHs in acetonitrile [51, 52]. The bimolecular rate constant for water addition to the naphthalene radical cation was found by Steenken to be $4 \times 10^4 \, M^{-1} s^{-1}$ [50]. These studies raise interesting questions concerning the mechanism of naphthalene photolysis in water with low intensity light sources. It is generally agreed that low intensity light sources cannot induce biphotonic events of the type reportedly observed by Steenken for naphthalene in water. On the other hand, naphthalene is not a good acceptor of $O_2(^1\Delta_g)$, which leaves one to question the mechanism of photolysis under low light flux conditions. Further resolution of the mechanism of naphthalene photodepletion in water requires additional transient and product studies.

The products of anthracene photolysis in water have been examined by Sigman et al. [53]. Photolysis in aerated water led to the formation of anthracene-9,10-endoperoxide and 9,10-anthraquinone in a 3:1 ratio at 10% conversion. At 70% conversion, two new products were identified, 9,10-dihydro-9,10-dihydroxyanthracene and 9-hydroxyanthrone. There was no evidence of anthracene-9,10-dimer formation, indicating that ground-state pairing was not occurring in water.

The low solubility of anthracene and short lifetime of $O_2(^1\Delta_g)$ in water prevented anthracene-9,10-endoperoxide formation by methylene blue sensitization of $O_2(^1\Delta_g)$. A $O_2(^1\Delta_g)$ mediated mechanism of endoperoxide formation in water must involve anthracene-$O_2(^1\Delta_g)$ geminate pair recombination, as discussed in Sect. 2.2.2. The wavelength-averaged quantum yield of anthracene photodepletion measured under aerated conditions, 1.1×10^{-3}, was similar to the value reported by Zepp [37]. Photolysis in deaerated solutions gave an 87% yield of the three isomers of 10,10'-dihydroxy-9,9',10,10'-tetrahydro-9,9'-bianthryl. The proposed mechanism of formation for these novel photoproducts was by initial single electron transfer from the anthracene excited state to water. In related work, Pagni and coworkers examined the photochemistry of anthracene in room temperature molten salts composed of mixtures of 1-ethyl-3-methyl-1H-imidazolium chloride (EMIC) and AlCl$_3$ [54]. In acidic mixes (EMIC:AlCl$_3$ <1) under degassed conditions, photolysis resulted in the formation of no less than 16 oxidized, neutral and reduced monomeric and dimeric anthracene-derived products as shown on the previous page. The observed photochemistry was attributed to electron transfer from anthracene to anthracenium ion, present at approximately 3% in the starting mix.

In the absence of water or other suitable nucleophiles, the observed chemistry is dominated by bimolecular electron transfer, hydrogen atom transfer and radical coupling reactions to give the observed products. This work clearly demonstrates how the presence of a small concentration of a good electron acceptor can channel the photochemistry away from more traditional pathways (i.e. anthracene photodimerization). In a study of the photochemistry of 9-methylanthracene in the same room temperature molten salt, Pagni and coworkers demonstrated similar chemistry arising from initial electron transfer to the 1-ethyl-3-methyl-1H-imidazolium cation (EMI$^+$) [55].

The photochemistry of acenaphthylene in water has been examined in laboratory samples [56]. Photodepletion of acenaphthylene in water was found to proceed much faster than in organic solvents. The major products from photolysis in water at concentrations above 5×10^{-5} M were the *cis* and *trans* dimers of acenaphthylene, and the *cis/trans* dimer ratio was shown to vary with conversion. Oxidation products formed included 1-acenaphthenone, 1,2-acenaphthenedione, 1,8-naphthalenedicarboxaldehyde and 1,8-naphthalic anhydride.

Perylene photooxidation has been studied by Guillet and coworkers in the presence of water-soluble polymeric photocatalysts [57]. This study shows that perylene, which is unreactive to $O_2(^1\Delta_g)$ in organic solvents but easily oxidized by electron acceptors such as dicyanobenzene, is readily oxidized in the aqueous solution of a water-soluble polymer containing 2-naphthyl and 4-phenylsulfonate pendant groups. The photooxidation was attributed to electron transfer to trace quantities of naphthoquinone pendant groups in the polymer. Light absorption by the polymer pendant groups and energy transfer along the polymer to the quinones was shown to play a significant role in accelerating the reaction. Two photoproducts, 1,2-perylenequinone and 3,10-perylenequinone, were identified in the reaction.

The quantum yields for the photodepletion of PAHs in pure water from the studies discussed above are summarized in Table 2. Where there is overlap between research groups, the values from different groups vary by more than the stated errors in the measured quantum yields. However, within a single research group, quantum yields measured at differing photolysis wavelengths are generally the same within experimental error (with the notable exception of fluoranthene). Wavelength independence is generally assumed when calculating quantum yields from Equation 1 for photodecomposition using sunlight or a broad-band irradiation source. The compiled data in Table 2 generally support this practice; however, the fluoranthene data highlights the need to experimentally verify wavelength independence. Correlation of the observed quantum yields with structural and chemical properties of the PAHs has been attempted by several authors, with varying degrees of success. From the limited number of product studies which have been done, it is clear that in some cases products can arise from $O_2(^1\Delta_g)$ addition. For example, anthracene photolysis yields the 9,10-anthracene endoperoxide; however, this product could equally arise from a single electron transfer route. In examples such as this, it is difficult to further delineate the mechanism due to the low solubility of the PAHs. Addition of singlet molecular oxygen quenchers will not affect a PAH-$O_2(^1\Delta_g)$ geminate pair recombination occurring within the solvent cage, as discussed above. Quenching the triplet state of the PAHs competitively with oxygen, however, would stop the reaction. Zepp's observation that PAH triplet quenchers did not alter the rate of photodepletion strongly suggests that the mechanism does not involve energy transfer and that the triplet state of the

PAHs is not accessed along the reaction path [37]. An electron transfer oxidation mechanism is further supported by the observed regiochemistry of the benzo[a]pyrene oxidation products identified by Mill [39]. Attempts by Paalme and Shevchuk to correlate PAHs photolysis rates with ($E_{OX} - E_T$) are also suggestive of a single electron transfer process [42, 43]. Additionally, the observed facile photodepletion quantum yield reported for naphthalene (a poor singlet molecular oxygen acceptor) in water also supports a mechanism which is not mediated by $O_2(^1\Delta_g)$, although the products from photolysis must be identified before a mechanism for their formation can be assigned. The best conclusion that can be drawn from the available data is that a continued effort in the field is required to better define the mechanism(s) which lead to PAHs photooxidation in aqueous media.

3.1.2
Heterogeneous Photochemistry of PAHs

A variety of organic phototransformations in heterogeneous media have been described in monographs, including fundamental considerations in dealing with photochemical excitation at interfaces [58–60]. Physical aspects of the photochemistry and radiation chemistry of molecules adsorbed on SiO_2, γ-Al_2O_3, zeolites and clays have also been reviewed recently by Thomas [61] and will not be repeated here. The emphasis of this review will be chemical in nature, dealing with kinetics and products of photodecomposition (of mostly unsubstituted PAHs) in heterogeneous media. Some studies have reported quantum yields of photolysis in heterogeneous media; however, due to the uncertainity in light loss by scattering at the solid/air interface, most investigators have chosen to report photolysis rates. Rate data taken from different research groups are difficult to compare directly due to differing light sources used by each group and variable light loss from scattering in reactors of different configurations. Therefore, no attempt will be made here to tabulate or correlate that data, but simply to report those studies and their most significant findings.

Katz and coworkers have studied the photochemistry of anthracene, benz[a]anthracene, pyrene, benzo[e]pyrene, benzo[a]pyrene, dibenz[a,c]-anthracene and dibenz[a,h]anthracene on cellulose [40]. Relative rates of photolysis were determined, and the presence of ozone (0.2 ppm) was shown to dramatically reduce the photolytic half-life; however, no photoproducts were identified. Blau and coworkers have determined the "relative" quantum yields for 10 PAHs on silica [62]. The "relative" quantum yields were converted to "real" quantum yields by comparison with the quantum yield of 0.085 determined for the photodepletion of anthracene in silica-cyclohexane slurry. Later work by Zingg et al. and Dabestani et al. has showed that in such slurries, anthracene partitions between the silica and the cyclohexane; this significantly affects the product distribution relative to that observed on dry silica [63–65].

König et al. have reported the relative rates of photodecomposition of anthracene, benz[a]anthracene, dibenz[a,c]anthracene and dibenz[a,h]anthracene on aluminum oxide and identified their photoproducts [66]. Half-lives of 0.12 h, 1.03 h, 2.42 h and 4.63 h, respectively, were reported for this series of structurally related PAHs. The quinones, 9,10-anthraquinone and 7,12-benz[a]-anthraquinone, were identified as photoproducts along with several dialdehydes derived from oxidative cleavage of one aromatic ring. The relative rates of reaction and sites of oxidation were seen to parallel the reaction rates of each PAH with maleic anhydride in Diels-Alder reactions. These results implicated $O_2(^1\Delta_g)$ in the oxidation mechanism.

Wehry, Mamantov and coworkers examined PAH susceptibility to photodecomposition when sorbed on fly ash and alumina [67–69]. Benzo[a]pyrene, pyrene and anthracene were found to be highly resistant to photodegradation when adsorbed on fly ash; however, both pyrene and benzo[a]pyrene were found to photolyze efficiently on alumina and in organic solvents. Phenanthrene and fluoranthene, on the other hand, were resistant to photochemical oxidation on fly ash and in solution. These results indicated that surface sorption could decrease the susceptibility to photochemical oxidation for some PAHs. In a more extensive study, Behymer and Hites investigated the photolytic behavior of 15 PAHs adsorbed on silica gel, alumina, and coal fly ash [70,71]. Photolytic rates were determined, but products were not identified. The photolytic half-lives of the PAHs were highly dependent on the nature of the substrate. Furthermore, half-lives for PAHs sorbed on silica and alumina were observed to vary significantly with PAH structure, while those on fly ash and carbon black were virtually independent of PAH structure. The half-lives were universally longer on carbon black and typically shortest on silica and alumina. The observed photolytic half-lives of eight PAHs on silica were correlated with the Dewar reactivity numbers (i.e. the energy required to remove a electron from a specific carbon center in a molecule). The authors concluded that fly ash and carbon black would stabilize some PAHs and facilitate their environmental transport.

In a more product and mechanism oriented work, Fatiadi has examined the photolysis of pyrene adsorbed on garden soil, silica gel G, precipitated silica gel, alumina (basic, acidic and neutral), silicic acid, alkali carbonates, Florisil (magnesium silicate), "industrial dust" and alkaline earth carbonates [72]. Photolysis on soil led to eight photoproducts, of which five were identified. The identifiable products were 1,1'-bipyrene, 1,6-, and 1-8-pyrenediones and 1,6- and 1,8-pyrenediols. Photolysis on soil in the presence of co-sorbed radical traps, acrylamide or 2,5-di-tert-amylhydroquinone, led to decreased yields of 1,1'-bipyrene; however, the yield of the two quinones was unaffected. The conclusion was that 1,1'-bipyrene was formed on the surface by a radical reaction. On supports other than garden soil, 1,1'-bipyrene was not formed.

Taskar et al. have also examined the photochemistry of pyrene on silica, alumina and carbon [73]. The rates of photochemical degradation of pyrene were found to be approximately two times faster on silica and alumina than on carbon. Both thermal (dark) and photochemical reactions were observed on silica and alumina and the major products were isomeric pyrenediols and 1,1-bipyrene, although the pyrenediols did not appear to form photochemically on alumina. The 1,6- and 1,8-pyrenediones were not observed in any significant quantity on these supports. The photochemistry of pyrene on γ-alumina and silica-alumina surfaces has also been investigated by Thomas and coworkers [74, 75]. The formation of the pyrene radical cation by both thermal and photochemical processes was confirmed by optical and ESR spectroscopic methods and oxygen was shown to be necessary for radical cation formation on the surface. The reaction of the radical cation with water vapor on the surface was also observed by optical spectroscopy and the presence or absence of oxygen during product isolation was shown to affect the yield of unidentified products. The only photoproducts identified by Thomas were mono- and dihydroxypyrene.

Eisenberg et al. have shown that the photolysis of anthracene, phenanthrene, 9,10-diphenylanthracene, fluoranthene, pyrene, benzo[a]anthracene, chrysene, benzo[a]pyrene and 9-fluorenone on Chromosorb 102 led to the formation of $O_2(^1\Delta_g)$ by monitoring the $O_2(^1\Delta_g)$ emission at 1,270 nm [76]. Direct photolysis of 9,10-diphenylanthracene was found to generate 9,10-diphenylanthracene endoperoxide as the only identifiable product, accounting for 78% of the consumed starting material. Addition of triethylamine, a potent $O_2(^1\Delta_g)$ quencher, to the system completely stopped endoperoxide formation. Similarly, the slow photodecomposition of chrysene on Chromosorb 102 was stopped by addition of triethylamine. The product mixture from chrysene was complex and speciation was not reported.

The rates and products of benzo[a]pyrene photodegradation have been examined on silica, carbon, TiO$_2$ and coal fly ash at submonolayer concentrations [77]. The half-lives were observed to range from 3 h on TiO$_2$ to in excess of 280 h on carbon. The 1,6-, 3,6- and 6,12-benzo[a]pyrenequinones were formed from photolysis on all four substrates along with proposed diol and tetraol products. These are the same quinones observed by Mill et al. from photolysis of benzo[a]pyrene in water [39]. Dark reactions were also observed by Daisey [77] on each substrate, with TiO$_2$ and coal fly ash giving the greatest extent of reaction. Energy transfer or electron transfer quenching and sensitization experiments were not reported although the mechanism was concluded to be complex on all substrates.

Dupont and coworkers have evaluated the soil-sorbed photolysis of a complicated mix of biorecalcitrant organic components of wood treating waste, including anthracene, fluorene and biphenyl [78]. Addition of methylene blue, riboflavin, hydrogen peroxide, peatmoss and diethylamine to the soil did not enhance the photolysis rates; however, the presence of anthracene in the soil appeared to enhance the photolytic decomposition of all components. Fluorenone and 9,10-anthraquinone were identified as the major products from fluorene and anthracene, respectively. The mass balance due to fluorene and fluorenone was 62–85%, while that for anthracene and 9,10-anthraquinone was 43–66%. Soil color was found to be the most significant factor affecting compound degradation rates in these systems. The systems were far too complicated for detailed mechanistic information to be obtained.

Anthracene has been widely studied at solid-liquid and solid-air interfaces. Addition of silica to a oxygen-free solution of anthracene in cyclohexane was shown by Sigman and coworkers to greatly enhance the rate of anthracene photolysis without altering the product composition from that formed upon photolysis in cyclohexane, where the photoproduct is anthracene-9,10-photo-dimer [63, 64]. In an oxygen-containing silica-cyclohexane slurry, the photo-dimer was formed along with anthracene-9,10-endoperoxide. The endoperoxide underwent thermal decomposition on the surface to give a complicated product set, including 9,10-anthraquinone, 9,10-dihydro-9,10-dihydroxyanthracene, bi-anthronyl, 9-hydroxyanthrone and anthrone. The mechanism of oxidation was ascribed to O$_2$($^1\Delta_g$) addition to anthracene. Anthracene photochemistry on dry silica, Cab-O-Sil (fumed silica) and alumina (neutral) at surface coverages less than 11% of a monolayer has been investigated by Dabestani et al [65]. Under deaerated conditions, photolysis on the silica surface gave anthracene-9,10-photodimer as the only product. The dimer was shown to originate from ground-state anthracene pairs forming at surface coverages exceeding 1% of a monolayer. In the presence of air, anthracene photolyzed much faster to give dimer as well as anthracene-9,10-endoperoxide and its thermal decomposition products, as observed in cyclohexane-silica slurries. Additionally, 10,10′-di-hydroxy-9,9′,10,10′-tetrahydro-9,9′bianthryl was also formed. Oxidation of anthracene on silica by $O_2(^1\Delta_g)$, sensitized with co-sorbed methylene blue, gave the same set of oxidation products as observed on direct photolysis. Direct photolysis of anthracene on silica in the presence of co-sorbed 2,5-dimethyl-furan, a good $O_2(^1\Delta_g)$ trap, gave only dimer. These results gave strong, although indirect, proof of the involvement of $O_2(^1\Delta_g)$ in the photooxidation of anthracene at the solid-air interface. Additional evidence for $O_2(^1\Delta_g)$ participation came from the direct observation of its emission at 1270 nm. Transient spectral data suggesting the formation of anthracene ground-state pairs on silica at low surface coverages have also been reported by Wilkinson et al [79].

Dabestani et al. have shown that the photochemistry of tetracene, a good acceptor of singlet molecular oxygen, at a silica-air interface closely resembles that of anthracene [80]. Photolysis of deaerated samples gaves the *syn-* and *anti-* tetracene-5,12-dimers. Photolysis at surface coverages below 1% monolayer, in the presence of oxygen, yielded mainly tetracene-5,12-endoperoxide and its subsequent thermal decomposition products. At higher surface loading, in the presence of air, direct photolysis led to greater yields of dimer. Singlet molecular oxygen sensitization (methylene blue) and quenching (2,5-dimethyl-furan) studies demonstrated that photooxidation resulted from singlet oxygen addition to tetracene at the silica-air interface.

Acenaphthylene photochemistry provides a good probe for medium effects due to the varied, often state-specific, reactions which are known for this PAH. Dimerization to give the *cis*-dimer can occur from both the singlet and triplet manifolds, and the *trans*-dimer is formed from only the triplet manifold. However, the ratio of *cis*- to *trans*-dimer resulting from within the triplet manifold is a function of the polarity of the medium where dimerization takes place [81]. An early study of acenaphthylene photochemistry on silica by Bauer et al. reported only the formation of *cis*- and *trans*-photodimers, and no oxidation products [82, 83]. A re-examination of the photochemistry of acenaphthylene on silica by Dabestani and coworkers revealed the formation of oxidation products as well as dimers [84]. Oxidation products formed from the photolysis of acenaphthylene at a silica-air interface (less than 10% monolayer) are similar to those formed in water and included 1,2-acenaphthenedione, 2-hydroxy-1-acenaphthenone, 1,8-naphthalenedicarboxaldehyde and 1,8-naphthalic anhydride. The oxidation products were shown to arise from $O_2(^1\Delta_g)$ addition to acenaphthylene, presumably giving a dioxetane or perepoxide intermediate, with subsequent thermal decomposition to give the isolated products. The yields of both dimers of acenaphthylene were shown to increase with monomer surface coverage and result from both excited singlet and triplet states of the PAH.

The photochemistry of phenanthrene has also been examined at a silica-air interface by Barbas et al [85]. The phenanthrene photoproducts were shown to result from addition of $O_2(^1\Delta_g)$ across the 9,10-position in phenanthrene with subsequent thermal decomposition to give 2,2'-biformylbiphenyl, 9,10-phenanthrenequinone, *cis*-9,10-dihydrodihydroxyphenanthrene, benzocoumarin, 2,2'-biphenyldicarboxylic acid, 2-formyl-2'-biphenylcarboxylic acid, 2-formyl-biphenyl, 1,2-naphthalenedicarboxylic acid and phthalic acid. The products accounted for 85–90% of the photolyzed phenanthrene. The formation of the *cis*-9,10-dihydrodihydroxyphenanthrene rather than the *trans* isomer revealed that $O_2(^1\Delta_g)$ added to phenanthrene on silica to give a dioxetane. Although the photochemical oxidation of phenanthrene on silica was determined to proceed entirely by a $O_2(^1\Delta_g)$ mechanism, the cation radical of phenanthrene can be generated on silica and alumina for study by EPR [86].

Naphthalene, which is a very poor acceptor of singlet oxygen, has been shown by Sigman and coworkers to undergo facile photolysis on silica and alumina, in the presence of air, to give phthalic acid as the only isolable product, accounting for less than 50% of the consumed naphthalene [87a]. Similarly, photolysis of 1-methylnaphthalene yielded 2-acetylbenzoic acid (35% yield) as the major product at a silica-air interface, with a small amount of 1-naphthalenecarboxaldehyde also being formed. Oxygen was required for these PAHs to react photochemially on the surface. Notably, 1-cyanonaphthalene did not undergo photolysis at the silica-air interface. Addition of singlet molecular oxygen, formed by methylene blue sensitization at the silica-air interface, did not result in the formation of any naphthalene photoproducts. This result makes a $O_2(^1\Delta_g)$ mediated oxidation mechanism unlikely. An electron transfer mechanism was suggested as the major reaction pathway on the basis of these results. Additionally, superoxide was observed by ESR at the silica-air interface during photolysis. Naphthalene photolysis has also been studied in the presence of a layer of water on TiO_2, Fe_2O_3, muscovite (a phyllosilicate mica), and fly ash [87b]. The results were similar to those reported for naphthalene on silica; however, on dry TiO_2, 1,4-naphthoquinone, 2-naphthol, phthalide, phthaldialdehyde, acetophenone, benzaldehyde and benzoic acid were found in addition to phthalic acid. Photoozonolysis of naphthalene at a silica-ozone interface (oxygen-free) has been shown by Li and coworkers to proceed several orders of magnitude faster than the corresponding photooxygneation reaction; however, photoproducts were not determined [88].

In an interesting case, 1-methoxynaphthalene has been observed by Sigman and coworkers to undergo photochemical oxidation at a silica-air interface by both $O_2(^1\Delta_g)$ mediated and electron transfer mechanisms [89]. Direct photolysis gives phthalic acid and monomethyl phthalate (analyzed as dimethyl phthalate), 1,4-naphthoquinone and 4-methoxy-1,2-naphthoquinone as the only detected photoproducts. Singlet molecular oxygen sensitization (by cosorbed methylene blue) at the silica air interface gave only the quinones, whereas direct photolysis in the presence of cosorbed 2,5-dimethylfuran gave only the acidic products.

In the studies of PAH photochemistry in heterogeneous systems discussed above, the mechanistic distinctions between electron transfer and $O_2(^1\Delta_g)$ oxidation pathways are better defined than in the aqueous PAHs studies. What remains unclear, however, are the factors controlling which oxidation mechanism will be active for a particular PAH on a given substrate and the exact role of the surface in facilitating one mechanism over the other. It appears that molecules which are good $O_2(^1\Delta_g)$ acceptors can react by an energy transfer mechanism upon photolysis, although electron transfer oxidation is also possible. In the case of 1-methoxynaphthalene on silica, both mechanisms are operative. Molecules such as naphthalene, which are also poor acceptors of $O_2(^1\Delta_g)$, readily photolyze on polar surfaces, such as silica. Molecules that are poor acceptors of $O_2(^1\Delta_g)$, and have high oxidation potentials, i.e. 1-cyanonaphthalene, do not undergo any appreciable photolysis, even on polar surfaces. The role of the surface in assisting these reactions may prove to be in stabilization of the radical cation formed; however, in these reactions electron transfer is not necessarily to Lewis acid or other defect sites in the surface. Transient studies by Thomas showed that oxygen was necessary for pyrene radical cation formation from pyrene on γ-alumina and silica-alumina, while it has also been shown that oxygen is necessary for photolysis of several PAHs on silica surfaces [74, 75, 87, 89]. Studies, notably those by Wehry and Hites [69, 70], have shown that dramatic differences in photolytic half-lives occur for a given PAH on different surfaces ($t_{1/2}$ ranging from <1 h to >1000 h), further underscoring the importance of the surface-PAH interaction.

Much information remains to be gained from simple model studies of PAH photolysis at solid-air and solid-water interfaces, although an expanded set of clean surface models needs to be addressed. The current state of product and mechanism studies does not allow for a priori prediction of either the photoproduct distribution or the mechanistic pathway of PAH degradation. Although experimentally difficult, additional product studies in "real" environmental systems would greatly assist in furthering this field of research, as would additional studies at the solid-water interface.

3.2
Photochemistry of PCBs

3.2.1
Introduction

Polychlorobiphenyls (PCBs) have found extensive use in industry because of their desirable electrical-insulating and thermal properties and their resistance to decomposition. Not surprisingly, PCBs have become widely distributed around the world and have found their way into the food chain and other living organisms. As a result, the physical, chemical and biological properties of these compounds have been widely studied. Numerous monographs on the subject have been written [90–96].

The photodegradation of PCBs by sunlight has been a topic of considerable interest for more than two decades. The topic is complex because there are 209 chlorinated biphenyls, often present in environmental settings in very low concentration, in a large number of environments, each of which is subject to many reaction variables. The controlled photodegradation of PCBs using artificial light sources has also become a subject of interest because of its potential utility in decontaminating hazardous environmental samples.

This section of the article will address these two types of photochemistry of PCBs in water and heterogeneous media. The photochemistry of PCBs and other haloaromatic compounds in solution, particularly in alkanes and alcohols, has been described in several places [97–103].

3.2.2
Background

Because all chlorobiphenyls are poorly soluble in water, early work on PCB photochemistry was carried out in alkanes and alcohols. Although these solvents, especially the alkanes, are poor mimics of water, the environmental liquid, a short discussion of this photochemistry is appropriate in order to put the photochemistry in water in perspective. More detailed discussions may be found in [97–103].

Chlorobiphenyls generally have weak absorption bands above 300 nm which match the shorter wavelengths of sunlight impinging on the surface of the earth. As with biphenyl itself, the chlorobiphenyls have non-coplanar benzene rings because of the unfavorable interactions of *ortho* hydrogen and chlorine atoms on adjacent rings. This deviation from planarity is largest for those PCBs with one or more "large" chlorine atoms at the *ortho* positions. Because the excited states prefer to be planar, this added strain may be responsible for the higher photoreactivity of *ortho*-substituted chlorobiphenyls.

The photochemistry of PCBs in alkanes is free radical in character because alkanes are non-polar [104] and incapable of solvating cations and anions. The photoreaction is initiated by homolysis of a carbon-chlorine bond to form a chlorine atom and a carbon-centered radical. Based on quenching experiments, especially with O_2, and the high efficiency of intersystem crossing from singlet to triplet excited state, the homolysis occurs in the triplet excited state of the chlorobiphenyl, even though the triplet excitation energy is likely less than the bond dissociation energy of the carbon-chlorine bond. This endothermicity may account for the low quantum yields of reaction. Reaction from an upper triplet state following intersystem crossing cannot be ruled out, however. This issue will be discussed at a later stage of the article. Once the carbon-centered radical and chlorine atom are produced, they react with the solvent to yield, respectively, a biphenyl with one less chlorine and HCl.

$$ArCl \xrightarrow{h\nu} {}^{1}ArCl*$$

$${}^{1}ArCl* \xrightarrow{ISC} {}^{3}ArCl*$$

$${}^{3}ArCl* \xrightarrow{ISC} Ar\bullet + Cl\bullet$$

$$Ar\bullet + RH \longrightarrow ArH + R\bullet$$

$$Cl\bullet + RH \longrightarrow HCl + R\bullet$$

Ar = biphenyl; RH = alkane, ISC = intersystem crossing

Low molecular weight alcohols are more waterlike solvents [104], capable of solvating cations and anions. Primary and secondary alcohols in particular have, nonetheless, weak carbon-hydrogen bonds capable of donating hydrogen atoms to carbon-centered radicals. The photochemistry of chlorobiphenyls in alcohols reflects these properties, often yielding products arising by both homolysis and heterolysis mechanisms. The pathways of heterolysis will be described in detail later.

$$ArCl \xrightarrow{h\nu} {}^{1}ArCl*$$

$${}^{1}ArCl* \xrightarrow{ISC} {}^{3}ArCl*$$

$${}^{3}ArCl* \longrightarrow Ar\bullet \; Cl\bullet$$

? electron transfer

$$Ar^{+} \; Cl^{-}$$

$$Ar\bullet + R{-}\underset{\underset{H}{|}}{\overset{\overset{R'}{|}}{C}}{-}OH \longrightarrow ArH + R{-}\underset{\underset{\bullet}{}}{\overset{\overset{R'}{|}}{C}}{-}OH$$

$$Ar^{+} + RR'CHOH \longrightarrow ArOCHRR' + H^{+}$$
$$\text{phenolic ether}$$

3.2.3
Aqueous Photochemistry of PCBs

The photochemistry of PCBs in water has, surprisingly, received scant attention in the last 25 years. Nonetheless, the published work is quite informative, suggesting pathways of photodecomposition in the environment. To comprehend these results, a short discussion of relevant properties and reactions of H_2O will be given.

Based on its dielectric constant and dipole moment, the empirically derived polarity parameters Ω and E_T, and the ionizing power [104], H_2O is a superior medium for ions and their associated reactions such as the S_N1 reaction. Thermally and photochemically induced heterolysis reaction should be facilitated in this solvent. Free radical reactions may likewise take place in water, but carbon-centered radicals will abstract hydrogen from H_2O slowly due to solvent's very large bond dissociation energy of 119 kcal mol^{-1}. Contrast this to the same hydrogen atom transfer reaction in cyclohexane which is exothermic and presumably fast. The reactions of the phenyl radical model this behavior nicely. Even though these enthalpies of reaction refer to the gas phase, the hydrogen atom transfers are also likely slow in liquid water, giving ionic reactions a chance to complete.

$$C_6H_5\bullet + H_2O \longrightarrow C_6H_6 + \bullet OH \qquad \Delta H = +9 \; Kcal/mol$$

$$C_6H_5\bullet + C_6H_{12} \longrightarrow C_6H_6 + C_6H_{11}\bullet \qquad \Delta H = -25 \; Kcal/mol$$

The heterolysis of a carbon-chlorine bond in a chlorobiphenyl, modeled here by C_6H_5Cl, is endothermic by over 200 kcal mol^{-1} (>9 ev) in the gas phase. Neither the singlet or triplet excitation energy of a chlorobiphenyl is sufficient to supply this energy. Solvation energies of the ions make up the difference, however. $\Delta H_{sol}(Cl^-)$ is about -90 kcal mol^{-1} [104] and $\Delta H_{sol}(C_6H_5^+)$ must be sizable as well. If the value for $C_6H_5^+$ is in the range of -50 to -60 kcal mol^{-1}, the heterolysis energy matches the excitation energy reasonably well. Any Coulombic attraction between the ions will reduce the enthalpy of the heterolysis as well.

$$C_6H_5Cl \longrightarrow C_6H_5^+ + Cl^- \quad \Delta H = +227 \; kcal/mol \; (gas \; phase)$$

$$\Delta H = \sim +80 \; kcal/mol \; (H_2O)$$

The heterolysis, which can take place in one step or in two via the $C_6H_5\bullet$ $Cl\bullet$ radical pair, yields Cl^- and $C_6H_5^+$, which will react rapidly with H_2O to form phenol. Thus, the hallmark of a heterolysis reaction of a chlorobiphenyl in water is the formation of a hydroxybiphenyl.

$$C_6H_5Cl \xrightarrow{h\nu} C_6H_5\bullet \; Cl\bullet$$

? | electron transfer

$$C_6H_5^+ \; Cl^-$$

$$C_6H_5\bullet + H_2O \longrightarrow C_6H_6 + HO\bullet \qquad endothermic$$

$$C_6H_5^+ + H_2O \longrightarrow C_6H_5OH + H^+ \qquad exothermic$$

Crosby and Moilanen investigated the photochemistry of 8 di-, tri-, and tetrachlorobiphenyls in aerated water containing small amounts of CH_3OH (0.3 – 2 mL L^{-1}) added to solubilize the substrates [105]. The behavior of 4,4'-dichlorobiphenyl was typical of the four substrates which reacted, yielding a mixture of reduced and hydroxylated products (yields not given).

Bunce reported interesting results on the photochemistry of 2-chlorobiphenyl in equimolar CH_3CN and H_2O [106]. Only biphenyl was produced in this organic rich medium. To elucidate where the added hydrogen in the product arose, 2-chlorobiphenyl was photolyzed in CH_3CN/D_2O and CD_3CN/H_2O and the deuterium content of the biphenyl deduced by mass spectrometry. Both components of the solvent contributed H/D to the BP. If one assumes that the isotope effect k_H/k_D for hydrogen atom abstraction from acetonitrile and water is 7, it is easy to show that k_H (water)/k_H(CH_3CN) (statistically corrected) is circa 1/3 to 1/5. This is in keeping with the large OH bond dissociation energy of water (119 kcal mol^{-1}) and the much smaller C-H bond dissociation energy of CH_3CN (94 kcal mol^{-1}).

Bunce has also shown that the photodecomposition ($\lambda = 300$ nm) of 2,4-di-, 2,4,6-tri-, and 2,2',5,5' tetrachlorobiphenyl (concentration $1 - 5 \times 10^{-3}$ M) in CH_3CN/H_2O (v:v 3:1) is inhibited by the relatively unreactive 4-chlorobiphenyl ($0 - 5 \times 10^{-3}$ M) [107]. The 4-chlorobiphenyl is likely acting as an internal filter, thus preventing the more reactive chlorobiphenyls from absorbing the light. Environmental samples of PCBs, which contain lower concentrations of chlorobiphenyls, should not show this internal filter effect.

Photolysis of 2-chlorobiphenyl in the more water rich CH_3CN/H_2O (10% CH_3CN by volume) afforded 2-hydroxybiphenyl, N-(2-biphenyl)acetamide, a product which can only arise from an aryl cation, and four products derived from 2-hydroxybiphenyl, with a quantum yield of reaction of 0.20 [108]. Humic acid had no effect on the reaction rate. 4-Chlorobiphenyl afforded 4-hydroxybiphenyl, exclusively, with a very low quantum yield of reaction of 2×10^{-3}. Analysis of the rate and spectral data suggested half-lives of decompositon of 18 years and 8.2 years for 2-chloro- and 4-chlorobiphenyl, respectively, when surface waters are exposed to sunlight at 40° latitude in the summer.

In pure aerated water 2-chlorobiphenyl (5×10^{-5} M) afforded 2-hydroxybiphenyl exclusively, with Φ, the quantum yield of reaction, being

2.1×10^{-2} [109]. 4-Chlorobiphenyl was much less reactive under the same conditions ($\Phi = 2.5 \times 10^{-4}$), yielding traces of 4-hydroxybiphenyl. These data translate to half-lives of 31 and 45 days, respectively. Sugiura et al. have also shown 4-chlorobiphenyl to be photochemically unreactive in water [110].

benzvalene

Photolysis of radiolabelled 4-chlorobiphenyl ($\sim 6.6 \times 10^{-9}$ M) in degassed water, when carried to completion, afforded equal amounts of 4-hydroxybiphenyl and 3-hydroxybiphenyl; no biphenyl was detected [111]. Owing to the sensitivity of the isotopic dilution method used for product analysis, one biphenyl would have been detected for every 10^6 4-chlorobiphenyls that had reacted. A variety of experiments proved that the 3-hydroxybiphenyl arose in two steps: (1) photoisomerization of 4-chlorobiphenyl to 3-chlorobiphenyl and (2) photohydrolysis of 3-chlorobiphenyl. Photoisomerization of chlorobenzenes had been reported earlier [112].

Experiments with radiolabelled 2-chloro- and 3-chlorobiphenyl (both $\sim 2 \times 10^{-5}$ M) in degassed water were also carried out [113]. 2-Chlorobiphenyl, which reacted more than five times faster than 3-chlorobiphenyl, yielded 2-hydroxybiphenyl, exclusively. 3-Chlorobiphenyl, on the other hand, afforded four products: 2-hydroxy- and 4-hydroxybiphenyl, likely formed by photohydrolysis of the corresponding chlorides, 3-hydroxybiphenyl and a small amount of biphenyl.

The photohydrolysis of 4-chlorobiphenyl in CH_3CN/H_2O (1:9 v:v) was suppressed by added HCl and NaCl [113]. Careful analysis of the data showed that the retardation was due to a common ion effect, the first ever seen for a photoreaction. This should not be surprising, however, because aryl cations are highly reactive with nucleophiles [114].

3.2.4
Micelles and Pseudo-Micelles

The photochemistry of PCBs in aqueous micellar solutions is important for two reasons: (1) The water soluble surfactants may mimic the behavior of naturally occurring water soluble organic compounds. (2) Micelles solubilize the hydrophobic PCBs. The photochemistry of PCBs in aqueous micellar solution thus represents a potential methodology for the destruction of PCBs. Surprisingly, little work has been carried out in either one of these areas.

Hashimoto and Thomas investigated the photoionization of 4-chlorobiphenyl from upper triplet states in anionic (sodium dodecylsulfate) and cationic (cetyltrimethyl-ammonium bromide) micellar solutions [115]. No carbon-chlorine bond cleavage was observed in this study. Photodechlorination of 6 chlorobiphenyls by $NaBH_4$ in anionic (sodium dioctylsulfosuccinate) and neutral (Brij 58; a polyether) micellar CH_3CN/H_2O (9:1 v:v) solutions occurred more rapidly and cleanly than in CH_3CN/H_2O alone [116, 117]. Experiments suggested that the photoreductions occurred by competing mechanisms: (1) direct attack of hydride on the chlorobiphenyl excited state and (2) photoinduced electron transfer from BH_4^- to the chlorobiphenyl excited state [116–118]. The photoreduction also occurred in CH_3CN/H_2O in the absence of detergents [116–118].

Pseudo-micelles are also good media for the destruction of PCBs. Guillet and coworkers have prepared a water soluble polymer, poly(sodium styrene-sulfonate-CO-2-vinylnaphthalene), which solubilizes PCBs in water and catalyzes their photodecomposition [119, 120]. Photolysis of 2,2′,3,3′,6,6′-hexachlorobiphenyl (HCB) (9.4×10^{-6} M) in water containing the polymer

($2 \, g \, dm^{-3}$) afforded several less chlorinated biphenyls [121]; no hydroxybiphenyls were reported. The source of added hydrogen in the reduced products is unknown. The photoreduction was faster than in pure water and was inhibited by added O_2. Based on quenching of the fluorescence of naphthalene (N) and the naphthalene in the polymer by HCB and the energetics of electron transfer, a photoinduced electron transfer mechanism for the reduction was proposed. Guillet and coworkers also showed that Aroclors 1254, 1248 and 1242 were also photoreduced in aqueous solutions of the polymer [122].

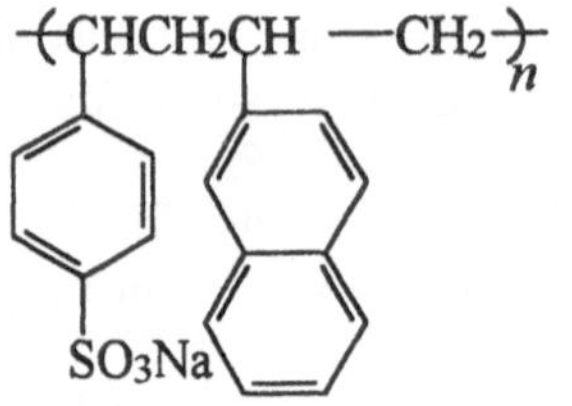

$$\left(\!\!\!\!\text{CHCH}_2\text{CH} \quad \text{---CH}_2\right)_{\!n}$$

SO$_3$Na

Polymer

$$N_0 \xrightarrow{h\nu} {}^1N^*$$

$${}^1N^* + HCB \longrightarrow \underset{\text{exciplex}}{[N\ ACB]} \longrightarrow N^{+\cdot}\ HCB^{\bar{\cdot}}$$

$$HCB^{\bar{\cdot}} \longrightarrow \longrightarrow \quad \text{Pentachlorobiphenyl}$$

3.2.5
Photoinduced Electron Transfer

Photoinduced electron transfer reactions have been studied extensively in recent years [123, 124] and can be applied to the photoreduction of PCBs. As these reactions generate ions, they should be particularly well suited to water and mixed water solvents. A few instances of this chemistry were described in the previous section in connection with micelles. Several additional examples will now be presented.

Epling has studied the photoreduction of 4-chloro- and 4,4'-dichlorobiphenyl in CH_3CN/H_2O (1:1 v:v) using triethylamine as electron donor and a dye, acriflavin, protoporphyrin X or methylene blue, as co-sensitizer [125]. This is an intriguing approach for the dechlorination of PCBs because it uses visible light.

The photoreduction of chlorobiphenyls in CH_3CN/H_2O [117] and CH_3CN/D_2O using alkyl sulfides as electron donors has also been reported [126]. Interestingly, the singlet excited state of the conjugate base of hydroquinone in isopropanol/H_2O (1:1 v:v) at pH = 7.0 also functions as an electron donor in the reduction of PCBs in Aroclor 1254 [127]. This reduction occurs in this manner because the singlet excited state of hydroquinone is more acidic (pK_a = 3.1) than hydroquinone itself.

$$\text{Dye} \xrightarrow{h\nu} \text{Dye}^*$$

$$\text{Dye}^* + \text{Et}_3\text{N} \longrightarrow \text{Dye}^{\overline{\cdot}} + \text{Et}_3\text{N}^{+\cdot}$$

$$\text{Dye}^{\overline{\cdot}} + \text{4-ClBP} \longrightarrow \text{Dye} + \text{4-ClBP}^{\overline{\cdot}}$$

$$^1\text{ArOH}^* + \text{H}_2\text{O} \longrightarrow \text{ArO}^{\overline{\cdot}*} + \text{H}_3\text{O}^+$$

$$^1\text{ArO}^{\overline{\cdot}*} + \text{Ar'Cl} \longrightarrow \text{ArO}^{\cdot} + \text{Ar'Cl}^{\overline{\cdot}} \longrightarrow \text{reduction}$$

3.2.6
Reactions with Photogenerated Hydroxyl Radicals

Fenton's reagent, consisting of a mixture of Fe^{+2} and H_2O_2 in water, thermally generates highly reactive hydroxyl radicals which have been used to degrade PCBs in water [128]. A photoFenton reagent – Fe^{+3}/H_2O_2/light (350 nm) – also generates hydroxyl radicals [129] which degrade PCBs in an Aroclor 1243/H_2O (pH = 2.75) slurry at 66 °C. No degradation occurred at room temperature and added surfactants had a detrimental effect on the degradation. As ferrous and ferric ions and H_2O_2 exist in surface waters, Fenton and photoFenton reagents may play a role in the removal of PCBs in the environment.

$$Fe^{+3}/H_2O_2 \xrightarrow[350\ nm]{h\nu} \cdot OH$$

$$PCB + OH\cdot \longrightarrow PCB\ degradation$$

The photoFenton reagent was also applied to the decomposition of 2-chloro-, 2,2′,5′-trichloro- and 2,2′,4,4′,5-pentachlorobiphenyl adsorbed on diatomaceous earth [130]. Only the first two biphenyls degraded. Because the pentachlorobiphenyl was too poorly soluble in water and the hydroxyl radicals were only active in water, the pentachlorobiphenyl did not react.

Interestingly, UV irradiation of ozone in seepage waters from landfills had little effect on PCB concentrations [131]. Irradiation of ozone in water is a likely source of hydroxy radicals. As we have just seen that hydroxyl radicals are highly reactive with PCBs, this lack of reactivity may be due to the fact that the seepage water contains PAHs which are more highly reactive with hydroxyl radicals than PCBs are.

3.2.7
Heterogeneous Photochemistry of PCBs

There are two major reasons why the photochemistry of PCBs in heterogeneous media has been investigated: (1) to lead to an understanding of the mechanisms by which PCBs adsorbed on soils and other solids photodegrade in the environment, and (2) to develop methods using solid semiconductor photocatalysts such as titanium dioxide to remove PCBs from contaminated water, soil and other solids. These two areas will be described in turn.

Korte and coworkers were undoubtedly the first to study the photochemistry of chlorinated biphenyls in an heterogeneous environment. In a series of papers these researchers demonstrated that PCBs adsorbed on silica gel in the presence of O_2 are mineralized by light [132–137]. Occhiucci and Patacchiola showed that the photodegradation of four chlorobiphenyls adsorbed on silica gel and montmorillonite, a clay mineral, was greatly enhanced (typically one to two orders of magnitude) by added triethylamine [138]. 4,4′-Dichlorobiphenyl, for example, afforded 4-chlorobiphenyl and biphenyl. These reactions are reminiscent of photoinduced electron transfer reactions occurring in solution.

2-Chlorobiphenyl and 4,4′-dichlorobiphenyl, as well as biphenyl, were subsequently shown to be hydroxylated when photolyzed on silica gel in the presence of air [139, 140]. 2-Chlorobiphenyl, for example, afforded dibenzofuran via the intermediate 2-chloro-2-hydroxybiphenyl. These hydroxylations are different than those seen in water where Cl is replaced by OH. Here the non-chlorine-containing carbons are hydroxylated.

Mao and Thomas explored further the mechanism by which the biphenyls are hydroxylated [141]. They examined the photochemistry of biphenyl, 4-chlorobiphenyl, and 4,4′-dichlorobiphenyl on γ-alumina and silica-alumina, a solid Brønsted acid [142], using electron spin resonance and diffuse reflectance absorption and fluorescence spectroscopy. All substrates afforded hydroxylated products when photolyzed in air at 254 nm. Based on the detection of radical cations of the substrates and charge transfer bands in certain cases, the authors formulated a mechanism involving electron transfer from the substrate excited state to O_2 on the surface (superoxide was not detected by ESR) followed by reaction of the radical cation with surface-bound water. It is interesting to note that O_2, in conjunction with the surface, functions as an electron acceptor in these cases, but as a triplet state quencher with the same substrates in solution.

$$4\text{-ClBP} \xrightarrow[254\text{ nm}]{h\nu} 4\text{-ClBP}^* \xrightarrow{O_2} 4\text{-ClBP}^{+\cdot} + O_2/\text{surface}^{\overline{\cdot}}$$

$$4\text{-ClBP}^{+\cdot} + H_2O \longrightarrow H^+ + 4\text{-ClBP(OH)}\cdot \longrightarrow \text{hydroxychlorobiphenyl}$$

Surprisingly few studies on the photochemistry of PCBs on soil have been reported. Biphenyl and other contaminants adsorbed on soil are known to photodegrade when exposed to sunlight [78]. PCBs (Aroclor 1248) likewise photodegrade on soil, but the reactions are slow and depend on soil type and the thickness of the soil sample [143]. It is possible to bypass this slow photoreactivity by removing the PCBs from the soil by extraction with biosurfactants in water and then photodegrading them in organic solvents [144].

Photocatalytic decompositon of pollutants on semiconductors such as titanium dioxide (TiO_2) has been studied extensively in recent years [145–147]. To see how this methodology applies to the degradation of PCBs consider what occurs when a semiconductor is irradiated. Photolysis leads to a promotion of an electron from the semiconductor's valence band (VB) to its conduction band (CB), creating a hole (h^+; oxidant) in the VB and an electron in the CB (reductant). If the semiconductor is in contact with water, which is usually the case in these experiments, h^+ reacts with water to produce H^+ and $\cdot OH$, the highly reactive radical primarily responsible for the destruction of the pollutant. Other radicals such as $\cdot OOH$ may also be generated. Jaeger and Bard, for example, have detected $\cdot OH$ and $\cdot OOH$ generated on TiO_2 by spin trapping experiments [148].

A large number of studies have been carried out on the photocatalytic degradation of individual chlorobiphenyls and PCB mixtures on semiconductors [149–160]. This has proven to be a rapid, very effective method for the removal of PCBs from water [149–154, 158–160], soils, slurries of clays or sediments and water [155–157] and hexane-water mixtures [152]. TiO_2 is generally the best semiconductor for the degradation [150], although Pt on TiO_2 [154], Pt/TiO_2 on SiO_2 [160] and TiO_2-zinc tetraphenylporphin on polyvinyl-pyridine in the presence of hexane-water mixtures [152] are also effective. Sunlight is sufficiently energetic to initiate the photodegradation on TiO_2 [155–156]. Although little work has been carried out on the mechanism of degradation, $\cdot OH$ appears to be the reagent responsible for the degradation. No degradation products have been identified in these photocatalytic decompositions. Less chlorinated PCBs are more reactive than more highly chlorinated PCBs [156–158]. The kinetics of the reaction do follow a Langmuir-Hinshelwood form [160], suggesting that the degradation occurs on the surface of the semiconductor; this may not be the case, however, because models of the degradation occurring in solution give the same kinetic form [161]. Added ultrasound accelerates the decompositon [154]. Added isopropanol reduces the rate of decomposition [151], probably because of its efficient reaction with hydroxyl radicals.

$$\cdot OH + CH_3\overset{\overset{\displaystyle OH}{|}}{\underset{\underset{\displaystyle H}{|}}{C}}CH_3 \longrightarrow H_2O + CH_3\overset{\overset{\displaystyle OH}{|}}{\underset{\displaystyle \cdot}{C}}CH_3$$

Acknowledgement. This research was sponsored by the Division of Chemical Sciences, Office of Basic Energy Sciences, US Department of Energy under contract DE-AC05-96OR22464 with Oak Ridge National Laboratory, managed by Lockheed Martin Energy Research Corp.

References

1. Callahan MA, Slimak MW, Gabelc NW, May IP, Fowler CF, Freed JR, Jennings P, Durfee RL, Whitmore FC, Maestri B, Mabey WR, Holt BR, Gould C (1979) EPA-440/4-79-029. US Environmental Protection Agency, Washington, DC
2. Mabey WR, Smith JH, Podoll RT, Johnson HL, Mill T, Chou TW, Gates J, Partridge IW, and Vandenberg D (1979) EPA-440/4-81-014. US Environmental Protection Agency, Washington, DC
3. Menzie CA, Potockim BB, Santodonato J (1992) Environ Sci Technol 26:1278
4. Harvey RG (1985) Polycyclic hydrocarbons and carcinogenesis. ACS symposium series 283. American Chemical Society, Washington DC
5. Jones KC, Stratford JA, Waterhouse KS, Furlong ET, Giger W, Hites RA, Schaffner, C, Johnston AE (1989) Environ Sci Technol 23:95

6. Broman D, Näf C, Rolff C, Zebühr Y (1991) Environ Sci Technol 25:1850
7. Gould IR (1987) Conventional light sources. In: Scaiano JC (ed) Handbook of organic photochemistry, vol II. CRC Press, Boca Raton, FL, p 155
8. Leifer A (1988) The kinetics of environmental aquatic photochemistry. Theory and practice. American Chemical Society, Washington, DC
9. Zepp RG, Cline DM (1977) Environ Sci Technol 11:359
10. Dulin D, Mill T (1982) Environ Sci Technol 16:815
11. Draper WM (1985) Chemosphere 14:1195
12. Draper WM (1987) Measurement of quantum yields in polychromatic light: Dinitroaniline herbicides. In: Zika RG, Cooper WJ (eds) Photochemistry of environmental aquatic systems. ACS symposium series 327. American Chemical Society, Washington, DC, p 268
13. Miller GC, Zepp RG (1979) Water Res 13:453
14. Plane JMC, Zika RG, Zepp RG, Burns LA (1987) Photochemical Modeling applied to natural waters. In: Zika RG, Cooper WJ (eds) Photochemistry of environmental aquatic systems. ACS symposium series 327. American Chemical Society, Washington, DC, p 250
15. Whitney LV (1941) J Opt Soc Am 31:714
16. Zepp RG, Wolfe NL (1987) Abiotic transformations of organic chemicals at the particle-water interface. In: Stumm W (ed) Aquatic surface chemistry: Chemical processes at the particle-water interface. John Wiley, New York, NY, p 423
17. Kearns DR (1971) Chem Rev 71:395
18. Lissi EA, Encinas MV, Lemp E, Rubio MA (1993) Chem Rev 93:699
19. Gorman AA, Rodgers MA (1989) Singlet Oxygen. In: Scaiano JC (ed) Handbook of organic photochemistry, vol II. CRC Press, Boca Raton, FL, p 229
20. Zepp RG, Wolfe NL, Baughman GL, Hollis RC (1977) Nature 267:421
21. Zepp RG, Baughman GL, Schlotzhauer PF (1981) Chemosphere 10:119
22. Grigor'ev EI, Myasnikov IA, Tsivenko VI (1982) Russ J Phys Chem 56:1059
23. Gohre K, Miller GC (1985) J Chem Soc Faraday Trans 1, 81:793
24. Stevens B, Ors JA, Pinsky ML (1974) Chem Phys Lett 27:157
25. Vohra KG, Chatha JPS, Arora PK, Raja N (1975) Gas phase studies with singlet molecular oxygen from a discharge-flow system. Proceedings of the Symposium on Singlet Molecular Oxygen, p 118
26. Williamson F, Brummer JG (1981) J Phys Chem Ref Data 10:809
27. Fischer AM, Winterele JS, Mill T (1987) Primary photochemical processes in photolysis mediated by humic substances. In: Zika RG, Cooper WJ (eds) Photochemistry of environmental aquatic systems. ACS symposium series 327. American Chemical Society, Washington, DC, p 141
28. Power J F, Sharma DK, Langford CH, Bonneau R, Joussot-Dubien J (1987) Laser flash photolytic studies of a well-characterized soil humic substance. In: Zika RG, Cooper WJ (eds) Photochemistry of environmental aquatic systems. ACS symposium series 327. American Chemical Society, Washington, DC, p 157
29. Zepp RG, Braun AM, Hoigné J, Leenheer JA (1987) Environ Sci Technol 21:485
30. Mill T (1980) Photooxidation in the Environment. In: Hutzinger O (ed) Handbook of Environmental Chemistry, vol 2, Part A. Springer, Berlin Heidelberg New York, p 77
31. Buxton GV, Greenstock CL, Helman WP, Ross AB (1988) J Phys Chem Ref Data 17:513
32. Kochany J, Maguire RJ (1994) Sci Total Environ 144:17
33. Payne JR, Phillips CR (1985) Environ Sci Technol 19:569
34. Larson RA, Berenbaum MR (1988) Environ Sci Technol 22:354
35. Tuveson RW, Wang G-R, Wang TP, Kagan J (1990) Photochem Photobiol 52:993
36. Kagan J, Kagan ED, Kagan IA, Kagan PA (1987) Do polycyclic aromatic hydrocarbons, acting as photosensitizers, partiipate in the toxic effects of acid rain? In: Zika RG, Cooper WJ (eds) Photochemistry of enviromental aquatic systems. ACS symposium series 327. American VChemical Society, Washington, DC p 191

37. Zepp RG, Schlotzhauer PF (1979) Photoreactivity of selected aromatic hydrocarbons in water. In: Jones PW, Leber P (eds.) Polynuclear aromatic hydorcarbons: Third international symposium on chemistry and biology- carcinogenesis and mutagenesis. Ann Arbor Science, Ann Arbor, MI, p 141

38. Picel KC, Stamoudis VC, Simmons MS (1985) Photolysis rates of selected polynuclear aromatic hydrocarbons in aqueous coal-oil systems. In: Cooke M, Bennis AJ (eds) Polynuclear aromatic hydrocarbons: Mechanisms, methods and metabolism, Battelle Press, Columbus, OH, p 1013

39. Mill T, Mabey WR, Lan BY, Baraze, A (1981) Chemosphere 10:1281

40. Katz M, Chan C, Tosine H, Sakuma T (1979) Relative rates of photochemical and biological oxidation of polynuclear aromatic hydrocarbons. In: Jones PW, Leber P (eds) Polynuclear aromatic hydrocarbons. Third international symposium on chemistry and biology- carcinogenesis and mutagenesis. Ann Arbor Science, Ann Arbor, MI, p 171

41. Zepp RG, Schlotzhauer PF (1983) Environ Sci Techmol 17:462

42. Paalme L, Uibopuu H, Rohtala I, Pahsapill J, Goubergrits M, Jacquignon, PC (1983) Reactivity of PAHs in UV- and γ-radiation initiated oxidation reactions. In: Cooke M, Dennis AJ (eds.) Polynuclear aromatic hydrocarbons: Formation, metabolism and measurement. Battelle Press, Columbus, OH, p 999

43. Shevchuk I (1986) Eesti NSV Tead Akad Toim, Keem 35:128

44. Rehm D, Weller A (1970) Israel J Chem 8:259

45. Syamala MS, Ramamurthy V (1986) J Org Chem 51:3712

46. Rideout DC, Breslow R (1980) J Am Chem Soc 102:7816

47. Wan P, Culshaw S, Yates K (1982) J Am Chem Soc 104:2509

48. Yates P, Wan P (1984) Rev Chemical Intermed 5:157

49. Fukuda K, Inagaki Y, Maruyama T, Kojima HI, Yoshida T (1988) Chemosphere 17:651

50. Steenken S, Warren CJ, Gilbert BC (1990) J Chem Soc Perkin Trans 2, 335

51. Vauthey E, Haselbach E, Suppan P (1987) Helv Chim Acta 70:347

52. Delcourt MO, Rossi MJ (1982) J Phys Chem 86:3233

53. Sigman ME, Zingg SP, Pagni RM, Burns JH (1991) Tetrahedron Letters 32:5737

54. Hondrogiannis G, Lee CW, Pagni RM, Mamantov G (1993) J Am Chem Soc 115:9828

55. Lee C, Winston T, Unni A, Pagni RM, Mamantov G (1996) J Am Chem Soc 118:4919

56. Sigman ME, Chevis EA, Brown A, Barbas JT, Dabestani R, Burch EL (1996) J Photochem Photobiol A: Chem 94:149

57. Burke NAD, Templin M, Guillet JE (1996) J Photochem Photobiol A: Chem 100:93

58. Fox MA (1985) Organic phototransformations in nonhomogeneous media, ACS symposium series 278. American Chemical Society, Washington, DC

59. Thomas JK (1984) The chemistry of excitation at interfaces. American Chemical Society, Washington, DC (Kalyanasundaram K (1987) Photochemistry in microheterogeneous systems. Academic Press, New York, NY

60. Anpo M, Matsuura T (1989) Studies in surface science and catalysis, Vol 47: Photochemistry on solid surfaces. Elsevier, New York NY

61. Thomas JK (1993) Chem Rev 93:301

62. Blau L, Güsten H (1982) Quantum yields of photodecomposition of polynuclear aromatic hydrocarbons adsorbed on silica gel. In: Cooke M, Dennis AJ, Fisher GL (eds) Polynuclear aromatic hydrocarbons. Physical and biological chemistry. Battelle, Richland, WA, p 133

63. Zingg SP, Sigman ME (1993) Photochem Photobiol 57:453

64. Sigman ME, Zingg SP (1994) Anthracene photochemistry in aqueous and heterogeneous media. In: Helz GR, Zepp RG, Crosby DG (eds) Aquatic and surface photochemistry. Lewis, Ann Arbor, MI, p 197

65. Dabestani R, Ellis KJ, Sigman ME (1995) J Photochem Photobiol A: Chem 86:231

66. König J, Balfanz E, Funcke W, Romanowski T (1985) Structure-reactivity relationships for the photooxidation of anthracene and its anellated homologues. In: Cooke M, Dennis AJ (eds) Polynuclear aromatic hydrocarbons: Mechanisms, methods and metabolism. Battelle Press, Richland, WA, p 739

67. Korfmacher WA, Wehry EL, Mamantov G, Natusch DFS (1980) Environ Sci Technol 14:1094
68. Korfmacher WA, Natusch DFS, Taylor DR, Mamantov G, Wehry EL (1980) Science 207:763
69. Wehry EL, Mamantov G (1994) Sorption and photochemical transformations of polycyclic aromatic compounds on coal stack ash particles. In: Helz GR, Zepp RG, Crosby DG (eds.) Aquatic and surface photochemistry. Lewis, Ann Arbor, MI, p 173
70. Behymer TD, Hites RA (1986) Photolysis of PAHs adsorbed on silica gel and fly ash. In: Cooke M, Dennis AJ (eds) Polynuclear aromatic hydrocarbons. Chemistry, characterization and carcinogenesis. Battelle, Richland, WA, p 65
71. Behymer TD, Hites RA (1985) Environ Sci Technol 19:1004
72. Fatiadi AJ (1967) Environ Sci Technol 1:570
73. Taskar PK, Solomon JJ, Daisey JM (1985) Rates and products of reaction of pyrene adsorbed on carbon, silica and alumina. In: Cooke M, Dennis AJ (eds) Polynuclear aromatic hydrocarbons: Mechanisms, methods and metabolism. Battelle Press, Richland, WA, p 1285
74. Mao Y, Thomas JK (1992) Langmuir 8:2501
75. Liu X, Iu K-K, Mao Y, Thomas JK (1994) Photoinduced reactions on clay and model systems. In: Helz GR, Zepp RG, Crosby DG (eds.) Aquatic and surface photochemistry. Lewis, Ann Arbor, MI, p 187–699
76. Eisenberg WC, Taylor K, Cunningham DLB, Murray RW (1985) Atmospheric fate of polycyclic organic material. In: Cooke M, Dennis AJ (eds) Polynuclear aromatic hydrocarbons: Mechanisms, methods and metabolism. Battelle Press, Richland, WA, p 395
77. Daisey JM, Boone PM (1986) Rates and products of reaction of benzo[a]pyrene adsorbed on various particles. Air pollution control association. Proceedings APCA 79th annual meeting, paper 86–77.3, p 1
78. Dupont RR, McLean JE, Hoff RH, Moore WM (1990) J Air Waste Manage Assoc 40:1257
79. Wilkinson F, Worrall DR, Williams SL (1995) J Phys Chem 99:6689
80. Dabestani R, Nelson M, Sigman ME (1996) Photochem Photobiol 64:80
81. Hartmann I-M, Hartmann W, Schenck GO (1967) Chem Ber 100:3146
82. Bauer RK, Borenstein R, de Mayo P, Okada K, Rafalska M, Ware WR, Wu KC (1982) J Am Chem Soc 104:4635
83. Bauer RK, de Mayo P, Okada K, Ware WR, Wu KC (1983) J Phys Chem 87:460
84. Barbas JT, Dabestani R, Sigman ME (1994) J Photochem Photobiol A: Chem 80:103
85. Barbas JT, Sigman ME, Dabestani R (1996) Environ Sci Technol 30:1776
86. Ueda H (1977) J Catalysis 47:284
87. (a) Barbas JT, Sigman ME, Buchanan AC III, Chevis EA (1993) Photochem Photobiol 58:155. (b) Guillard C, Delprat H, Hoang-Van C, Pichat P (1993) J Atmos Chem 16:47
88. Schutt WS, Sigman ME, Li Y (1996) Analytica Chim Acta 319:369
89. Sigman ME, Barbas JT, Chevis EA, Dabestani R (1996) New J Chem 20:243
90. PCB in Water A Bibliography Vol 1–3 (1978) Dept of Interior, Washington
91. Nesbit ICT (1976) Criteria Document for PCBs. Mass Audubon Soc, Lincoln, Mass
92. D'Itri FM, Kamrin MA (eds) (1983) PCBs: Human and Environmental Hazards. Butterworths, Boston
93. Ackerman DG, Scinto LL, Bukshi PS, Delamyea RG, Johnson RJ, Richard G, Takata AM Sworzyu EM (1983) Destruction and Disposal of PCBs by Thermal and Non-thermal Methods. Noyes Data Corp, Park Ridge NJ
94. PCBs and the Environment vol 1–3 (1986) Ward JS (ed) CRC Press, Boca Raton FL
95. Hazards, Decontamination and Replacemnt of PCB (1988) Crine JP (ed) Plenum, New York
96. Polychlorinated Biphenyls and Terphenyls, 2nd edn. (1993) vol 140 of Environmental Health Criteria World Health Org, Geneva
97. Hutzinger O, Safe S, Zitko V (1974) Chemistry of PCBs. CRC Press, Cleveland, chap 6
98. Safe S, Bunce NJ, Chittim B, Hutzinger O, Ruzo, LO (1976) In: Keith LH (ed) Identification and Analysis of Organic Pollutants in Water. Ann Arbor Science, Ann Arbor

99. Grimshaw J (1981) Chem Soc Rev 10:181
100. Bunce NJ (1982) Chemosphere 11:701
101. Davidson RS, Goodin JW, Kemp G (1984) Adv Phys Org Chem 20:191
102. Choudhry GG, Webster GRB, Hutzinger O (1988) Toxicol Environ Chem 17:267
103. Malkin J (1992) Photophysical and Photochemical Properties of Aromatic Compounds CRC Press, Boca Raton FL
104. Reichardt C (1990) Solvents and Solvent Effects in Organic Chemistry. 2nd edn. VCH,Weinheim
105. Crosby DG, Moilanen KW (1973) Bull Environ Contam Toxicol 10:372
106. Bunce NJ (1978) Chemosphere 7:653
107. Bunce NJ, Kumar Y, Brownlee BG (1978) Chemosphere 7:155
108. Dulin D, Drossman H, Mill T (1986) Environ Sci Technol 20:72
109. Mansour M, Fecht E, Méallier P (1989) Toxicol Environ Chem 20–21:139
110. Sugiura K, Aoki M, Kaneko S, Daisaku I, Komatsu Y, Shibuya H, Suzuki H, Goto M (1984) Arch Environ Contam Toxicol 13:745
111. Moore T, Pagni RM (1987) J Org Chem 52:770
112. Choudry GC, Root AAM, Hutzinger O (1979) Tetrahedron Lett 20:2059
113. Orvis J, Weiss J, Pagni RM (1991) J Org Chem 56:1851
114. Scaiano JC, Kim-Thuan N (1983) J Photochem 23:269
115. Hashimoto S, Thomas JK (1991) J Photochem Photobiol A: Chem 55:377
116. Epling GA, Florio E (1986) Tetrahedron Lett 27:675
117. Epling GA, Florio EM, Bourque AJ, Qian, X-H, Stuart JD (1988) Environ Sci Technol 22:952
118. Tsujimoto K, Tasaka S, Ohashi K (1975) J Chem Soc Chem Commun 758
119. Guillet JE, Nowakowska M (1999) Chemistry in Britain 327
120. Guillet JE (1991) Pure Appl Chem 63:917
121. Nowakowska M, Sustar E, Guillet JE (1991) J Am Chem Soc 113:253
122. Sustar E, Nawakowska M, Guillet JE (1992) J Photochem Photobiol A Chem 63:357
123. Kavarnos GJ, Turro NJ (1986) Chem Rev 86:401
124. Kavarnos GJ (1993) Fundametals of Photoinduced Electron Transfer, VCH, New York
125. Epling GA, Wang Q, Qui Q (1991) Chemosphere 22:959
126. Davidson RS, Goodin JW, Pratt JE (1982) Tetrahedron Lett 23:2225
127. Chaudhary SK, Mitchell RH, West PR (1984) Chemosphere 13:1113
128. Sediak DL, Andrew AW (1991) Environ Sci Technol 25:1419
129. Pignatello, JJ, Chapa G (1994) Environ Toxicol Chem 13:423
130. Sediak DL, Andrew AW (1994) Water Res 28:1207
131. Vollmuth S, Wenzel A, Niessner R (1995) Proc SPIE-Int Soc Opt Eng 2504:520
132. Gäb S, Nitz S, Parlar H, Korte F (1975) Chemosphere 4:251
133. Gäb S, Schmitzer J, Thamm HW, Parlar H, Korte F (1977) Nature 270:331
134. Kotzias D, Klein W, Korte F (1977) Chemosphere 6:99
135. Kotzias D, Klein W, Lotz F, Nitz S, Korte F (1979) Chemosphere 8:301
136. Lotz F, Nitz S, Korte F (1979) Chemosphere 8:763
137. Kotzias D, Nitz S, Korte F (1981) Chemosphere 10:415
138. Occhiucci G, Patacchiola A (1982) Chemosphere 11:255
139. Kotzias D, Hustert K, Parlar H, Korte F (1983) Naturwissenschaften 70:413
140. Kotzias D, Herrmann M, Parlar H, Korte F (1984) Chemosphere 13:623
141. Mao Y, Thomas JK (1992) J Chem Soc Faraday Trans 88:3079
142. Tanabe K, Misono M, Ono Y, Hattori H (1989) New Solid Acids and Bases. Kodansha, Tokyo
143. (1994) EPA Document DPA/540/F-94/502
144. Samson R, Cseh T, Hawari J, Greer CW, Zaloum R (1994) Sci Tech Eau 23:15
145. Legrini G, Oliveros E, Braun AM (1993) Chem Rev 93:671
146. Ollis DF (1985) Environ Sci Technol 19:480
147. Ollis DF, Pelizzetti E, Serpone N (1991) Environ Sci Technol 25:1523
148. Jaeger CD, Bard AJ (1979) J Phys Chem 83:3146

149. Carey JH, Lawrence J, Tosine HM (1976) Bull Environ Contam Toxicol 16:697
150. Carey JH, Oliver BG (1980) Water Poll Res J Can 15:157
151. Tunesi S, Anderson MA (1987) Chemosphere 16:1447
152. Ménassa PE, Mak MKS, Langford CH (1988) Environ Tech Lett 9:825
153. Pelizzetti E, Borgarello E, Serpone N (1988) Chemosphere 17:499
154. Johnston AJ, Hocking P (1993) In: Tedder DW, Pohland FG (eds) Emerging Technologies in Hazard Waste III. 3:106
155. Zhang P, Scrudato RJ, Pagano JJ, Roberts RN (1993) Photocatalytic decomposition of PCBs in aqueous systems with solar light. In: Ollis DF, Al-Elcabi H (eds) Photocatalytic Purification and Treatment of Water and Air. Elsevier, Amsterdam, p. 1213
156. Zhang P-C, Scrudato RJ, Pagano JJ, Roberts RN (1993) Chemosphere 26:1993
157. Chiarenzelli J, Scrudato R, Wunderlich M, Rafferty D, Jensen K, Oenga G, Roberts R, Pagano J (1995) Chemosphere 31:3259
158. Huang I-W, Hong C-S, Bush B (1996) Chemosphere 32:1869
159. Ogawa S, Nozawa K, Hanasaki Y, Hirayama T (1996) Jpn J Toxicol Environ Health 42:44
160. Zhang Y, Crittenden JC, Hand DW, Perram DL (1996) J Solar Energy Engineering 118:123
161. Turch C, Ollis DF (1990) J Catal 122:178

7 Phototransformations Induced in Aquatic Media by NO$_3^-$/NO$_2^-$, Fe^{III} and Humic Substances

P. Boule, M. Bolte and C. Richard

Laboratoire de Photochimie Moléculaire et Macromoléculaire Université Blaise Pascal, UMR CNRS 6505 (Clermont-Ferrand) F-63177 Aubière Cedex, France.
E-mail: boule@cicsun.univ-bpclermont.fr

This review describes the reaction mechanisms of three typical classes of compounds which play a role in the photochemical processes occurring in natural waters, namely nitrate ions, Fe^{III} salts and humic substances.

The UV excitation of nitrate ions leads to the formation of hydroxyl radicals, atomic oxygen and nitrogen dioxide. Hydroxyl radicals induce the oxidation of organic substrates. Nitrogen dioxide disproportionates into nitrite and nitrate ions but, in some cases, it can induce nitration and nitrosation. Nitrite ions have a similar behaviour. In both cases the presence of oxygen and an ˙OH quencher plays an important role in the orientation of the reaction.

The photoinductive role of iron (III) depends on the nature of the organic pollutant. If the substrate is able to complex iron (III), the resulting iron salt undergoes an intramolecular redox process giving rise to iron (II) and the oxidized ligand. If the substrate does not present any complexing feature, its degradation is initiated by ˙OH radicals arising from iron (III) aquocomplexes in the excited state.

Upon excitation of humic substances, several reactive species are produced: singlet oxygen, solvated electrons, hydroxyl radicals, HO$_2$˙/O$_2$˙⁻. They oxidize or reduce a large variety of organic compounds. Triplet excited states of humic substances also seem capable of reacting with substrates through energy, hydrogen or electron transfer processes. Examples of phototransformations are given.

Keywords: nitrate ions, iron (III) complexes, humic substances, induced photooxidations, photonitration.

Contents

The Handbook of Environmental Chemistry Vol. 2 Part L
Environmental Photochemistry (ed. by P. Boule)
© Springer-Verlag Berlin Heidelberg 1999

1
Introduction

Phototransformation plays an important role in the degradation of organic substances present in natural waters, particularly in the case of slightly bio-degradable pollutants. Direct photolysis takes place with substrates absorbing solar-light, i.e. wavelengths longer than ca. 300 nm. Nevertheless with substances absorbing shorter wavelengths photodegradation can be induced by compounds such as transition metals, some inorganic ions and natural organic molecules containing carbonyl or phenolic groups. These substances produce radical or oxidizing species. The present study is focussed on the mechanisms of reactions of three typical classes of compounds which play a major role in surface waters: nitrate ions, salts of Fe^{III} and humic substances.

2
Reactions Induced by Excitation of Nitrate and Nitrite Ions

2.1
Occurrence of Nitrates and Nitrites in Natural Waters. Their Observed or Expected Photochemical Incidence

Nitrate ions are usually present in natural waters, including cloud water. Their concentration in sea water changes with latitude, most probably because it depends on biological activity. The concentration is about 30 µmol l^{-1} near the Antarctic Peninsula [1] and about 10 µmol l^{-1} in the Sargasso Sea [2b]. In fresh waters, it is highly dependent on geographic location and agricultural activities. The concentration is usually lower than 10^{-3} mol l^{-1}. Concentration of nitrites is generally very low (usually lower than 2 µM in sea water) [2a]. The photolysis of nitrate ions contributes to their formation. They are chemically stable but photochemically unstable in sea water and the photolysis can be considered as a non-trivial sink for nitrogen and a source of OH radicals in surface waters [2a]. Consequently photochemical reactions induced by both ions cannot be dissociated. It was observed that the excitation of nitrate or nitrite ions induced the oxidation of organic compounds present in the solution. This was attributed to the formation of hydroxyl radicals [3–5]. It was indeed demonstrated by Zafiriou and True that the excitation of nitrite and nitrate ions is a significant source of hydroxyl radicals [2].

Russi et al. estimated the order of magnitude of the steady state concentration of OH radicals at 5×10^{-16} mol l^{-1} in different aquatic environments and noted that it depends on nitrate concentration [5]. Haag and Hoigné used butyl chloride as a probe molecule for the formation of hydroxyl radicals in lake water (the Greifensee lake in Switzerland) irradiated in UV light [6]. Their kinetic study shows that the formation of OH-radicals cannot be explained by the photolysis of H$_2$O$_2$ but is consistent with the excitation of nitrate ions. Atomic oxygen is also formed by the excitation of nitrate ions, but they are more likely to form ozone by reaction with dioxygen [4]. So from this point of view, nitrates have some depolluting effect.

On the other hand, it was reported in the 1980s by Suzuki et al. that the UV irradiation of solutions containing aromatic derivatives and nitrite or nitrate ions leads to the formation of mutagens [7–13]. A similar occurrence was observed with nitrite and aminoacids, especially with tryptophan in acidic or neutral solution [14]. The excitation of nitrite ions induces the transformation of dimethylamine into nitrosodimethylamine, a well-known carcinogen, and it was suggested that this reaction may occur in environmental conditions [15]. So a more detailed analytical and mechanistic study is useful for a better understanding of these reactions. It has been the subject of several recent publications. The main results are presented in the subsequent sections. The knowledge of mechanisms is useful to assess photochemical transformation under environmental conditions.

2.2
Analytical Study

2.2.1
Reactions Induced by Nitrate Ions

According to the conditions, the excitation of nitrate ions can induce several kinds of reactions: oxidations, nitrations or nitrosations. Analytical studies were mainly developed with aromatic derivatives, but it can be assumed that oxidations occur with most organic substances since hydroxyl radicals are involved. Most often, concentrations of nitrate ions used were in the range $10^{-3} - 5 \times 10^{-2}$ M $(60 - 3100$ mg l^{-1} in NO$^-_3)$ i.e. higher than under common environmental conditions. Suzuki et al. reported the formation of hydroxynitrobiphenyls by UV irradiation of a solution of sodium nitrate (NO$_3$-N 1650 ppm) in the presence of biphenyl coated on silica, glass or kaolinite used as carriers [8, 16]. They suggested a reaction in two stages, the first one being the formation of hydroxybiphenyls. Bunce et al. confirmed the formation of these products by irradiating biphenyl and nitrate ions in aqueous methanol at 254 nm[17]. It is noteworthy that their experiments are not really revelant to environmental conditions and that biphenyl is directly excited at 254 nm. Sarakha et al. also obtained hydroxynitrobiphenyls by exciting nitrate ions in the presence of 2-hydroxybiphenyl [18]. Several products of oxidation, namely phenylhydroquinone, phenylbenzoquinone and 2-hydroxydibenzofuran were also identified.

The three types of reactions (oxidation, nitration, nitrosation) were observed with phenol and resorcinol. In the case of phenol, the main initial reactions reported by Niessen et al. are the formations of pyrocatechol and 2-nitrophenol. However other products (hydroquinone, resorcinol, benzoquinone, hydroxybenzoquinone) result from oxidation. The formation of 4-nitrophenol and 4-nitrosophenol corresponds to a minor pathway. 4-Nitropyrocatechol appears as a secondary product formed from pyrocatechol [19]. Guillaume et al. also obtained resorcinol, hydroquinone, benzoquinone and nitrophenol, but nitrohydroquinone as well (confirmed by the authors) [20]. The formation of nitrosophenol was not reported, but dihydroxybiphenyls were detected by GC-MS [20]. More recently, the quantum yield of phenol conversion induced by excitation of nitrate ions was evaluated at ca 4×10^{-3} for phenol concentration in the range $10^{-4} - 5 \times 10^{-3}$ mol l^{-1}, whatever the nitrate concentration between 4×10^{-3} mol l^{-1} and 10^{-1} mol l^{-1} [21]. It was also reported that oxygen has only a slight influence on oxidation, but it reduces nitration and nitrosation. Both reactions are favoured by decreasing pH. Formate ions used as ·OH quencher favour nitration in air-saturated solution and inhibit it in deoxygenated medium. This effect will be discussed in the section on mechanisms. Resorcinol has similar reactivity [21, 22]. The quantum yield of induced phototransformation was evaluated at 2.8×10^{-3} with resorcinol 5×10^{-4} mol l^{-1} and 10^{-3} with resorcinol 10^{-4} mol l^{-1}. However, the 2-nitroderivative was a minor product and the formation of 4-nitroso is higher than the formation of 4-nitroresorcinol. It was observed that formate ions have the same influence as with phenol.

The main photoproducts obtained with pyrocatechol were hydroxybenzoquinone and 4-nitropyrocatechol. No nitroso derivative was detected.

Usui et al. reported that 4-hydroxy-3-nitrobenzoate is the main photoproduct obtained by irradiation at 313 nm of a solution of sodium nitrate 0.2 mol l^{-1} and 4-hydroxybenzoate 0.5×10^{-4} mol l^{-1}. The addition of an OH scavenger inhibits the reaction but not completely (the quantum yield was depressed by 57% with KI 10^{-2} mol l^{-1} but only by 1% with NaBr 10^{-2} mol l^{-1}) [23].

The excitation of nitrate ions in the presence of nitrophenols [24] or chlorophenols [25] only induces hydroxylations. The reaction is more efficient in the *ortho* or *para* position with respect to phenol function. With chlorophenols, the

Table 1. Main reactions induced on aromatic compounds by excitation of nitrate ions in aqueous solution

Substrate	concentration (mmol l^{-1})		Oxida-tion	Nitration	Nitrosa-tion	References
	substrate	nitrate				
Phenol	0.5–1	1.6–10	x	x	x	19, 21
	0.13–1.6	1–8	x	x		20
Resorcinol	0.5	50	x	x	x	21, 22
Pyrocatechol	0.5	50	x	x		21
Hydroquinone	0.5	50	x	x_m		21[a]
Biphenyl	0.6[b]	118	x	x		11
	4	10–100	x			17[m]
	0.3[b]	–1.2	x	x		8
	0.3[b]	1.2	x	x		16
Hydroxybiphenyls	formed in situ	10–100		x		17[m]
	0.3[b]	1.2	x	x		13
	0.025–0.25	100	x	x		18
Hydroxybenzoates	0.01–0.1	50–200		x		23
Nitrophenols	1–2	20–40	x			24
Chlorophenols	1	10–50	x			25

x_m: minor reaction.
[a] The reaction may result from the excitation of nitrate ions or from the excitation of hydro-quinone since both species absorb in the same range.
[b] Coated on suspended silica.
[m] In aqueous methanol.

reaction can occur with or without dechlorination, and nitration was observed with dechlorinated products. The formation of hydroquinone from 2-chloro-phenol can be attributed to the intermediate formation of chlorohydroquinone followed by its direct photolysis. Induced reactions are more difficult to observe with nitrophenols since it is not possible to excite nitrate ions selectively [24].

Results obtained with aromatic compounds are summarized in Table 1. It appears that the orientation of the reaction is highly dependent on the substituents on the ring. Oxidation, mainly hydroxylation, occurs in all cases whereas nitration and nitrosation are inhibited by electro-withdrawing substituents such as $-NO_2$ and $-Cl$. Nitrosation was mainly observed with phenol and resorcinol. It is localized *para* to the OH group. The formation of a dinitrosoresorcinol is probably possible but its structure was not definitively established

2.2.2
Reactions Induced by Nitrite Ions

Ohta et al. reported that the irradiation in sunlight or artificial UV light of sodium nitrite in aqueous solution containing dimethylamine leads to the formation of nitrosodimethylamine [15]. Nitrite concentration was in the range

Table 2. Quantum yield of induced phototransformation of resorcinol initial concentration $(5 \times 10^{-4} \, \text{mol} \, l^{-1})$ and formation of 4-nitrosoresorcinol in an air saturated solution irradiated at 366 nm [29]

NO$_2^-$ mol l^{-1}	ϕ disp. $\times 10^2$	yield nitrosoresorcinol
5×10^{-4}	1.21 ± 0.15	0.44
1×10^{-3}	1.15 ± 0.15	0.65
5×10^{-3}	1.05 ± 0.15	0.89

$2 \times 10^{-3} - 5 \times 10^{-2} \, \text{mol} \, l^{-1}$ and it was noted that the reaction is favoured by alkaline pH. However, most of the studies were focussed on aromatic derivatives. Nitration of 4-hydroxybenzoic acid was observed by excitation of nitrite ions in air-saturated solution [23]. Suzuki et al. reported that the excitation of nitrite ions in the presence of pyrene dispersed in solution on silicagel used as carrier can lead to the formation of nitropyrene and pyrene-quinone. The mutagenicity of the ether extract, mainly due to nitropyrene, increases with nitrite concentration, irradiation time and with lower pH. No formation of nitroso derivatives was observed [12]. It was also reported by Kochany and Choudhry that the irradiation of a mixture of sodium nitrite $(1.4 \times 10^{-1} \, \text{mol} \, l^{-1})$ and bromoxynil (3,5-dibromo 4-hydroxybenzonitrile) induces the formation of nitro derivatives [26].

In contrast the main product formed from phenol is 4-nitrosophenol. The efficiency of the reaction is enhanced by increasing nitrite concentration, by deoxygenation [21, 27] and by increasing pH [27]. It is inhibited by OH quenchers such as thiocyanate [27] and formate ions [21]. In air-saturated solution nitrosation stays the main pathway compared to oxidation and it is not completely eliminated by formate 5×10^{-2} mol l^{-1}.

Similar reactivity was observed with resorcinol [28]. As it appears in Table 2, the quantum yield of induced transformation in air-saturated solution decreases, and the specificity of *para* nitrosation increases when the concentration of nitrite increases. The formation of dinitroresorcinol (probably 2,4-dinitroso) was also observed when the conversion rate exceeded 20%. Oxidation into trihydroxybenzenes was not observed.

Nitrosation was not observed with pyrene, biphenyl, pyrocatechol and hydroquinone. With pyrene and biphenyl it was attributed to the instability of nitrosoderivative [12]. Hydroxybenzoquinone and 4-nitropyrocatechol are the main products formed from a solution of pyrocatechol $5 \times 10^{-4} \, \text{mol} \, l^{-1}$ and nitrite ions $10^{-3} \, \text{mol} \, l^{-1}$ [21]. It was reported that the formation of 4-nitropyrocatechol increased about 3.5-fold and the formation of hydroxybenzoquinone was reduced approximately in the same proportion when the concentration of nitrite was $5 \times 10^{-3} \, \text{mol} \, l^{-1}$. Formate ions 5×10^{-2} mol l^{-1} have a higher inhibiting effect on oxidation than on nitration [21].

With hydroquinone the main pathway is oxidation into benzoquinone and hydroxybenzoquinone, but a minor formation of nitrohydroquinone was also observed [21]. Reactions photoinduced by excitation of nitrite ions are summarized in Table 3.

Table 3. Main reactions induced on organic substances by excitation of nitrite ions in aqueous solution

Substrate	concentration (mmol L⁻¹)		Oxidation	Nitration	Nitrosation	References
	substrate	nitrite				
Phenol	0.5	1	x		x	21
	1.1	1–12			x	27
Resorcinol	0.5	1			x	28
Pyrocatechol	0.5	1–5	x	x		21
Hydroquinone	0.4	10	x	x_m		21
Biphenyl	0.3[b]	1.2	x	x		16
Hydroxybiphenyls	0.3[b]	1.2		x		13
4-Hydroxybenzoate				x		23
Pyrene	0.25[b]	0–1.2	x	x		12
Bromoxynil	0.0078	0–25		x		26
Nitrophenols	1–2	1–2	x			24
Dimethylamine	5	3–50			x	15

[a] minor reaction; [b] coated on suspended silica.

2.3
Mechanisms

2.3.1
Direct Photolysis of Nitrite and Nitrate Ions

The UV spectra of NO_3^- and NO_2^- are given in Fig. 1. The characteristics of the first weak absorption band of both ions are shown in Table 4. The stronger band at shorter wavelengths is not involved in phototransformations by sunlight in natural waters.

The shoulder near 290 nm in the UV spectrum of NO_2^- is probably due to a second $n \rightarrow \pi^*$ transition and has some importance in sunlight photolysis. In acidic media, nitrites are partly as nitrous acid since the pKa of the latter is 3.37. Nitrous acid absorbs in the same range as nitrite ions, but the UV spectrum is structured and the molar absorption coefficient is 2–3 times

Table 4. Wavelength of maximum absorption of NO_3^- and NO_2^- in aqueous solution. The molar absorption coefficient in l mol⁻¹ is given in parentheses

NO_3^-	References	NO_2^-	References
304 nm (7.1)	40	355 (22.5)	33
300 nm (7.4)	31	354 (24)	34
302 nm (7.2)	22, 30	352 (22)	21
≈300 nm (7.3)	32		

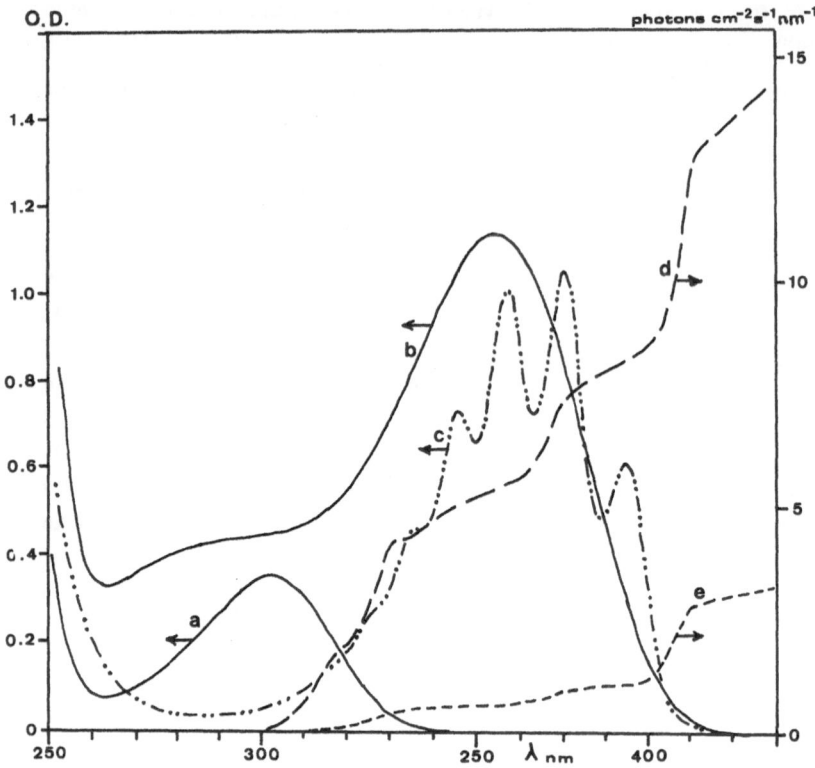

Fig. 1a–e. UV absorption of NO_3^-, NO_2^- and HNO_2. **a** nitrate ions 5×10^{-2} M, **b** nitrite ions 5×10^{-2} M, **c** nitrous acid 2×10^{-2} M. Sunlight spectrum in Central Europe according to ref. 29, **d** 15 June **e** 15 December

higher. It appears in Fig. 1 that nitrous acid, nitrite and nitrate ions absorb much more sunlight in summer than in winter.

It is well-established that the photolysis of nitrite ions leads to the formation of hydroxyl radicals and nitrogen monoxide [35,36]:

$$NO_2^- \xrightarrow{h\upsilon} NO^\bullet + O^{\bullet-}$$

$$O^{\bullet-} + H_2O \rightarrow {}^\bullet OH + OH^- \text{ (pKa of }^\bullet OH: 11.9 \pm 0.2 \text{ [37])}$$

The quantum yield ϕ is higher at shorter wavelengths ($\phi = 0.06 - 0.07$ between 298 and 340 nm [34, 38] than at longer wavelengths ($\phi_{355\,nm} = 0.025$, $\phi_{371\,nm} = 0.015$ [34], $\phi_{355\,nm} = 0.046$ [38]). A similar effect was reported by Harrison et al. [39]. This phenomenon may be related to the presence of a shoulder on the UV spectrum. The photolysis of HNO_2 is more efficient since the quantum yield was evaluated at 0.45 ± 0.1 [24].

The photolysis of nitrate ions excited near 300 nm is a little more complex since two primary processes are involved [40, 41].

$$NO_3^- \xrightarrow{h\upsilon} NO_2^- + O$$

$$NO_3^- \xrightarrow{h\upsilon} NO_2^{\bullet} + O^{\bullet -}$$

$$O^{\bullet -} + H_2O \rightarrow {}^{\bullet}OH + OH^-$$

According to Warneck and Wurzinger the first process is less efficient than the second one at 305 nm, the quantum yields being evaluated at $(1.1 \pm 0.1) \times 10^{-3}$ and $(9.2 \pm 0.4) \times 10^{-3}$ respectively. Zepp et al. obtained a value a little higher for the quantum yield of ${}^{\bullet}OH$ formation at 313 nm, for pH between 6.2 and 8.2 [4]:

$$\phi_{\bullet OH} = (13 \pm 2) \times 10^{-3} \text{ at } 20\,°C \quad \phi_{\bullet OH} = (17 \pm 3) \times 10^{-3} \text{ at } 30\,°C$$

The last value was confirmed by Zellner et al. at 308 nm [38]. Bayliss and Bucat proved experimentally that in an alkaline solution the quantum yield is much higher at short wavelength and that in these conditions the main primary process is the formation of atomic oxygen. However, it does not really correspond to environmental conditions [32].

Nitrogen dioxide formed in the photolysis of nitrate ions disproportionates into nitrite and nitrate

$$2NO_2^{\bullet} \xrightleftharpoons{k_1} N_2O_4 \xrightarrow[H_2O]{k_2} NO_2^- + NO_3^- + 2H^+$$

recombination constant $k_1 = 4.5 \times 10^8\, l\, mol^{-1}\, s^{-1}$ [42]
equilibrium constant $K = 1.53 \times 10^{-5}\, mol\, l^{-1}$ [42]
and $k_2 = 1000\, s^{-1}$ [42].
So the photolysis of nitrite must not to be neglected in environmental photochemistry.

2.3.2
Induced Photoreactions

Oxidation of organic substances is mainly attributed to hydroxyl radicals since atomic oxygen $(O\,^3P)$ is less efficient than ${}^{\bullet}OH$. Frequently hydroxyl radicals generate an organic radical by abstraction of a hydrogen atom. With aromatic derivatives a radical adduct $Ar\ldots{}^{\bullet}OH$ is often involved. Further steps of the oxidation most often involve dioxygen or other oxidant present in the solution. Recombination or disproportionation of organic radicals can also be observed.

The mechanisms of nitration and nitrosation are not so clear. It has been suggested that the nitration and the nitrosation of aromatic derivatives result from the addition of $NO_2^{\bullet}$ or $NO^{\bullet}$ to the organic radical or to the adduct Substrate $\ldots{}^{\bullet}OH$ [12, 19, 23, 27]. In the presence of oxygen nitration and nitrosation compete with oxidation. These mechanisms are consistent with reactions observed with phenol [27] and 4-hydroxybenzoate [23].

$PhOH + {}^{\bullet}OH \rightarrow PhO^{\bullet} + H_2O$ PhOH = phenol

$PhO^{\bullet} + NO^{\bullet} \rightarrow$ 4-nitrosophenol

$HBA^- + {}^{\bullet}OH \rightarrow HO\text{-}HBA^-$ HBA$^-$ = anionic form of 4-hydroxy-
                                                     3-nitrobenzoic acid

$HO\text{-}HBA^- + NO_2^{\bullet} \rightarrow$ 4-hydroxy-3-nitrobenzoate

However the following features were reported in the case of the transformation of resorcinol photoinduced by excitation of nitrite and nitrate ions: i) when nitrite ions are excited in the presence of resorcinol the quantum yield of nitrosation increases with increasing concentration of nitrite ions in spite of the well known $^\bullet$OH quenching property of nitrite ions [28]; ii) the nitration and the nitrosation of resorcinol induced by nitrate ions in air-saturated solution is enhanced by formate ions used as $^\bullet$OH quenchers [21, 22]. So a mechanism involving an electrophilic substitution of nitrogen oxides NO$^\bullet$ and $NO_2^\bullet$ (perhaps also N_2O_3 and N_2O_4) on the aromatic ring has been suggested. It is in good agreement with the fact that nitration is unfavoured by the presence of electrowithdrawing substituents on the ring (chlorophenols [25], nitrophenols [24]).

– with nitrite ions: $NO_2^- + {}^\bullet OH \rightarrow NO_2^\bullet + OH^-$

$$ArH + NO^\bullet + NO_2^\bullet \text{ (or } N_2O_3) \rightarrow ArNO + NO_2^- + H^+$$

ArH   = resorcinol
ArNO = 4-nitrosoresorcinol

This reaction competes with the regeneration of nitrite ions:

$NO_2^\bullet + NO^\bullet \rightarrow N_2O_3$

$N_2O_3 + H_2O \xrightarrow{k_3} 2NO_2^- + 2H^+$ $k_3 = 530 \text{ s}^{-1}$ [43]

Moreover in aerated solution NO$^\bullet$ may be oxidized.

– with nitrate ions: $ArH + 2NO_2^\bullet \text{ (or } N_2O_4) \rightarrow ArNO_2 + NO_2^- + H^+$

$ArNO_2$ = 4-nitroresorcinol

The formation of nitrosoresorcinol was tentatively explained by the following reaction:

$$ArH + 2NO_2^\bullet \text{ (or } N_2O_4) \rightarrow ArNO + NO_3^- + H^+$$

Nitration and nitrosation compete with the formation of nitrite ions:

$2NO_2^\bullet \rightarrow N_2O_4$

$N_2O_4 + H_2O \xrightarrow{k_2} NO_2^- + NO_3^- + 2H^+$ $k_2 = 1000 \text{ s}^{-1}$ [42]

In air-saturated solution formate ions enhance nitration because they inhibit the recombination $NO_2^\bullet + {}^\bullet OH$ but in the absence of oxygen $CO_2^{\bullet-}$ formed from formate ions may play a role in the reaction.

The formation of mutagens with aminoacids (mainly with tryptophan) is probably related with a nitrosation, but the mechanism has not been reported.

2.4
Assessment of Oxidation, Nitration and Nitrosation in Environmental Conditions

It is well established that nitrate and nitrite ions are a source of hydroxyl radicals in natural waters. They induce the photochemical oxidation of most of organic substrates. The quantum yield of formation of oxidizing species from nitrate ions is not affected by their dilution. With nitrite ions the orientation of the reaction depends on their concentration since they are good •OH quenchers: oxidation is unfavoured compared to nitration and nitrosation by increasing concentration. Actually under common environmental conditions the concentration of nitrite is low compared to the concentration of organic matter and this quenching effect plays a minor role. Oxidation of pollutants by •OH competes with the oxidation of natural organic matter present in the aquatic medium. Thus the photo-depolluting effect of nitrate ions is more efficient at low concentration of natural organic matter and low concentration of nitrite ions. This conclusion is also valid for the other sources of •OH such as iron salts.

The mechanisms of photoinduced nitration and nitrosation of aromatic derivatives are not completely understood but it is sure that they involve nitrogen oxides NO$_2^•$ or NO•. Nitration and nitrosation compete with the regeneration of nitrite and nitrate ions by oxidation of NO•, recombination and hydrolysis of nitrogen oxides. Consequently the occurrence of nitration and nitrosation is highly dependent on the conditions and their assessment in environmental conditions is quite difficult. It is noteworthy that nitro and nitroso products absorb sunlight and they are involved in further photochemical reactions. So they reach a maximum concentration that depends on the concentration of photoinducing ions. Nitration and nitrosation are expected to be minor reactions in aerated aquatic medium at low concentration of ions and low concentration of substrate. In contrast they occur in waters containing relatively high concentration of organic matter (natural organic matter and aromatic derivatives, and nitrate ions (NO$_3^-$ > ca. 50 ppm), or nitrite ions (NO$_2^-$ > ca. 3 ppm); fortunately it is uncommon in natural waters. Nitrosation photoinduced by nitrate ions is a minor reaction except with resorcinol; nitration of aromatic derivatives photoinduced by these ions are unfavoured by electrowithdrawing substituents on the ring and, in pure water, by the presence of oxygen. For a more quantitative assessment, further studies would be useful. The following points have to be clarified in order to extrapolate laboratory studies to environmental conditions: reactivity of nitrogen oxides toward phenolic compounds, influence of dissolved organic matter present in natural waters, influence of the concentration of nitrate and nitrite ions on the orientation of the reaction and on the accumulation of nitro and nitroso products.

3
Transformation Photoinduced by Iron (III)

Iron is the most abundant transition metal in the soil and the photochemical reaction of iron (III) in the aquatic compartment plays an important role in the geochemical cycles of various elements. In most natural waters, iron (III) is probably complexed with hydroxide ion and organic ligands. The photoredox process that occurs with iron (III) species gives rise to iron (II) and accordingly it strongly affects the chemical and biological processes which are highly sensitive to the speciation of iron in surface waters.

The photochemical process involving iron (III) as a photoinitiator of degradation and formation of iron (II) depends on the nature of the organic substrate:

1. the organic matter or the pollutant is able to complex iron (III) and the resulting complex absorbs solar radiation and undergoes a direct photolysis;
2. iron (III) is only complexed by hydroxide ion and the degradation is achieved by ·OH radicals formed by irradiation of iron (III) aquocomplexes ($\lambda > 300$ nm).

In both cases iron (II) is formed.

The detailed analysis of photoinduced degradation of some organic compounds is given as examples to illustrate these two classes of mechanisms.

3.1
Photodegradation of Iron (III) – Organic Substrates Complexes

The photoredox cycling between reduced or oxidized iron species and particles or dissolved organic species is of particular importance. It can represent a significant sink of humic substances and an efficient route for the degradation of dissolved organic matter [44]. The photoreaction between humic substances and iron (III) is similar to that observed with iron (III) carboxylates: upon irradiation in a LMCT (ligand to metal charge transfer) transition, a decarboxylation is observed together with the formation of an organic radical. Simultaneously iron (III) is reduced in iron (II).

Aminopolycarboxylic acids present analogous behaviour. They are widely used as sequestering agents in industrial cleaning and household detergents because of their powerful chelating properties. Among them, ethylenediaminotetraacetic acid EDTA and nitrilotriacetic acid NTA are the most commonly used. Because of its poor biological degradation in sewage treatment plants [45] EDTA is found in considerable concentrations in European rivers [46] and light-induced transformation can be of interest. Accordingly, photochemical degradation of iron (III) EDTA or NTA have been studied extensively. Both are electron donors and a photoredox process was observed in iron (III) complexes.

EDTA   NTA

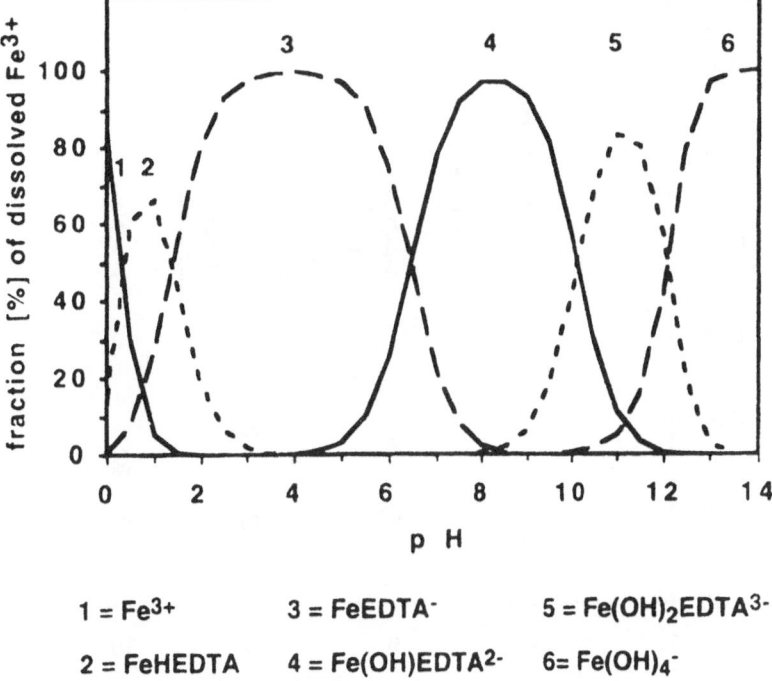

1 = Fe^{3+} 3 = $FeEDTA^-$ 5 = $Fe(OH)_2EDTA^{3-}$

2 = $FeHEDTA$ 4 = $Fe(OH)EDTA^{2-}$ 6= $Fe(OH)_4^-$

Fig. 2. Dissolved iron-EDTA species in the system $[Fe^{3+}$ dissolved$] = 10^{-6} M$ $[EDTA] = 10^{-6} M$ and I (ionic strength) $= 0.1$ M. (From Günter KF et al. [47], with permission)

3.1.1
Photochemical Degradation of Iron (III) EDTA Complex

The photochemical rate of iron (III) EDTA degradation depends upon the various hydrolytic forms of the complex. The different species of iron (III) EDTA according to the pH are shown in Fig. 2 [47]. In the pH range 4–8, $FeEDTA^-$ and $Fe(OH)EDTA^{2-}$ represent approximately 100% of EDTA speciation.

Values of the photochemical quantum yield Φ have been reported by several authors and a large scattering is observed: the values range from 0.002 to 0.35 [47–52]. Part of the discrepancy can be attributed to the formation of FeEDTA dimers. Neither the UV spectrum nor the overlap with solar emission were strongly affected when the pH was varied from 4 to 8 and the quantum yield appears to be roughly pH independent (Fig. 3) [47].

Most authors reported the dependence of Φ on the irradiation wavelength: the shorter the excitation wavelength, the higher the quantum yield. The first stable photoproduct is iron (II) ethylenediaminotriacetic acid

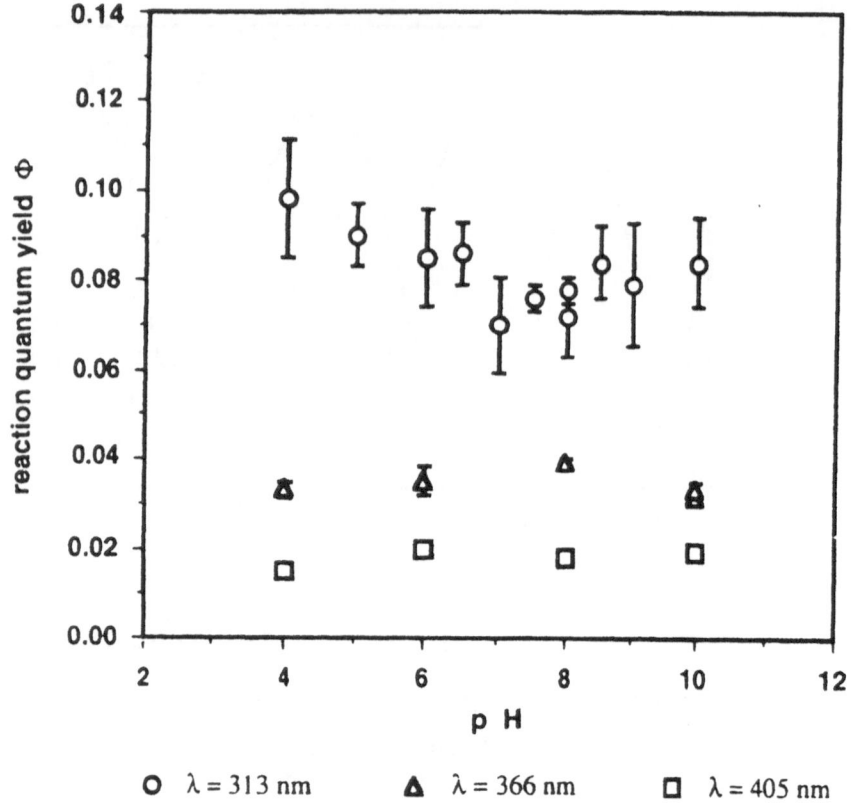

Fig. 3. Reaction quantum yield of Fe (III)-EDTA in phosphate buffer solution (1 mM) as a function of pH and irradiation wavelength. (From Günter KF et al. [47], with permission)

Table 5. Ratios of quantum yields from irradiation of FeEDTA complex

[FeEDTA]	λ_{exc} nm	pH	$\Phi Fe^{2+}/\Phi CO_2$	$\Phi Fe^{2+}/\Phi HCHO$	References
2×10^{-3} M	254	neutral	1.71		[48]
5×10^{-2} M	254	neutral	2.23		[48]
5×10^{-2} M	254	neutral	1.71		[48]
3×10^{-3} M	254	neutral	2.00		[48]
5×10^{-2} M	254	neutral	2.10		[48]
6×10^{-3} M	350	9		1.8	[50]
8×10^{-3} M	350	9		1.7	[50]

(ED3A) together with CO_2, formaldehyde and iron (II) probably complexed by EDTA [49].

Some authors mentioned ratios of roughly 2:1:1 between $\Phi Fe^{2+}:\Phi(CO_2):$ $\Phi(HCHO)$ [48, 50] (Table 5). So the mechanism can be summarized in the following way:

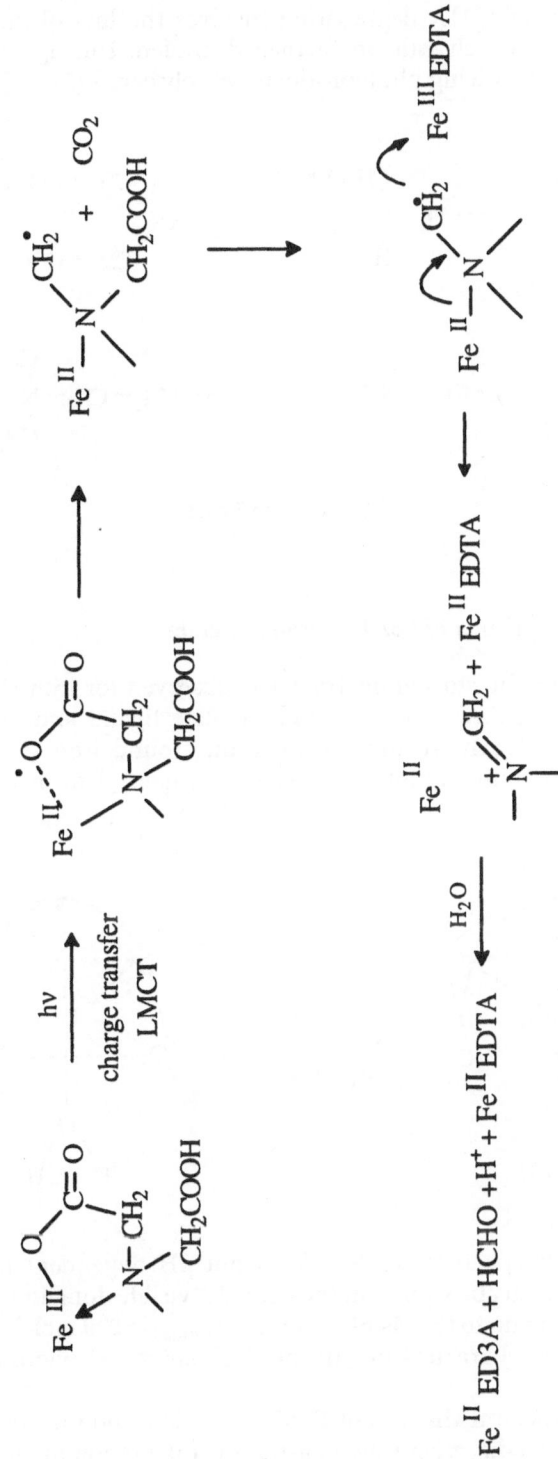

The first step of EDTA degradation involves the loss of an acetate group leading to ED3A which can be further degraded. During the process, the formation of the following photoproducts was observed [51, 53].

$$
\begin{array}{cc}
\underset{\underset{\text{HOOC}-\text{CH}_2}{|}}{\overset{\overset{\text{H}}{|}}{\text{N}}}-\text{CH}_2-\text{CH}_2-\underset{\underset{\text{H}}{|}}{\overset{\overset{\text{CH}_2\text{COOH}}{|}}{\text{N}}}
&
\text{HN}\underset{\text{CH}_2-\text{COOH}}{\overset{\text{CH}_2-\text{COOH}}{\diagdown}}
\end{array}
$$

$$
\begin{array}{cc}
\underset{\underset{\text{HOOC}-\text{CH}_2}{|}}{\overset{\overset{\text{H}}{|}}{\text{N}}}-\text{CH}_2-\text{CH}_2-\text{NH}_2
&
\text{H}_2\text{N}-\text{CH}_2-\text{CH}_2-\underset{\underset{\text{CH}_2-\text{COOH}}{|}}{\overset{\overset{\text{CH}_2-\text{COOH}}{|}}{\text{N}}}
\end{array}
$$

$$
\text{NH}_2-\text{CH}_2-\text{COOH}
$$

3.1.2
Photochemical Degradation of Iron(III)-Nitrilotriacetate

In dilute solution, the stoechiometry 1:1 is observed for iron (III) nitrilotriacetate complex (FeNTA). NTA is a four dentate ligand and as a result, two molecules of water achieve the coordination around iron (III) [54]. FeNTA undergoes a protolytic equilibrium between a neutral form and the monohydroxy anion.

The UV-visible spectrum of FeNTA is not pH dependent in the domain $4 < \text{pH} \leq 7$: there is an absorption in the near UV-visible domain with a shoulder at 260 nm and a tail up to the visible domain ($\varepsilon_{260\,\text{nm}} = 6000 \text{ mol}^{-1}\text{l cm}^{-1}$). In contrast to the spectral characteristics, the photochemical behaviour is strongly pH dependent [55].

In Fig. 4 the spectral changes of FeNTA solution upon irradiation are presented: in acidic media, where the neutral form is predominant, a continuous

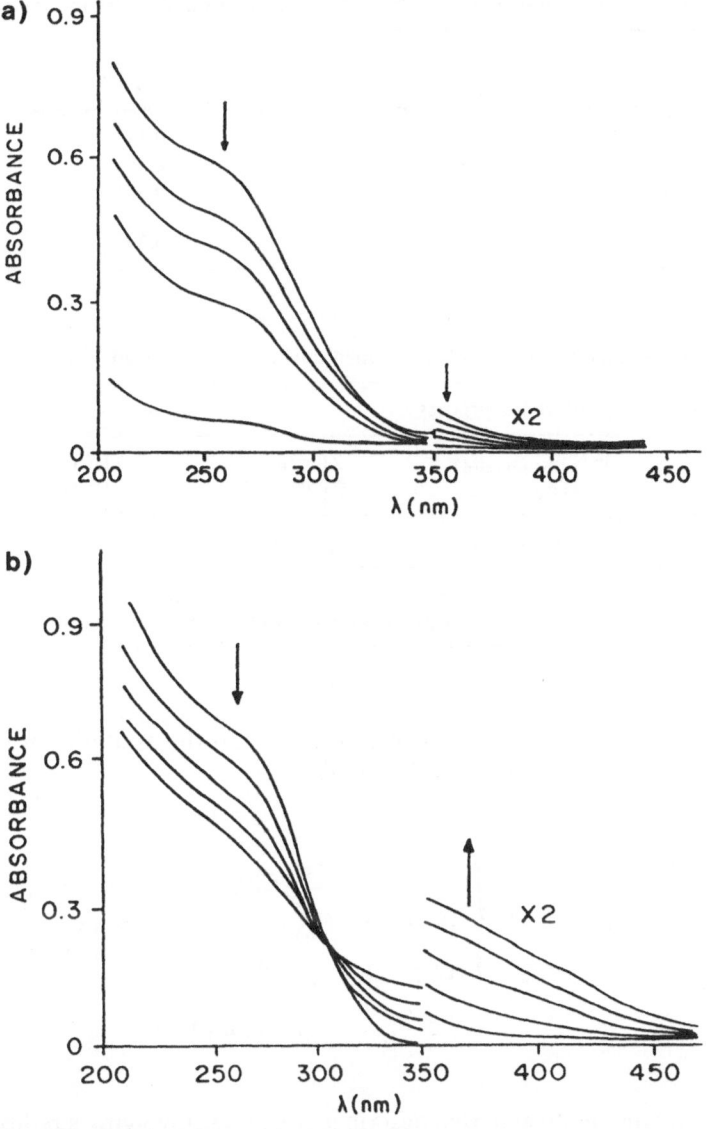

Fig. 4a, b. Change in the spectrum of a solution FeNTA 10^{-4}M as a function of irradiation time at 365 nm. **a** pH = 4, **b** pH = 7

decrease appeared whereas an isosbestic point was observed upon irradiation of the monohydroxyanion present at pH 7. It is well defined ($\lambda_i = 305$ nm) at $\lambda_{exc} = 365$ nm and not precise at $\lambda_{exc} = 254$ nm. As already mentioned for FeEDTA, the quantum yield of FeNTA disappearance is strongly wavelength dependent: it increases when the excitation wavelength decreases (Table 6). The

Table 6. Quantum yields of FeNTA disappearance as a function of excitation wavelength [55]

λ_{exc} nm	Φ pH = 4	Φ pH = 7
254	0.50	0.19
296	0.40	0.18
313	0.30	0.08
334	0.22	–
365	0.18	0.09

Table 7. Stoichiometry of photoproducts formed during FeNTA degradation

acidic solution $\Rightarrow$ photoredox process

254 nm	$2\,FeNTA_{disappeared}$	$2\,Fe^{2+}_{formed}$	$2\,HCHO_{formed}$
365 nm	$2\,FeNTA_{disappeared}$	$2\,Fe^{2+}_{formed}$	$1\,HCHO_{formed}$

neutral solution $\Rightarrow$ photosolvation then redox process

254 nm	$2\,Fe^{2+}_{formed}$	$2\,HCHO_{formed}$
365 nm	(the redox process only represents a few percent)	

photoproducts are analogous whatever the pH and the excitation wavelength, but the stoichiometry is different (Table 7):

$$FeNTA \longrightarrow \begin{cases} Fe^{2+} \\ HCHO \\ CO_2 \end{cases} + HN \begin{array}{c} CH_2COOH \\ \\ CH_2COOH \end{array} \quad (IDA)$$

[51],[55]-[58]

A mechanism identical to that described for EDTA, involving a redox process between iron (III) and the carboxylate group can be assigned to FeNTA photodegradation upon irradiation at 365 nm in acidic medium. A second redox process between iron (III) and the OH group giving rise to iron (II) and $^\bullet$OH accounts for the stoichiometry observed upon irradiation at 254 nm. Hydroxyl radicals abstract a hydrogen from CH_2 group of NTA likely in a cage phenomenon [55]. In neutral solution, a photosolvation with the release of NTA and aquocomplex of iron (III) is observed prior to the redox process. Upon irradiation at 254 nm, the oxidation of the organic moiety is complete, whereas the redox process only represents a few percent of FeNTA transformation when $\lambda_{exc} = 365$ nm [55].

The degradation of IDA, the major photoproduct, is only efficient at pH < 3.5 [59]. Actually, IDA is not able to complex iron (III) in the pH range usually found in natural waters. Accordingly, at constant NTA input, some accumulation of IDA can occur and the formation of its carcinogenic nitrosoderivative can be observed.

The intramolecular photoredox process occuring in iron (III)-organic pollutant complexes represents an efficient route to degrade the pollutant. However the complete degradation of the pollutant depends on the capability of the photoproducts to complex iron (III).

3.2
Degradation of Organic Substrates Photoinduced by Iron (III) Aquocomplexes

In addition to iron (III) polycarboxylates, many other iron (III) species with ligands such as Cl^- or HO_2^- [60], HSO_3^- [61], H_2O [62], OH^- [63–65] absorb at $\lambda > 300$ nm.

The aquocomplexes play an important role in natural waters since the photoredox process occuring upon excitation is a major source of $^{\bullet}OH$ radicals.

$$\diagup\!\!\!\!\!\diagdown Fe^{III}OH \xrightarrow{\ h\nu\ } Fe^{2+} + HO^{\bullet} \qquad (62–65]$$

$^{\bullet}OH$ radicals, known as being powerful oxidizing agents, cause the degradation of most organic substrates.

The quantum yield of the reaction depends on the excitation wavelength and on the iron (III) species [64–66]; $FeOH^{2+}$ the species present at 3 < pH < 5 is the most photoactive one (Table 8) [65]. In ferric solutions, the percentage of $FeOH^{2+}$ strongly depends on the starting concentration and on the age of the solution (before and after dilution). Using the HQSA method [65] it is possible to determine the concentration of $FeOH^{2+}$ in solution. In dilute solutions (C < 0.3 mM), the dimeric form $Fe_2(OH)_2^{4+}$ can be neglected.

Two examples of organic substrates, the transformation of which is photoinduced by iron (III) are given: 2,6-dimethylphenol taken as a model molecule of phenol derivatives and diuron a herbicide widely used in agriculture.

Table 8. Quantum yields of iron (II) formation

Species	λ_{exc} (nm)	Φ_{Fe}^{2+}	References
$Fe_2(OH)_2^{4+}$	350	0.010	[66]
$FeOH^{2+}$	313	0.14	[65]
	360	0.017	
Iron (III) solution 0.1 mM (90% $FeOH^{2+}$)	313	0.08	[67]
	365	0.055	
Iron (III) solution 0.1 mM (10% $FeOH^{2+}$)	313	0.02	[67]
	365	0.008	

3.2.1
Degradation of 2,6-Dimethylphenol (DMP) [67]

DMP

The substitution of the hydrogen atoms by methyl groups in the two *ortho* positions of DMP is assumed to make the overall mechanism simpler since the only mesomeric form of the expected phenoxyl radical corresponds to the *para* position.

The mixture of DMP/iron (III) is thermally stable provided the ratio $R = [FeOH^{2+}]/[DMP]$ is ≤ 1. At higher R values, a fast thermal redox process leading to the disappearance of DMP is observed.

There is no detectable complexation between iron (III) and DMP in the ground state and the mixture is stable for a few days ; there is neither a redox reaction nor precipitation. An irradiation at 365 nm, at which the light is only absorbed by iron (III) species, causes the disappearance of DMP.

The rate of DMP disappearance and of iron (II) formation is very sensitive to the percentage of monomeric species: the higher the percentage, the faster the degradation is. The initial quantum yields are collected in Table 8. Whatever the percentage of monomeric species, there is no significant influence of oxygen on the process.

The influence of isopropanol used as a scavenger of •OH radicals is strongly dependent on the percentage of monomeric species: at low percentage ($< 5\%$) no influence is observed whereas with 75% of the monomeric form, the rate of DMP disappearance is reduced by a factor of 2.5. It can be concluded that •OH radicals are only involved at high percentages of monomeric species.

ESR spectroscopy permits us to confirm the presence of radicals during the process; it was demonstrated by spin trapping experiments with α-phenyl,*N*-

Table 9. Quantum yields of iron (II) formation and DMP disappearance (aerated solution, $[iron (III)] = [DMP] = 0.1$ mM pH = 3.5) [67]

Conditions	% FeOH^{2+}	Φ_{DMP}	Φ_{Fe}^{2+}
	75	5.0×10^{-3}	5.9×10^{-2}
	50	3.7×10^{-3}	2.2×10^{-2}
	< 5	8.0×10^{-5}	2.0×10^{-4}
with isopropanol	75	1.8×10^{-3}	5.5×10^{-2}
	50	1.4×10^{-3}	–
	< 5	8.0×10^{-5}	2.0×10^{-4}

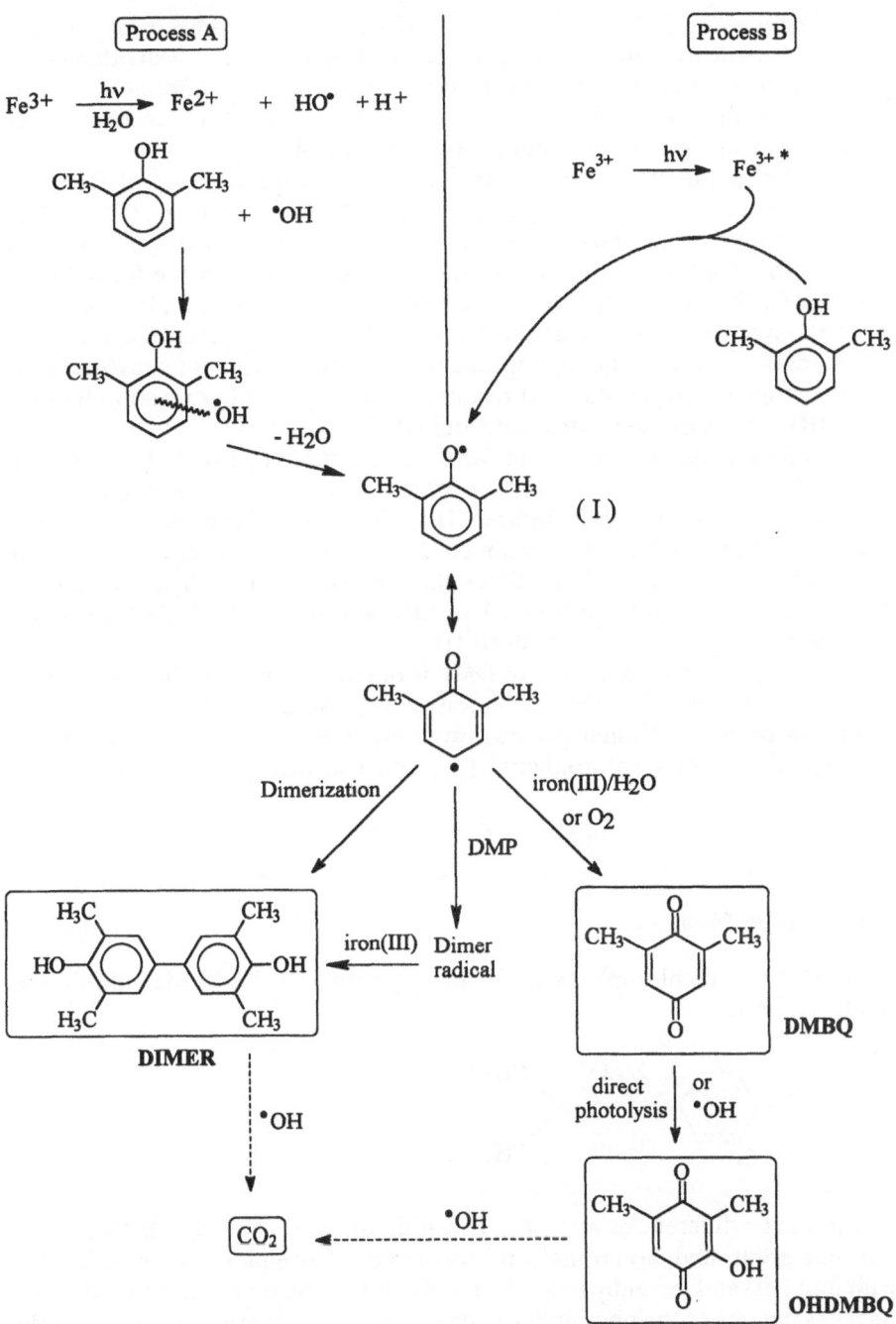

Scheme 1

tert-butylnitrone (PBN) that the effect of isopropanol addition depends on [FeOH^{2+}] again. At high percentage (90%) the signal intensity is divided by two whereas there is only a very weak influence in the case of solutions containing a low percentage of FeOH^{2+}. ESR spectroscopy confirms the presence of two radicals, one of them being inhibited by isopropanol.

The photoproducts of the reactions are the *para-para* dimer of DMP, 2,6-dimethylbenzoquinone and 2-hydroxy-3,5-dimethylbenzoquinone (HODMBQ).

The complete mechanism is described in Scheme 1. Process A involves ˙OH radicals that lead to 2,6-dimethylphenoxyl radicals through the formation of the ˙OH-DMP adduct [68] and the subsequent water elimination [69]. A second process involving a radical unaffected by the presence of isopropanol is also involved: the nature of the photoproducts supports the direct formation of the 2,6-dimethylphenoxyl radical (I) resulting from a quenching reaction between iron (III) species in the excited state and DMP.

Accordingly, the degradation of DMP photoinduced by iron (III) is the result either of the attack by ˙OH radicals and/or a charge transfer reaction between excited iron (III) and DMP. Radical (I) is the common intermediate of both processes. The major photoproduct 2-hydroxy-3,5-dimethylbenzoquinone (HODMBQ) comes either from direct photolysis of 2,6-dimethylbenzoquinone (DMBQ) or from a subsequent attack of ˙OH radicals on DMBQ [70]. The dimer is formed by coupling of two radicals (I).

The complete mineralization of DMP is obtained upon irradiation of aquo-complexes of iron (III), the formation of CO$_2$ occurs without a significant induction period. A similar process involving ˙OH radicals was described for the degradation of 2-chlorophenol [71] and triazines [72] photoinduced by iron (III).

3.2.2
Degradation of Diuron

Diuron (3-(3,4-dichlorophenyl)-1,1-dimethylurea is a herbicide inhibitor of photosynthesis.

Two major differencies appear between diuron and DMP: firstly there is no phenolic group and, accordingly, no possibility of complexation, even weakly, with iron (III) and secondly two chlorine atoms are present on the aromatic ring. As far as the rate of the photoinduced degradation is concerned, the same conlusions are drawn. The higher the percentage of FeOH^{2+} the higher the rate is [73].

Two major paths are described for diuron degradation but both involve the reaction with ˙OH radicals arising from iron (III) species in the excited state (scheme 2).

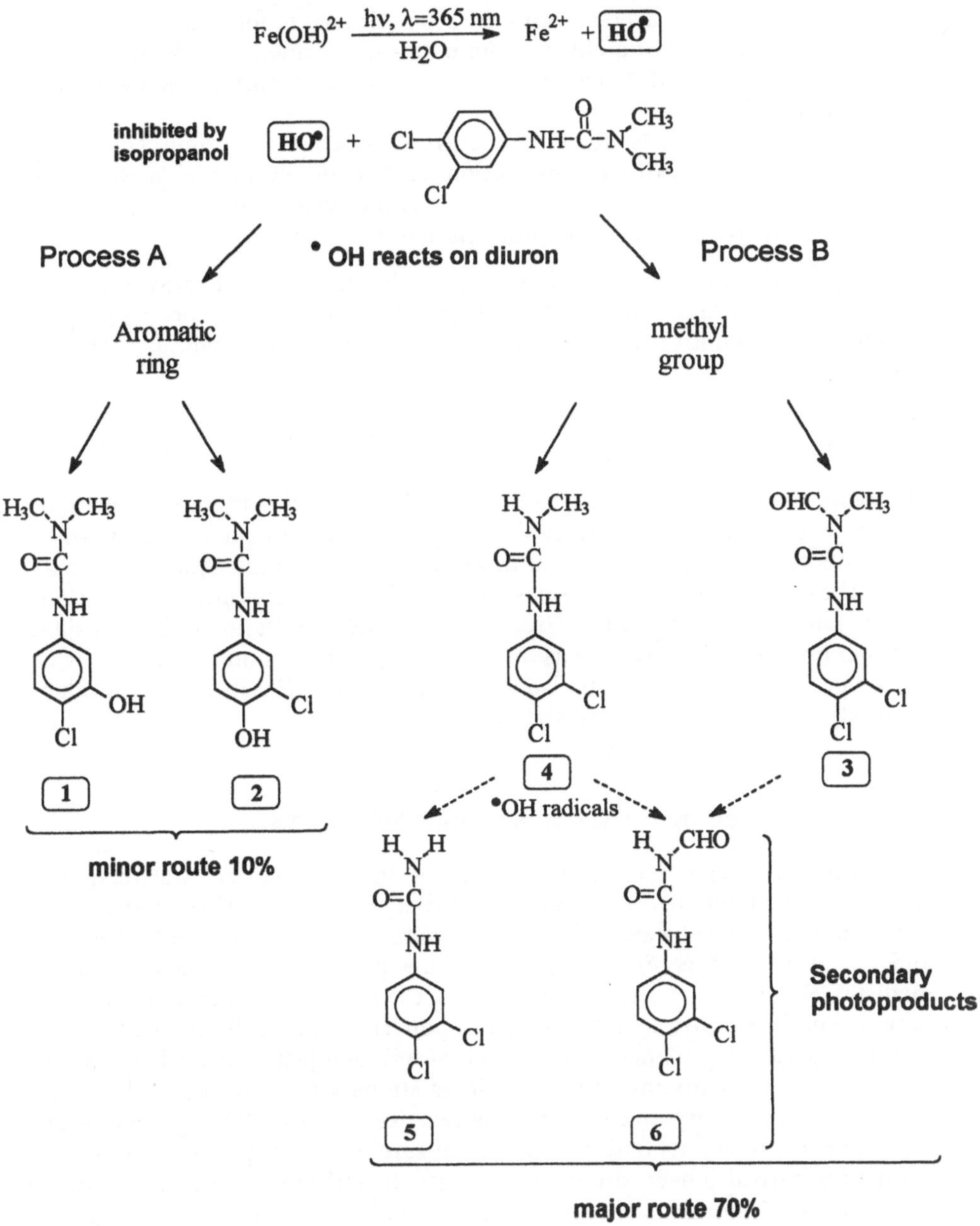

$$Fe(OH)^{2+} \xrightarrow[H_2O]{h\nu,\ \lambda=365\ nm} Fe^{2+} + \boxed{HO^{\bullet}}$$

inhibited by isopropanol $\boxed{HO^{\bullet}} +$ [chemical structure: Cl-substituted phenyl ring with $-NH-\overset{O}{\underset{\parallel}{C}}-N\overset{CH_3}{\underset{CH_3}{}}$]

Process A ⟋ $^{\bullet}OH$ **reacts on diuron** ⟍ **Process B**

Aromatic ring   methyl group

1   **2**

minor route 10%

$^{\bullet}OH$ radicals

4   **3**

5   **6**

Secondary photoproducts

major route 70%

Scheme 2

1. process A: •OH radicals react with the aromatic ring to form an adduct with the subsequent loss of a chlorine atom and the formation of hydroxylated photoproducts 1 and 2. This way only represents a minor way of diuron degradation ;
2. process B: •OH radicals attack the methyl of dimethylurea group. The alkyl-radical formed by hydrogen abstraction reacts with oxygen and gives rise to photoproducts 3 and 4. Photoproducts 5 and 6 result from the subsequent attack of OH radicals on photoproducts 3 and 4.

The complete mineralization of diuron is obtained upon irradiation of a mixture iron (III)-diuron in aqueous solution. A similar route (process B) is put forward to account for the degradation of tetraacetylethylenediamine [74].

3.3
Conclusion

The photoredox process that occurs in iron (III) species present in natural waters is an important source of oxidizing agents. The subsequent reaction with pollutants can represent an efficient way of degradation in the aquatic environment. The complete mineralization is obtained due to the continuous formation of •OH radicals resulting either directly from $FeOH^{2+}$ excitation or from a side reaction between H_2O_2 formed as an intermediate species and iron (III) by a Fenton process [64]. The results also indicate that iron (III) salts may be of value in the clean up of contaminated waters.

4
Reactions Induced by Excitation of Humic Substances

Among the great variety of photoreactions occurring in surface waters, many result from the light absorption by dissolved organic matter. These substances found in most natural waters absorb solar radiation in the range 300–500 nm and act as sensitizers or precursors for the production of reactive intermediates. Numerous studies were devoted in the last twenty years to the determination of the reactive species produced upon irradiation of humic acids (HA) and fulvic acids (FA) extracted from different soils or aquatic media. Laser-flash photolysis experiments and continuous irradiations with probe molecules were undertaken to identify and titrate these reactive species. The degradation of many organic pollutants under the action of sunlight in the presence of humic substances have also been studied to quantify the efficiency of the humic substances as photoinductors (or photosensitizers) and to predict the fate of these pollutants in the environment.

4.1
Primary Transient Species Produced by Excitation of Humic Substances

Laser-flash photolysis studies ($\lambda_{exc} = 355$ nm) were carried out in order to observe and characterize the phototransients produced upon excitation of natural water samples or humic substances solutions [75, 76]. The formation of three transients was revealed by transient absorption spectroscopy. The solvated electron was observed 20 ps after the pulse, its lifetime is about 1 µs at pH 7. These species are produced in a monophotonic process. The second transient ($\lambda_{max} = 475$ nm) was assigned to the radical cation, arising from an excited state of the humic material in the same process as the solvated electron.

$$HS^* \rightarrow HS^{\cdot+} + e_{aq}^-$$

The third component with a lifetime of about 100 µs was tentatively assigned to the triplet states. The formation of solvated electrons was confirmed by laser excitation of humic substances of various origins [77]. The primary quantum yields ranged from 0.005 to 0.008. Steady-state irradiations were carried out in the presence of 2-chloroethanol used as electron scavenger. The quantum yields measured in these conditions were about 2 orders of magnitude lower than the primary quantum yields [77]. This discrepancy between laser-flash experiments and steady-state irradiations was explained by a decrease of electrons-radical cations recombination at high light intensity. It can be concluded that the role of solvated electrons in sunlight-irradiated natural surface waters is likely to be minor. In deoxygenated media these species can reduce halogenated compounds. In aerated solutions they are mainly trapped by oxygen. Hydrogen peroxyl radicals and superoxide anions are further formed. These oxygenated species are not very reactive. Their disproportionation leads to the production of hydrogen peroxide that is a source of the hydroxyl radicals [78, 79]. These highly reactive species can be further formed by photolysis of hydrogen peroxide or by its reaction with ferrous ions (Fenton reaction).

Time resolved photoacoustic spectroscopy in conjunction with magnetic circular dichroism spectroscopy was used to determine energies and quantum yields for the formation of triplet states of fulvic acid in aqueous solution [80, 81]. The energy of the triplet states was estimated as 170 kJ mol^{-1}, the intersystem crossing yields as 0.85–0.35 and the lifetime of triplet states was deduced to be longer than 2 µs. The values of the quantum yields of triplet formation are much higher than those deduced from steady-state irradiations [78]. Thus, results from photophysics and primary photochemistry of humic substances cannot be used to evaluate the kinetics of the photoreations occurring in sunlight conditions.

4.2
Main Pathways of Organic Pollutant Phototransformation

Excited triplet states of humic substances are the main species responsible for the photoinduced degradation of aquatic pollutants. Humic substances are known to photosensitize oxygenations through a singlet oxygen pathway. Direct reactions between excited triplet states of humic substances and organic substrates are also likely to occur. They can be described as energy, electron or hydrogen atom transfer reactions.

4.2.1
Photosensitized Oxygenations

It is now established that singlet oxygen is formed by energy transfer between the triplet excited states of humic substances and ground state molecular oxygen.

$$^3HS^* \rightarrow HS + {}^1O_2$$

Many studies have been undertaken in order to evaluate the ability of humic substances to photosensitize the production of singlet oxygen [82–91]. The efficiency of this process was determined by using furan derivatives (dimethylfuran [82, 83], furfuryl alcohol [84, 88] or furoin [90]) as trapping agents

The quantum yield of singlet oxygen production depends on several parameters, mainly the irradiation wavelength and the nature of humic substances. Values in the range 1%–3% were reported by excitation at 365 nm [87]. The quantum yield was shown to decrease as the wavelength increased [83, 85, 86, 91]. It is about two times lower at 434 nm than at 365 nm. Quantum yields vary with the nature of the humic substances [86]. However, so far, no relationship with any structural characteristics has been found. It was shown that synthetic humic substances obtained by photo-oxidation of phenolic compounds do not sensitize the production of singlet oxygen [91].

4.2.2
Reactions of Pollutants with the Triplet States

Excited triplets of humic substances can transfer their energy to ground state organic molecules. This reaction is likely to occur if the energy level of the triplet state of the acceptor is lower than that of humic substances.

$$^3HS^* + A \rightarrow HS + {}^3A^*$$

Very few examples of energy transfer reactions have been reported in the literature.

The *cis-trans* photoisomerization of 1,3-pentadiene was shown to be sensitized by various humic substances [82]. This reaction is reversible and, as a consequence, the ratio *cis-trans* reaches a photostationary state at long irradiation times. Using the fact that the value of the ratio depends on the energy level of the donor, it was estimated that up to half of the triplets produced in sunlight-irradiated humus solutions have energies higher than or equal to 250 kJ mol⁻¹ [85]. The *E-Z* isomerization of cinnamic esters was photosensitized by humic acids [92]. The degree of association of these hydrophobic compounds with humic substances appeared to have no significant influence on the efficiency of the energy transfer process.

Reactive excited triplet states of humic substances also seem able to abstract hydrogen atoms or electrons from oxidizable organic molecules. Aniline [83] and phenolic compounds [93–96] are likely to be oxidized via this type of reaction. The phototransformation of a series of methyl and methoxyphenols was studied by using dissolved natural organic materials and the aromatic ketones benzophenone, 3'-methoxyacetophenone and 2-acetonaphtone as sensitizers [94]. The apparent first-order rate constants of phenols disappearance were measured by monitoring the exponential decays in natural and synthetic sensitizer solutions as a function of the irradiation time. The values relative to 2,4,6-trimethylphenol lay within the range 1–0.022 with all the humic substances studied. A similar selectivity was found with the synthetic sensitizers. By contrast, peroxyl radicals exhibit a much higher selectivity toward phenols. It was thus concluded that excited triplet states rather than peroxyl radicals play an important role in the phototransformation of phenols in natural waters. Experiments were performed in D_2O and H_2O in the presence and in the absence of sodium azide, a singlet oxygen scavenger [94]. In this way, it was possible to determine the contribution of singlet oxygen in the phenol degradation using the fact that the lifetime of singlet oxygen is 13 times longer in D_2O than in H_2O. It was also possible to point out a possible isotope effect, the energy of the bond $\Phi O–H$ being smaller than that of $\Phi O–D$. The results showed that singlet oxygen plays a minor role in the phototransformation of phenol in H_2O with humic substances and aromatic ketones. A very small isotope effect was observed with the humic substances, benzophenone and 3'-methoxy-acetophenone suggesting than an electron transfer is more likely than a hydrogen atom transfer. In the case of 2-acetophenone, a significant isotope effect was found in favour of a hydrogen atom transfer mechanism. The cycle of reactions (Scheme 3) was proposed to explain the experimental results [94]:

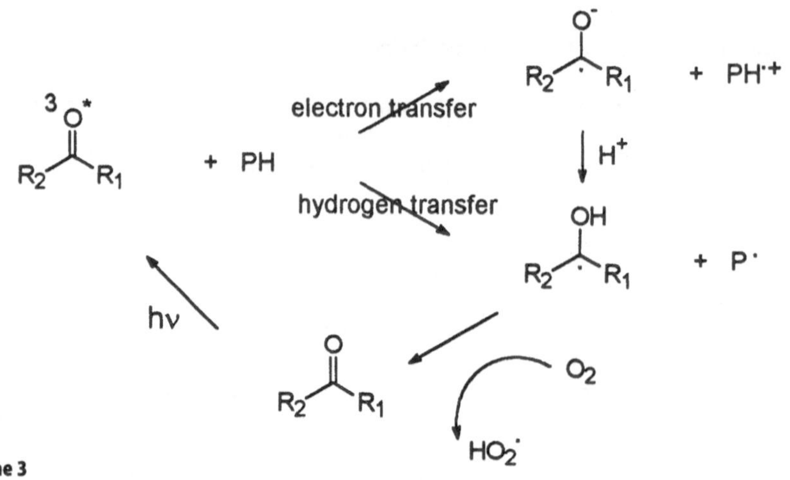

$$3 \overset{O^*}{\underset{R_2 \quad R_1}{\bigwedge}}$$

electron transfer

+ PH

hydrogen transfer

hv

$$\overset{O}{\underset{R_2 \quad R_1}{\bigwedge}}$$

$$\overset{O^-}{\underset{R_2 \quad \cdot R_1}{\bigwedge}} \quad + \quad PH^{\cdot +}$$

$$\downarrow H^+$$

$$\overset{OH}{\underset{R_2 \quad \cdot R_1}{\bigwedge}} \quad + \quad P^\cdot$$

O_2

$HO_2^\cdot$

Scheme 3

In the presence of oxygen, carbonyl groups may be regenerated with formation of hydroperoxyl radicals. The phenoxyl radicals $P^\cdot$ are further oxidized. The influence of oxygen concentration on the reaction was also studied [94]. In the absence of oxygen, the rates of phenols consumption are very low. The apparent rate constants of reaction are higher in air-saturated than in oxygen-saturated solutions with model sensitizers. This result was expected because of the deactivation of triplet excited states by oxygen. With humic substances, the rate constants measured in oxygen-saturated solutions sometimes equalled those measured in air-saturated solutions. This lack of oxygen effect in the range 2.6×10^{-4} M and 1.2×10^{-3} M was explained by an enhancement of intramolecular triplet states deactivation of the sensitizer moieties that are embedded in the humic substances with respect to a free sensitizer [94].

It was observed that the apparent first-order rate constants for photosensitized oxidation of several phenolic derivatives decrease as the phenol concentrations increase [93–95] and it was deduced from these results that long-lived photo-oxidants ($\tau > 100$ μs) are also probably involved in the transformation of phenols. However, they are much less reactive than the triplet states and react only with substrates having a very low oxidation potential such as methoxyphenols. Their role in the photosensitized transformation of pollutants is likely to be minor except at very low substrate concentrations. The moderately reactive radicals probably include the peroxyl radicals which were thought to be the main species responsible for the transformation of phenolic derivatives [97, 98].

4.2.3
Other Reactions

Humic substances form very stable complexes with iron. On irradiation, oxido-reduction reactions are likely to occur creating ferrous iron and a free radical from the organic matter. In neutral media, the ferrous ion is reoxidized into ferric ion by oxygen [99]. In acidic solutions, this reaction is very slow and oxidation by hydrogen peroxide can occur leading to the formation of hydroxyl radicals.

$$Fe^{3+}-L \xrightarrow{\quad h\nu \quad} Fe^{2+} + L^{\cdot+}$$

$$\text{neutral medium} \diagdown_{O_2} \quad H_2O_2 \diagdown \text{acidic medium}$$

$$Fe^{3+} + O_2^{-\cdot} \qquad\qquad Fe^{3+} + \,^{\cdot}OH + OH^-$$

In accordance with the involvement of hydroxyl radicals in acidic media, the rate of the photoinduced transformation of 2,4,6-trimethylphenol was found to increase when solutions were acidified to pH 3.6 [96]. Moreover, this increase was suppressed upon addition of isopropanol used as a hydroxyl radical scavenger [96]. When carboxylic groups of dissolved humic substances are esterified, the oxygen consumption rates decrease by 50% showing that these functional groups are involved in the ligand-to-charge transfer process [99].

4.3
Examples of Photoinduced Transformations

When the irradiations are performed at 254 nm, it is generally observed that humic substances decrease the rate of pollutant degradation [100–102]. Due to competitive light absorption by coloured humic substances the direct photolysis of pollutants is inhibited. Atrazine [100] and prometryn [2-(methylthio)-4,6-bis(isopropylamino)-s-triazine] [101] undergo dealkylation when they are irradiated at 254 nm in the presence of humic substances. The formation of these dealkylated products probably results from the oxidation of the starting materials by hydroxyl radicals, these species being generated during excitation of humic substances at short wavelengths. The two hydroxylated products 3-(4-hydroxyphenyl)-1,1-dimethylurea and 3-(2-hydroxyphenyl)-1,1-dimethylurea were found when 3-phenyl-1,1-dimethylurea (fenuron) was irradiated in the presence of humic acids at 254 nm [103] whereas the direct photolysis of fenuron yields o- and p-amino-N,N-dimethylbenzamide by a photo-Fries mechanism.

The photodegradation of numerous chemicals in sunlight was shown to be accelerated by humic substances or dissolved organic matter contained in natural waters: ethylenethiourea [104], trinitrotoluene [105], nitroaromatics [106], 2,4-dichlorophenol [107], 2,4-D esters [108], sustar [109], atrazine [110] and several other pesticides (such as carbamates, ureas, phosphates, pyrethoid or triazole derivatives) [111].

ethylene thiourea   2,4-D esters   "sustar"

Contradictory results were reported concerning the effect of humic acids on the transformation rate of methoxychlor. Its rate of degradation was found to be faster in natural waters than in pure water [112], whereas, according to other workers, the concentration of methoxychlor remains constant upon irradiation for several hours in the presence of humic substances [113].

methoxychlor

The rate of (2,4,5-trichlorophenoxy) acetic acid (245-T) photodegradation is higher in natural water samples or in fulvic acid solutions than in distilled water; 2,4,5-trichlorophenol is the main product of the humic-induced photoreactions. It was proposed that this product is formed by an electron transfer from the carboxylate group to reactive species of the humic substances [114]. A minor enhancing effect on the rate of photolysis of 4-amino-3,5,6-trichloropicolinic acid (picloram), an other carboxylic acid, was observed.

(2,4,5-trichlorophenoxy)acetic acid   picloram

Humic substances photoinduce the transformations of 3,4-dichloroaniline [115, 116] and metalaxyl [117] in solar light. Metalaxyl (or N-(2,6-dimethyl-phenyl)-N-(methoxyacetyl)-alanine methyl ester), is used as a herbicide.

metalaxyl   7   8

Phenanthrene was irradiated at $\lambda > 290$ nm (Suntest apparatus) in the presence of hydrogen peroxide and five fulvic acids from different origins [118]. The rate of phenanthrene degradation was greatly accelerated by H_2O_2. Two fulvic acids enhanced the rate of phenanthrene transformation with an efficiency comparable to that of H_2O_2. The other fulvic acids retarded the photodegradation of phenanthrene. Whatever the photoinductor used, the same photoproducts were detected: 9,10-phenanthrenequinone, 9-hydroxy-phenanthrene, 2,2'-biphenyldialdehyde, 2,2'-biphenyldicarboxylic acid and 2-phenylbenzaldehyde.

Humic substances sensitized the phototransformation of bromacil, their sensitizing effect being lower than that of riboflavin or methylene blue [119]. The photodegradation only occured in the presence of oxygen and the reaction was attributed to singlet oxygen.

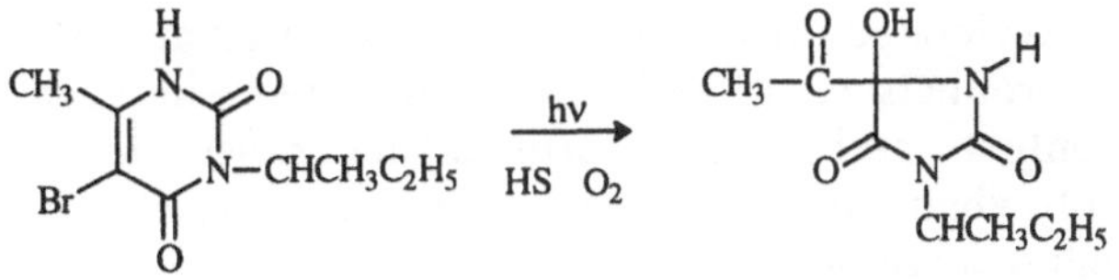

bromacil   9

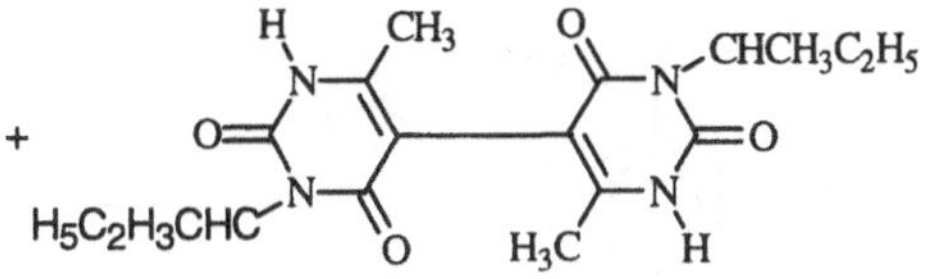

10

Nonylphenol and polyethoxylates derivatives are photodegraded when they are solubilized in filtered lake water and exposed to sunlight [120]. The photochemical oxidation of nonylphenol polyethoxylates is lower than that of nonylphenol itself. The transformations are mainly due to sensitized photolyses, direct photolyses being comparatively low. Singlet oxygen was shown to play a minor role in these photoreactions.

Fenuron was the subject of a detailed study. The irradiation at 365 nm of neutral and oxygenated solutions containing humic substances leads to the disappearance of fenuron and to the formation of the following main reaction products [121]:

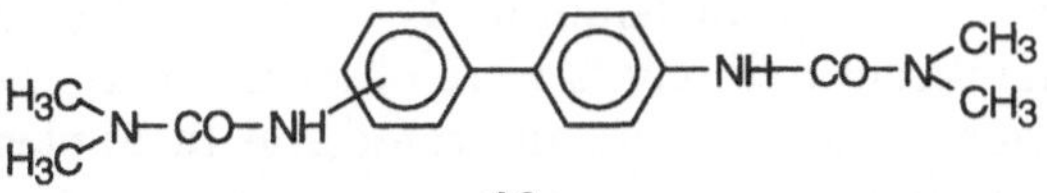

12

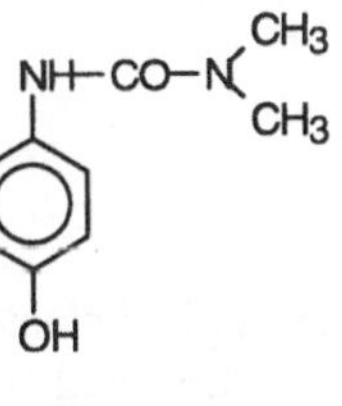

11

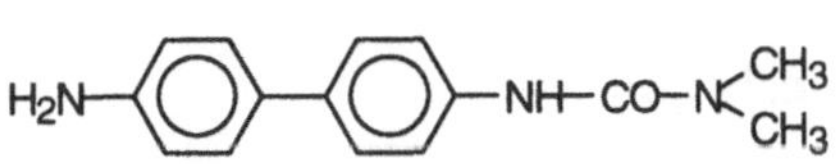

13

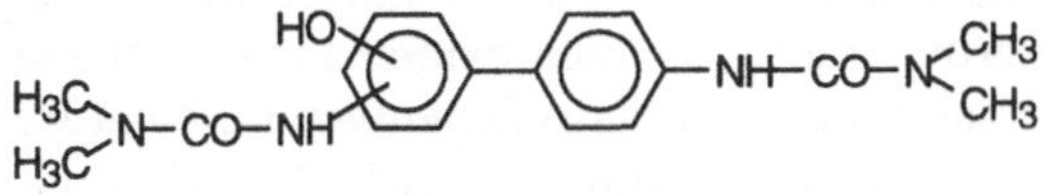

14

The proportion of biphenyls was found to increase with the initial fenuron concentration.

Additional experiments were carried out in order to determine what type of reaction is involved in the transformation of fenuron [121]. No degradation of fenuron was observed when mixtures containing rose Bengal and fenuron were irradiated at 546 nm, showing that singlet oxygen reacts very slowly with the substrate. On the other hand, a photosensitizer with a carbonyl group such as acetophenone strongly accelerated the disappearance of fenuron at 313 nm. In air-saturated solutions, products 11, 12 and 13 were detected whereas in de-oxygenated solutions only 12 and 13 were formed. The triplet state photo-sensitizer most probably abstracts a hydrogen atom (or an electron) from fenuron yielding the following radicals:

The products observed result from the rearrangement of this radical. Addition on another starting molecule leads to 12 and 13 and addition of oxygen in *para* position followed by a reduction process yields 11. The same mechanism is probably involved when humic substances are used as sensitizers. Apparent quantum yields of fenuron disappearance (Φ_a) were measured with commercial Aldrich humic acid upon irradiation at 365 nm. In neutral media, Φ_a is equal to 2.3×10^{-5}. It is about 2 times higher in acidic than in neutral media. This difference most probably results from the additional involvement of hydroxyl radicals at low pH. In accordance with this assumption, new reaction products were observed, which seem due to the oxidation of methyl groups. Humic and fulvic acids of various origins were used to photosensitize the oxidation of fenuron [121]. Qualitative results were the same. However, significant differences were observed in Φ_a values. The humic and the fulvic acids extracted from the same soil (a Ranker soil) both exhibited the best photoinductive properties.

5
Conclusion

Nitrate ions, salts of Fe^{III} and humic substances are three typical classes of com-pounds which play an important role by inducing photochemical reactions in natural waters. All these substances are able to produce reactive species such as hydroxyl radicals which oxidize most of the organic compounds. However, the overall mechanism also depends on the nature of the photoinductive species.

The main reaction induced by excitation of nitrate in dilute solution is oxidation: nitrate is a significant source of hydroxyl radicals in natural waters.

However, under excitation, nitrate ions are transformed into nitrogen dioxide which disproportionates into nitrite and nitrate ions but can, in some cases, lead to nitration or nitrosation and therefore generate carcinogenic or mutagenic compounds. Nitrite ions have similar inductive properties but the orientation of the reaction depends on their concentration since they behave as both a source of •OH and •OH quenchers. Nitration and nitrosation are favoured by increasing the concentration of nitrite and most likely with increasing concentration of substrate. It is likely, but not experimentally proved, that these two reactions play a minor role in natural waters. More research would be useful to extrapolate laboratory results to environmental conditions.

With iron (III), two different photochemical behaviours can occur. If the pollutant is able to complex iron (III), the resulting complex that absorbs solar light undergoes an intramolecular photoredox process between the metallic centre and the ligands of the inner sphere of solvation. When there is no complexation, aquocomplexes of iron (III) absorb solar radiation giving rise to •OH radicals that attack the pollutant. In both cases, iron (III) is reduced to iron (II) which can be reoxidized to iron (III) in the presence of oxygen; this reaction confers an interesting photocatalytic aspect to the process.

By means of laser-flash photolysis and probe molecules technique, it was possible to characterize the reactive species produced upon excitation of humic substances: namely $^1O_2$, •OH, solvated electrons. Kinetic studies using substrates with a low energy triplet state or oxidable substrates provided evidence that direct reactions with the triplet excited states of humic substances are possible through hydrogen, electron or energy transfer processes. It was experimentally proved that the degradation of many substances is accelerated by irradiation of humic substances in sunlight. However, in most cases further research would be necessary to elucidate mechanisms.

References

1. Goeyens L, Sörensson F, Tréguer P, Morvan J, Panouse M, Dehairs F (1991) Mar Ecol Prog Ser 77:7
2. Zafiriou OC, True MB (1979) (a) Marine Chem 8:9 (b) Marine Chem 8:33
3. Kotzias D, Parlar H, Korte F (1982) Naturwissenschaften 69:444
4. Zepp RG, Hoigné J, Bader H (1987) Environ Sci Technol 21:443
5. Russi H, Kotzias D, Korte F (1982)) Chemosphere 11:1041
6. Haag WR, Hoigné J (1985) Chemosphere 14:1659
7. Suzuki J, Okazaki H, Nishi Y, Suzuki S (1982) Bull Environ Contam Toxicol 29:511
8. Suzuki J, Okazaki H, Sato T, Suzuki S (1982) Chemosphere 11:437
9. Suzuki J, Hagino T, Ueki T, Nishi Y, Suzuki S (1983) Bull Environ Contam Toxicol 31:79
10. Suzuki J, Watanabe T, Suzuki S (1988) Chem Pharm Bull 36:2204
11. Suzuki J, Sato T, Suzuki S (1985) Chem Pharm Bull 33:2507
12. Suzuki J, Hagino T, Suzuki S (1987) Chemosphere 16:859
13. Suzuki J, Sato T, Ito A, Suzuki S (1990) Bull Environ Contam Toxicol 45:516
14. Suzuki J, Ueki T, Shimizu S, Uesugi K, Suzuki S (1985) Chemosphere 14:493
15. Ohta T, Suzuki J, Iwano Y, Suzuki S (1982) Chemosphere 11:797
16. Suzuki J, Sato T, Ito A, Suzuki S (1987) Chemosphere 16:1289
17. Bunce NJ, Cater SR, Willson JM (1985) J Chem Soc Perkin Trans II 1985:2013
18. Sarakha M, Boule P, Lenoir D (1993) J Photochem Photobiol 75:61
19. Niessen R, Lenoir D, Boule P (1988) Chemosphere 17:1977

20. Guillaume D, Morvan J, Martin G (1989) Environ Technol Letters 10:491
21. Machado F, Boule P (1995) J Photochem Photobiol A-86:73
22. Machado F, Boule P (1994) Toxicol Environ Chem 42:165
23. Usui Y, Takebayashi S, Takeuchi M (1992) Bull Chem Soc Jpn 65:3183
24. Alif A, Boule P (1991) J Photochem Photobiol A-59:357
25. Schedel G, Lenoir D, Boule P (1991) Chemosphere 22:1063
26. Kochany J, Choudhry GG (1990) Toxicol Environ Chem 27:225
27. Suzuki J, Yagi N, Suzuki S (1984) Chem Pharm Bull 32:2803
28. Machado F, Boule P (1994) Toxicol Environ Chem 42:155
29. Frank R, Klöpffer W (1988) Chemosphere 17:985
30. Meyerstein D, Treinin A (1961) Trans Faraday Soc 57:2104
31. Wagner I, Strehlow H, Busse G (1980) Z Phys Chem 123:1
32. Bayliss NS, Bucat RB (1975) Aust J Chem 28:1865
33. Strickler SJ, Kasha M (1963) J Am Chem Soc 85:2899
34. Zafiriou OC, Bonneau R (1987) Photochem Photobiol 45:723
35. Treinin A, Hayon E (1970) J Am Chem Soc 92:5821
36. Strehlow H, Wagner I (1982) Z Phys Chem 132:151
37. Rabani J, Matheson MS (1964) J Am Chem Soc 86:3175
38. Zellner R, Exner M, Herrmann H (1990) J Atmos Chem 10:411
39. Harrison CC, Malati MA, Smetham NB (1995) J Photochem Photobiol A Chem 89:215
40. Daniels M, Meyers RV, Belarto EV (1968) J Phys Chem 72:389
41. Warneck P, Wurzinger C (1988) J Phys Chem 92:6278
42. Grätzel von M, Henglein A, Lilie J, Beck G (1969) Ber Bunsenges Physik Chem 73:646
43. Grätzel von M, Taniguchi S, Henglein A (1970) Ber Bunsenges Physik Chem 74:488
44. Voelker B, Morel FMH, Sulzberger B (1997) Environ Sci Technol 31:1004
45. Alder AC, Siegrist H, Guser W, Giger W (1990) Water Res 24:733
46. Frimmel FH (1989) GWF Gas Wasserfach: Wasser/Abwasser 130:106
47. Günter KF, Hilger S, Canonica S (1995) Environ Sci Technol 29:1008
48. Natarajan P, Endicott JF (1973) J Phys Chem 77:2049
49. Klöpffer W, Kohl EG, Kaufmann G (1988) Batelle Institut, Draft report PUG/WKI0
50. Carey JH, Langford CH (1973) Can J Chem 51:3665; (1975) Can J Chem 53:2436
51. Svenson A, Kaj L, Björndal H (1989) Chemosphere 18:1805
52. Frank R, Rau H (1990) Ecotox Environ Saf 19:55
53. Lockhart HB Jr, Blakeley RV (1975) Environ Sci Technol 9:1035
54. Gustavson RL, Martell AE (1963) J Phys Chem 67:576
55. Andrianirinaharivelo SL, Pilichowski JF, Bolte M (1993) Transition Metal Chem 18:37
56. Trott T, Hanwood R, Langford CH (1972) Environ Sci Technol 9:367
57. Stolzberg R, Hume D (1975) Environ Sci Technol 9:654
58. Andrianirinaharivelo SL, Mailhot G, Bolte M (1995) Solar Energy Materials and Solar Cells 38:459
59. Mailhot G, Bordes AL, Bolte M (1995) Chemosphere 9:1729
60. Evans MG, George P, Uri N (1949) Trans Faraday Soc 34:230
61. Faust BC, Hoffmann MR (1986) Environ Sci Technol 20:943
62. Langford CH, Carey JH (1975) Can J Chem 53:2430
63. Weschler CJ, Mandich ML, Grädel TE (1986) J Geophys Res 91:5189
64. Benkelberg HJ, Warneck P (1995) J Phys Chem 99:5214
65. Faust BC, Hoigné J (1990) Atmos Environ 24A:79
66. Knight RJ, Sylva RN (1975) Inorg Nucl Chem 37:779
67. Mazellier P, Mailhot G, Bolte M (1997) New J Chem 21:389
68. Sehested K, Corfitzen H, Christensen HC, Hart EJ (1975) J Phys Chem 75:310
69. Terzian R, Serpone N, Fox M (1995) J Photochem Photobiol A:Chem 90:125
70. Mazellier P, Bolte M (1996) J Photochem Photobiol A:Chem 98:141
71. Kawaguchi H, Inagaki A (1994) Chemosphere 28:57
72. Larson RA, Schlauch MB, Marley KA (1991) J Agric Food Chem 39:2057
73. Mazellier P, Jirkovský J, Bolte M (1997) Pestic Sci 49:259

74. Brand N, Mailhot G, Bolte M (1997) Chemosphere 34:2637
75. Power JF, Sharma K, Langford CH, Bonneau R, Joussot-Dubien J (1986) Photochem Photobiol 44:11
76. Fischer A M, Winterle J S, Mill T (1987) ACS Symposium Series 327:141
77. Zepp RG, Braun AM, Hoigné J, Leenheer JA (1987) Environ Sci Technol 21:485
78. Cooper WJ, Zika RG, Pestane RG, Fischer A M (1989) ACS Symposium Series 219:333
79. Cooper WJ, Zika RG (1983) Science 220:711
80. Bruccoleri A, Langford CH, Arbour C (1990) Environ Technol 11:169
81. Bruccoleri A, Pant BC, Sharma K, Langford CH (1993) Environ Sci Technol 27:889
82. Zepp RG, Baughman GL, Scholtzhauer PF (1981) Chemosphere 10:109
83. Zepp RG, Baughman GL, Scholtzhauer PF (1981) Chemosphere 10:119
84. Haag WR, Hoigné J, Gassman E, Braun AM (1984) Chemosphere 13:631
85. Zepp RG, Scholtzhauer PF, Sink RM (1985) Environ Sci Technol 19:74
86. Haag WR, Hoigné J (1986) Environ Sci Technol 20:341
87. Frimmel FH, Bauer H, Putzien J, Murasecco P, Braun AM (1987) Environ Sci Technol 21:541
88. Scully FS, Hoigné J (1987) Chemosphere 16:681
89. Tratnyek PG, Hoigné (1991) J Environ Sci Technol 25:1596
90. Aguer JP, Richard C (1993) Toxicol Environ Chem 39:217
91. Aguer JP, Richard C (1994) J Photochem Photobiol A 84:69
92. Van Hoort P, Lammers R, Verboom H, Wondergem E (1988) Chemosphere 17:35
93. Kawaguchi H (1993) Chemosphere 27:2177
94. Canonica S, Jans U, Stemmler K, Hoigné J (1995) Environ Sci Technol 29:1822
95. Canonica S, Hoigné J (1995) Chemosphere 30:2365
96. Aguer JP, Richard C (1996) J Photochem Photobiol A 93:193
97. Faust BC, Hoigné J (1987) J Environ Sci Technol 21:957
98. Hoigné J, Faust BC, Haag WR, Scully FE, Zepp RG (1989) ACS Symposium Series 219:363
99. Miles CJ, Brezonik PL (1981) Environ Sci Technol 15:1089
100. Khan SU, Schnitzer M (1978) J Environ Sci Health, B13:299
101. Khan SU, Gamble DS (1983) J Agric Food Chem 31:1099
102. Frimmel FH, Hessler DP (1994) In: Helz H, Zepp RG, Crosby DG (eds) Aquatic and Surface Photochemistry, CRC Press, p 137
103. Aguer JP, Richard C (1996) Pestic Sci 46:151
104. Ross RD, Crosby DG (1973) J Agric Food Chem 21:335
105. Mabey WR, Tse D, Baraze A, Mill T (1983) Chemosphere 12:3
106. Simmons M, Zepp R (1986) Wat Res 20:899
107. Hwang HM, Hodson RE, Lee RF (1986) Environ Sci Technol 20:1002
108. Zepp RG, Wolfe NL, Gordon JA, Baughman GL (1975) Environ Sci Technol 9:1144
109. Miller GC, Crosby DG (1978) J Agric Food Chem 26:1316
110. Minero C, Pramauro E, Pelizzetti E, Dolci M, Marchesini A (1992) Chemosphere 24:1597
111. Jensen-Korte U, Anderson C, Spiteller M (1987) Sci Total Environ 62:335
112. Zepp RG, Wolfe NL, Gordon JA, Fincher RC (1976) J Agric Food Chem 24:727
113. Van Noort P, Smit R, Zwaan E, Zijlstra J (1988) Chemosphere 17:395
114. Skurlatov Y I, Zepp RG, Baughman GL (1983) J Agric Food Chem 31:1065
115. Miller GC, Zisook R, Zeep R (1980) J Agric Food Chem 28:1053
116. Mill M J, Crosby DG (1983) Marine Chem 14:111
117. Sukul P, Moza PN, Hustert K, Kettrup A (1992) J Agric Food Chem 40:2488
118. Wang CX, Yediler A, Peng A, Kettrup A (1995) Chemosphere 30:501
119. Acher AJ, Saltzman S (1980) J Environ Qual 9:190
120. Ahel M, Scully FE, Hoigné J, Giger W (1994) Chemosphere 28:1361
121. Aguer JP, Richard C, Andreux F (1992) J Photochem Photobiol A 103:163

8 Mechanism of Phototransformation of Phenol and Derivatives in Aqueous Solution

Claire Richard[1] and Gottfried Grabner[2]

[1] Laboratoire de Photochimie Moléculaire et Macromoléculaire, UMR n°6505, Université Blaise Pascal, F-63177 Aubière Cedex. *E-mail: richardc@cicsun.univ-bpclermont.fr*
[2] Institut für Theoretische Chemie und Strahlenchemie, Althanstrasse 14, A-1090 Wien

This article reviews recent research on the mechanism of the photolysis of phenolic compounds. Relevant photophysical studies are also included. The main emphasis is on photodegradation of compounds of actual environmental relevance or structurally related to such compounds, such as halogen-, nitro-, or cyano-substituted phenols in aqueous solutions, and on the photosensitizing action of phenolic compounds.

Keywords: phenols, photophysics, photolysis, photodehalogenation, photosensitization.

Contents

The Handbook of Environmental Chemistry Vol. 2 Part L
Environmental Photochemistry (ed. by P. Boule)
© Springer-Verlag Berlin Heidelberg 1999

1
Introduction

Phenol and its derivatives are present in natural waters as a consequence of their use as domestic products or in industrial activities. These compounds are also systematically introduced in soil or in aquatic media through the intensive application of fungicides, insecticides or herbicides. The contamination of the environment – and especially of ground waters – by these pollutants represents a toxicological hazard because many of them were found to be highly toxic. Biological transformations are often an efficient degradation path for these compounds. However, photochemistry constitutes an alternative path for their degradation.

Earlier work on photophysics and photochemistry of phenol and some of its derivatives, in particular the amino acid tyrosine, has been summarized in a review by Creed [1]. The photophysical properties and primary photoreactions of phenol in aqueous solution were well understood by then, with the exception of triplet state properties which were not well characterized. Phenol and tyrosine are moderately fluorescent, with fluorescence quantum yields of the order of 0.2. The main photoreaction in terms of quantum yield is electron ejection to form a hydrated electron. The efficiency of this process depends on excitation energy and temperature and is greatly enhanced when the molecule is present as an anion (the pK values of many phenols range between 9 and 10).

These earlier studies were mainly motivated by the relevance of photo-processes of phenols to photochemistry and photobiology, in particular to the mechanism of protein photodegradation. A new perspective came to light concerning the role that these photoprocesses might play in the environment. Accordingly, the interest shifted to the study of derivatives which were known or assumed to be of environmental relevance, with halogen-substituted phenols in the front row. Among these, 4-chlorophenol constitutes one of the most frequently employed substrates in studies of photocatalyzed pollutant degradation (see the corresponding chapter).

This review is limited to reactions occurring after direct absorption of light by the phenolic compounds. Photosensitized transformations also take place in environmental conditions. These reactions are described in other chapters. The first part will focus on photophysical properties of phenol and its derivatives, in particular alkylphenols and dihydroxybenzenes. The second part will report on photochemical reaction mechanisms described in the literature for variously substituted phenols, with an emphasis on dehalogenation of halogen-substituted phenols.

2
Photophysical Properties

A detailed account of earlier work on the photophysics of phenol, alkylphenols, and in particular tyrosine can be found in Creed's review [1]. A large number of more recent studies on fluorescence and phosphorescence spectroscopy of these compounds have been motivated by efforts to understand the photophysics of proteins and will not be considered here; a comprehensive review of the fluorescence of tyrosine in proteins and polypeptides appeared a few years ago [2].

The influence of substitution and solvent environment on the photophysical properties of phenolic compounds has been the subject of a few studies in the last few years. One of these is an inquiry into the effect of complexation with β-cyclodextrin on phenol and methylated phenols by means of absorption, circular dichroism, fluorescence, and transient absorption spectroscopies [3]. Cyclodextrins are cyclic oligosaccharides forming inclusion complexes with suitable host molecules; these complexes constitute one of the prototypes of host-guest systems. The fluorescence properties of poly-methylated phenols were found to be strongly affected by complexation, with both quantum yield and lifetime showing a marked increase; taking 2,4,6-tri-methylphenol as an example, τ_F was found to increase from 0.6 to 3.65 ns and ϕ_F from 0.025 to 0.12 when going from the aqueous solution to the cyclodextrin complex. This was shown to be mainly due to a decrease of the rate constant for nonradiative decay induced by the alteration of the OH group environment in the complex.

The photophysical behavior of di- and trihydroxybenzenes in aqueous solution was the subject of a recent study [4]. Room-temperature ($T = 22-24\,°C$) fluorescence lifetimes and quantum yields are distinctly smaller for the dihydroxybenzenes when compared to phenol and become very small ($\phi_F < 0.001$) for the trihydroxybenzenes. The *ortho*-disubstituted compound (pyrocatechol) is less fluorescent ($\phi_F = 0.033$, $\tau_F = 0.5$ ns) than the *meta*- and *para*-disubstituted isomers (resorcinol: $\phi_F = 0.077$, $\tau_F = 1.6$ ns; hydroquinone: $\phi_F = 0.09$, $\tau_F = 1.27$ ns). The measurement of the dependence of ϕ_F on the wavelength of excitation showed that the fluorescence quantum yield is constant throughout the first absorption band. The same result was obtained for the quantum yield of electron ejection (see below). The triplet-triplet absorption spectra of pyrocatechol ($\lambda_{max} = 410$ nm) and of resorcinol ($\lambda_{max} = 375$ nm) were characterized. The corresponding spectrum for hydroquinone ($\lambda_{max} = 430$ nm) had been measured in an earlier study [5]. The intersystem crossing quantum yields ($T = 22\,°C$) in aqueous solution were determined for phenol ($\phi_{isc} = 0.66$), pyrocatechol ($\phi_{isc} = 0.17$), resorcinol ($\phi_{isc} = 0.46$) and hydroquinone ($\phi_{isc} = 0.46$; in reasonable agreement with $\phi_{isc} = 0.39$ as previously determined by nitrate quenching [5]). These values show that intersystem crossing is the quantitatively main photophysical pathway for these compounds with the exception of pyrocatechol where other nonradiative processes dominate. The triplet lifetimes as determined by transient absorption spectroscopy are of the order of $1-10\,\mu s$. These triplet properties indicate that phenol and the dihydroxybenzenes are potentially efficient photosensitizers (see below).

Triplet quenching data have been obtained for aqueous hydroquinone using transient absorption spectroscopy [5]. Acrylamide and O_2 were found to react with triplet hydroquinone at diffusion-controlled rates, while reaction with nitrate was one order of magnitude slower ($k = 3.1 \times 10^8\,M^{-1}\,s^{-1}$). The extent of electron transfer in the course of the quenching by O_2 and nitrate was estimated by monitoring the formation of semiquinone anions. The reaction with nitrate was found to proceed quantitatively by charge transfer, that with O_2 only partially so, but in both cases the radical yield was significantly enhanced in comparison with solutions in which the triplet quencher was absent.

Knowledge about singlet oxygen formation by excited phenolic compounds is scarce. Indications of a self-sensitization mechanism were found in a study of the reactivity of dihydroxybenzenes with singlet oxygen ($^1\Delta_g$) [6]. Self-sensitization quantum yields of 0.21 for hydroquinone and 0.12 for resorcinol were measured in dioxane, but no self-sensitization was observed in D_2O. A possible fast self-quenching reaction of singlet oxygen was invoked to explain this result. It was concluded that photosensitization by aqueous dihydroxybenzenes via a singlet oxygen mechanism might be of relevance if very efficient singlet oxygen quenchers are present. Indications of both energy- and electron-transfer triplet photoreactions were also found in a study of the photochemistry of aqueous tyrosine [7] (see below).

3
Photoreaction Mechanisms of Phenol Derivatives in Dilute Solutions

3.1
OH Bond Rupture

The photoinduced breaking of an OH bond producing a H atom and a phenoxyl radical, as straightforward as it may seem, is of minor importance only in the photochemistry of phenols in aqueous solution [1]. However, it constitutes the main primary photochemical step in nonpolar medium. This was shown in a detailed study of the photophysics and photochemistry of phenol and methylated phenols in nonpolar solvents [8]. In n-hexane, the photodissociation quantum yields upon excitation in the first singlet range from 0.08 (for p-cresol) to 0.21 (for 2,6-dimethylphenol); the value for unsubstituted phenol itself is 0.13. It seems appropriate to emphasize the possibility of this photoprocess as it might play a role in other media of lower polarity in heterogeneous environments. No studies of such systems have been carried out so far.

A further parameter influencing the efficiency of OH bond rupture is the energy of excitation. The quantum yield has long been known to increase dramatically when aqueous phenol is excited in its S_2 state [1]. This effect was also noted in a more recent study of the photochemistry of hydroquinone in aqueous solution [5], where a quantum yield of 0.13 was found at an excitation wavelength of 266 nm.

3.2
Photoionization

The study of hydrated electron formation by photoexcitation of dihydroxybenzenes [4] confirmed earlier results [1] showing that photoionization is the main primary photoprocess in unsubstituted phenol as well as in alkylphenols. The quantum yields for e_{aq}^- formation in neutral aqueous solution (22 °C), as determined by scavenger techniques as well as by transient absorption spectroscopy, are 0.028 for pyrocatechol, 0.014 for resorcinol, and 0.009 for hydroquinone. These yields refer to a one-photon process. It should be mentioned that when high-fluence pulsed lasers are used for excitation, two-photon electron ejection generally predominates. The efficiency of this process is negligible in conditions of steady-state irradiation.

The one-photon electron ejection efficiency is constant when the excitation wavelength lies within the first absorption band (S_1) of the compound and increases at lower excitation wavelengths. This behavior is opposed to that of the fluorescence quantum yield. The same antagonism is observed in the temperature dependence: e_{aq}^- formation increases with increasing temperature, whereas the fluorescence yield decreases. These results are rationalized by a model describing electron ejection and fluorescence as competing processes in the singlet manifold [4].

The quantum yield of e_{aq}^- formation increases markedly when the phenols are present as phenolate anions [1]. It can assume substantial values even in cases where there is no electron ejection from the neutral molecule. In 4-chlorophenolate for instance, electron ejection competes with dehalogenation [9].

Electron ejection produces phenoxyl radicals, or, in the case of dihydroxybenzenes, semiquinone radicals as primary photochemical products. The kinetics of the reactions of these radicals play a key role in determining the photoproduct distribution. The mechanism of the thermal reactions of phenoxyl radicals has been described in detail [10-14]. The study of the kinetics of recombination of phenoxyl and semiquinone radicals with other radicals continues to be of interest [15-17].

3.3
Dehalogenation

As a general rule, the direct photolysis of halogenophenols in water results in the heterolytic cleavage of the C–X bond and HX is formed [18]. The nature of the organic photoproducts is highly dependent on the position of the halogen on the aromatic ring.

3.3.1
Monohalogenophenols

The excitation of 2-halogenophenols in neutral or acidic medium yields cyclopentadiene carboxylic acids and pyrocatechol. The quantum yields of photolysis of 2-chlorophenol was found to be in the range 0.03–0.068 [19–21] (see Table 1). Upon irradiation of the anion, cyclopentadienic carboxylate ions are produced with higher yield ($\lambda = 0.28-0.30$) [19–21]. The mechanism was recently investigated by means of laser flash photolysis and transient absorption spectroscopy [21]. The transient detected at the end of the pulse ($\lambda_{max} = 260$ nm) was assigned to the ketene arising from HX elimination and ring contraction. This species is subsequently converted into fulvene-6,6-diol

Table 1. Quantum yields of phototransformation of substituted phenols in aqueous solution

Substrate	ϕ	Experimental conditions[a] (λ_{exc}, pH)	Reference
2-F-phenol	0.14 ± 0.02	270 nm, neutral	21
2-F-phenol	0.44 ± 0.04	280 nm, pH=11	21
2-Cl-phenol	0.068 ± 0.008	280 nm, neutral	21
2-Cl-phenol	0.28 ± 0.03	280 nm, pH=11	19, 21
2-Br-phenol	0.085 ± 0.009	280 nm, neutral	21
2-Br-phenol	0.29 ± 0.03	280 nm, pH=11	21
3-Cl-phenol	0.09 ±0.01	280 nm, pH=5	20
4-Cl-phenol	0.68 ± 0.07	280 nm, neutral	9
4-Cl-resorcinol	0.08 ± 0.02	280 nm, neutral	47
4-Cl-resorcinol	0.28 ± 0.03	280 nm, pH=11	47
2-nitrophenol	$(2.3 ± 0.2) \times 10^{-6}$	365 nm, pH=8.2	57
3-nitrophenol	$(1.5 ± 0.2) \times 10^{-6}$	365 nm, pH=8.2	56
4-nitrophenol	$(1.7 ± 0.4) \times 10^{-5}$	365 nm, pH=8.2	55
4-nitrosophenol	0.0042 ± 0.0007	365 nm, pH=3	59
4-hydroxybenzonitrile	0.009 ± 0.0002	254 nm, pH=5.6, deoxygenated solution	62
hydroquinone	0.002 ± 0.0004	296 nm, neutral, deoxygenated solution	5
hydroquinone	0.048 ± 0.005	296 nm, neutral, O_2-saturated solution	5
3,5-dichlorophenol	0.03 ± 0.01		
3,4-dichlorophenol	0.02 ± 0.005	280 nm, pH = 5	39
2,4-dichlorophenol	0.01 ± 0.005		
3,5-dichlorophenol			
3,4-dichlorophenol	0.11 ± 0.02	296 nm, pH=11	39
2,4-dichlorophenol			

[a] air-saturated solutions if not specified otherwise.

(λ_{max} = 293 nm) with a rate constant of the order of 10^6 s^{-1} (Fig. 1), in agreement with earlier work on 2-bromophenolate [22]. Cyclopentadiene carboxylic acids are finally produced after H-atom migration. The reaction sequence is outlined in Scheme 1.

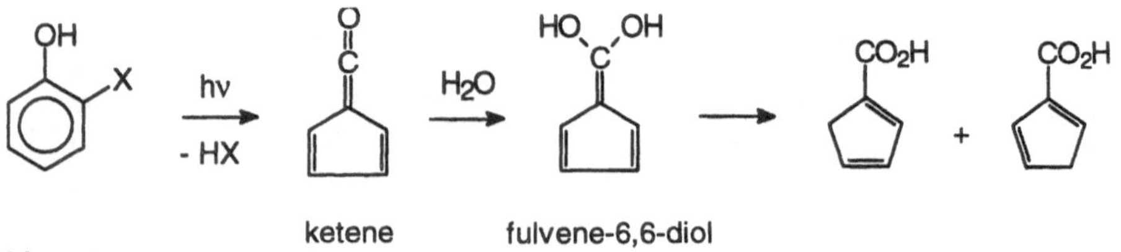

ketene   fulvene-6,6-diol

Scheme 1

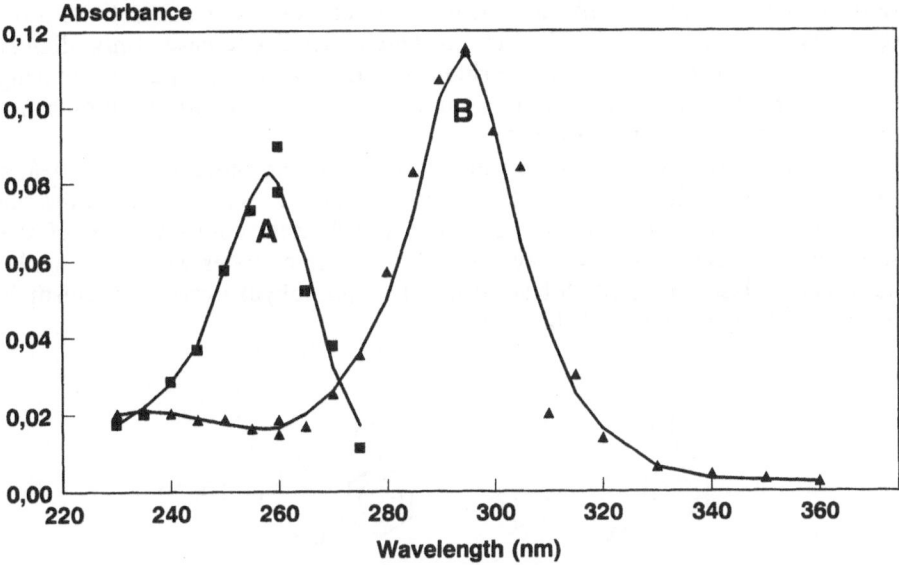

Fig. 1. Transients observed upon nanosecond laser flash photolysis (120 mJ cm^{-2}) of a neutral aqueous solution of 2-chlorophenol (5×10^{-4} M). Spectrum A: measured at pulse end, assigned to the ketene; spectrum B: measured 8 μs after pulse end, assigned to the fulvene diol

Such a ring contraction is also observed in the Wolff rearrangement (Scheme 2), which has been proposed to involve an intermediate α-ketocarbene [23].

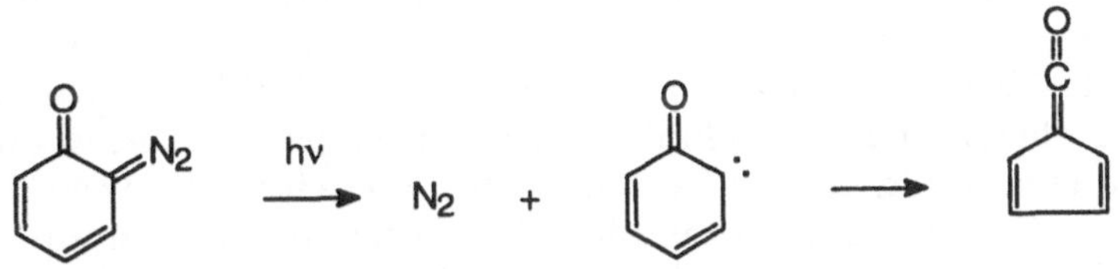

Scheme 2

By analogy to this reaction, it was proposed that an α-ketocarbene might be formed in the photolysis of the 2-halogenophenols after release of HX [19]. However, this carbene has a triplet ground state, and it is therefore not evident that it participates in the Wolff rearrangement which is generally assumed to proceed on the singlet surface [24]. A concerted mechanism of HX elimination from the singlet excited state of the 2-halogenophenols must therefore be considered as an alternative. Transient absorption measurements [21] have indeed shown that the carbene can be trapped in a neutral aqueous solution of 2-chlorophenol by reaction with O_2 or aliphatic alcohols to form, respectively, a quinone oxide and phenoxyl radicals, but with a yield ($\phi = 0.004$) that is one order of magnitude smaller than that of the ketene ($\phi = 0.036$). The phenoxyl radical yield increases with the alcohol content of water-alcohol mixtures ($\phi = 0.045$ in neat 2-propanol), but the ketene yield decreases only slightly ($\phi = 0.026$ in neat 2-propanol). Carbene reaction with 2-propanol and ring contraction thus do not seem to be competitive, i.e., ring contraction does not involve the carbene as intermediate.

In the photolysis of 3-halogenophenol, resorcinol accounts for over 80% of the starting material converted [25]. This reaction is selective whatever the experimental conditions, be it excitation wavelength, concentration of the substrate or concentration of oxygen [26]. No intermediates were detected by nanosecond laser flash photolysis [9]. A fast photohydrolysis mechanism is therefore likely to operate (Scheme 3).

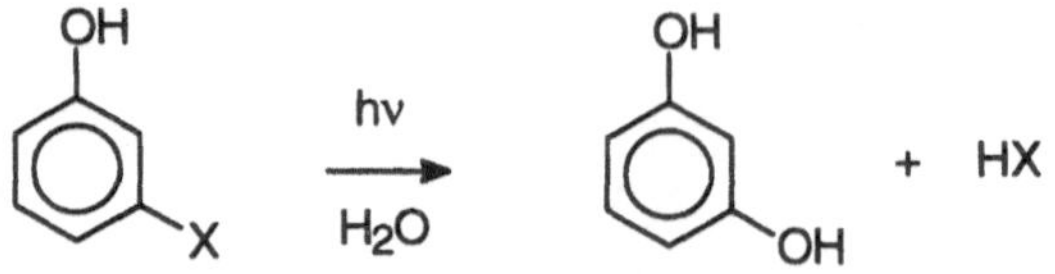

Scheme 3

In the last ten years, several studies have been devoted to the analysis of photoproducts formed upon irradiation of 4-chlorophenol in water [27–32]. Due to its apparent complexity and to some contradictory reports about the effect of oxygen on the formation of *p*-benzoquinone, the mechanism of the reaction remained unclear. It was only recently possible to elucidate it with the help of laser flash photolysis experiments [9]. The findings were confirmed by further studies [33, 34]. The distribution of products depends on the experimental conditions. In the absence of oxygen, hydroquinone and 5-chloro-2,4'-dihydroxybiphenyl are formed. The yield of hydroquinone decreases whereas that of 5-chloro-2,4'-dihydroxybiphenyl increases when the substrate concentration is increased. In the presence of oxygen, *p*-benzo-quinone is the major primary product, but formation of 5-chloro-2,4'-dihy-droxybiphenyl is also observed at high substrate concentration. Contrary to some reports [28, 32], *p*-benzoquinone is not formed at low conversion extent in the absence of oxygen. Several other products such as phenol, 4,4'-and 2,4'-dihydroxybiphenyl, and 2,5,4'-trihydroxybiphenyl, as well as oligomers were identified at high conversion extent or in concentrated solutions.

Nanosecond laser flash photolysis experiments [9] revealed that the carbene, 4-oxocyclohexa-2,5-dienylidene ($\lambda_{max} = 384$ and 370 nm) (Fig. 2, spectrum A) is formed after elimination of HCl from excited 4-chlorophenol. Once produced, this intermediate reacts through different ways. The reaction with water leading to hydroquinone is slow ($k = 1.5 \times 10^3 \, M^{-1} s^{-1}$) because of the triplet character of the carbene and the high barrier for O-H rupture in the molecule of water. In constrast, addition of oxygen is easy ($k = 3.5 \times 10^9 \, M^{-1} s^{-1}$) resulting in the production of benzoquinone-O-oxide ($\lambda_{max} = 460$ nm) (Fig. 2, spectrum B).

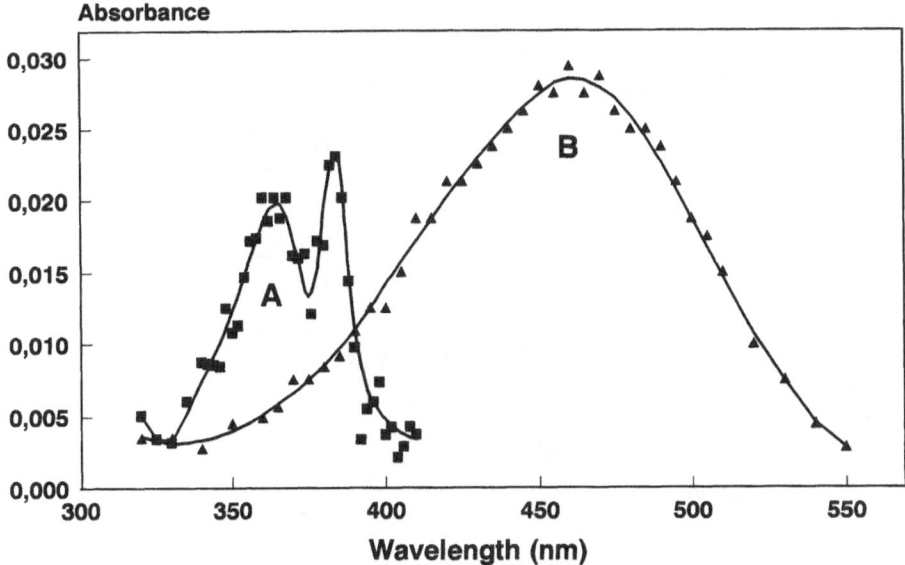

Fig. 2. Transients observed upon nanosecond laser flash photolysis (12 mJ cm^{-2}) of a neutral aqueous solution of 4-chlorophenol (1.5 × 10^{-3} M). Spectrum A: measured at pulse end in argon-saturated solution, assigned to 4-oxo-cyclohexa-2,5-dienylidene; spectrum B: measured 1 μs after pulse end in O$_2$-saturated solution, assigned to benzoquinone-O-oxide

This transient is further converted into benzoquinone with production of hydrogen peroxide. The carbene also reacts with 4-chlorophenol itself ($k = 1.5 \times 10^8$ M^{-1} s^{-1}). This process becomes major at high substrate concentration and yields the coupling product 5-chloro-2,4′-dihydroxybiphenyl. Further reactions with other 4-chlorophenol molecules lead to the formation of oligomers. The carbene can also be reduced by H-donor molecules such as alcohols ($k = 1.7 \times 10^7$ M^{-1} s^{-1} with 2-propanol) into the phenoxyl radical which is further converted into phenol [9]. An overview of the variety of possible carbene reactions is shown in Scheme 4.

The intermediate formation of 4-oxocyclohexa-2,5-dienylidene in the photolysis of 4-chlorophenol was recently confirmed by the detection, using Fourier-transform electron paramagnetic resonance, of a correlated pair of a 2-propanoyl and a phenoxyl radical in a solution of 4-chlorophenol in 2-propanol [33].

Due to the high reactivity of the carbene and to the high quantum yields of phototransformation of some primary products, secondary processes occur early in the reaction and the mechanism becomes quickly very complex in the conditions of steady-state irradiation. However, the phenomena are now well elucidated at low conversion extent and the primary photoproducts are clearly identified. This allows the rationalization of a few reports of 4-halophenol photolysis products [35–37] which, according to current knowledge, are due to secondary reactions. In one of these studies [37], EPR experiments were undertaken in order to detect the free radicals produced upon excitation of 4-bromo-benzoquinone-O-oxide

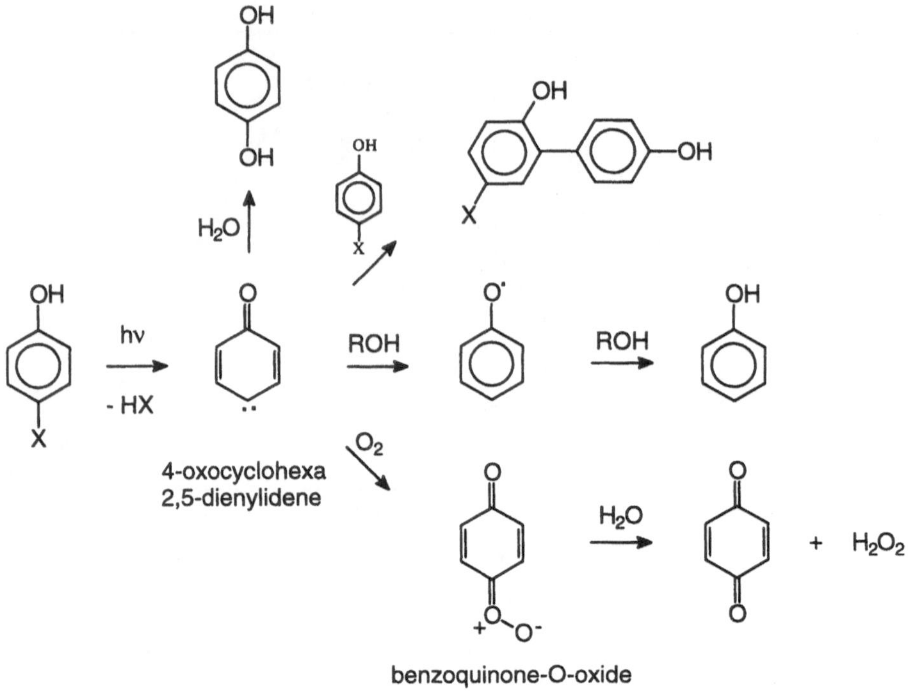

Scheme 4

phenol in slightly basic medium. DMPO (5,5-dimethylpyrroline-N-oxide) was used as a spin trap. Four distinct species were observed during irradiation with a mercury-xenon lamp. The signals were assigned to the DMPO-H, DMPO-aryl and DMPO-OH adducts and to the p-benzo-semiquinone anion. The aryl free radicals are most probably formed by reductive debromination by hydrated electrons. This assumption is supported by the observation that one-photon ionization of 4-chlorophenolate anions is a quite efficient process ($\phi = 0.095$) [9]. The hydroxyl radicals and p-benzosemiquinone anions detected by EPR are probably secondary products, resulting from the photochemical decomposition of p-benzoquinone [38]. Another group reported the production of 4-chloro-catechol and hydroquinone in oxygen-saturated solutions [35, 36]. The formation of 4-chlorocatechol is probably due to the reaction of 4-chlorophenol with hydroxyl radicals, these species being produced in small amount in the photolysis of p-benzoquinone [38]. Hydroquinone, on the other hand, is one of the main photoproducts of the phototransformation of aqueous p-benzo-quinone.

3.3.2
Polyhalogenophenols

The quantum yield of photolysis of polyhalogenated phenols decreases as the number of halogen atoms on the ring increases (see Table 1). With 3,4-halogenophenols, hydroxylation occurs at the *meta* position leading to the corresponding 3-hydroxy derivative. The effect of a chlorine atom on *ortho* position is prevalent: ring contraction is the main route of phototransformation of 2,4-dihalogenophenol [39, 40].

Pentachlorophenol is a highly toxic compound commonly used as a fungicide and bactericide. Several studies were undertaken to clarify the mechanism of its photodegradation [41–44]. Irradiation of pentachlorophenol with sunlight or UV lamps results in a rather rapid degradation at pH = 7.3. Total disappearance under simulated sunlight is reached after approximatively 20 hours. At pH = 3.3, the photodegradation is ten times slower. As the pK_a is equal to 4.7, it can be deduced that the phenolate ion is more photoreactive than the molecular form. Three major types of photoproducts such as lower chlorinated phenols (photoreduction), tetrachlorodihydroxybenzenes (replacement of Cl^- ions by OH^-), and non-aromatic fragments were isolated. Tetrachlorohydroquinone is subsequently oxidized even in the dark into chloranil, 2-hydroxy-3,5,6-trichlorobenzoquinone, 2,5-dichloro-3,6-dihydroxybenzoquinone and dichloromaleic acid (Scheme 5). It was also reported that highly toxic polychlorinated dibenzo-*p*-dioxins and polychlorinated dibenzofurans are formed during the photolysis of pentachlorophenol in water [42, 43].

The formation of 2-methyl-4,5,6,7-tetrachlorobenzoxazole was observed when pentachlorophenol is irradiated in a water-acetonitrile mixture

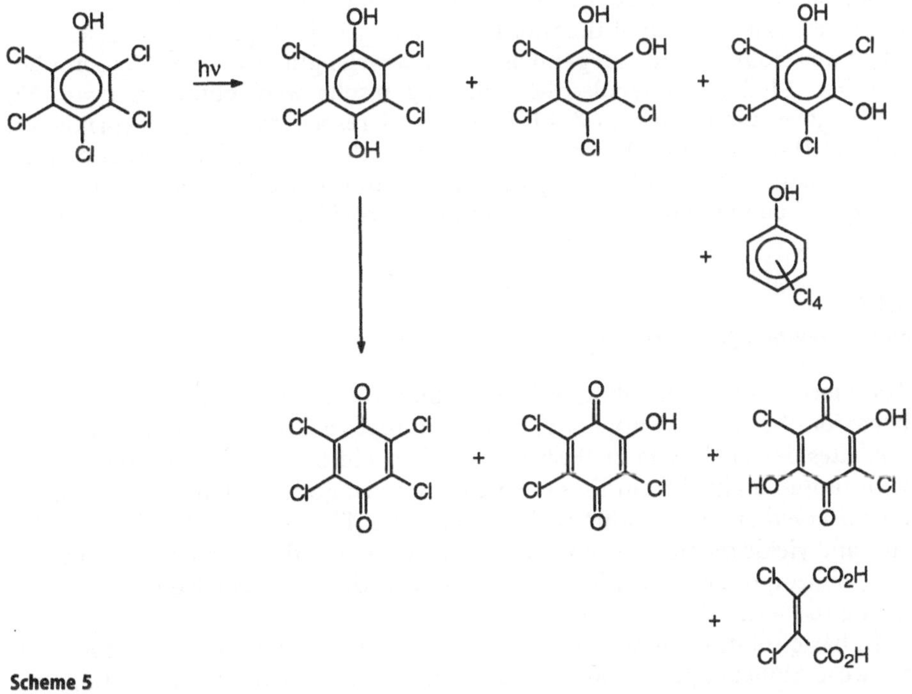

Scheme 5

(Scheme 6). It was suggested that this product results from the addition of acetonitrile on the 3,4,5,6-tetrachlorocyclohexa-3,5-diene-1,2-ketocarbene [44, 45].

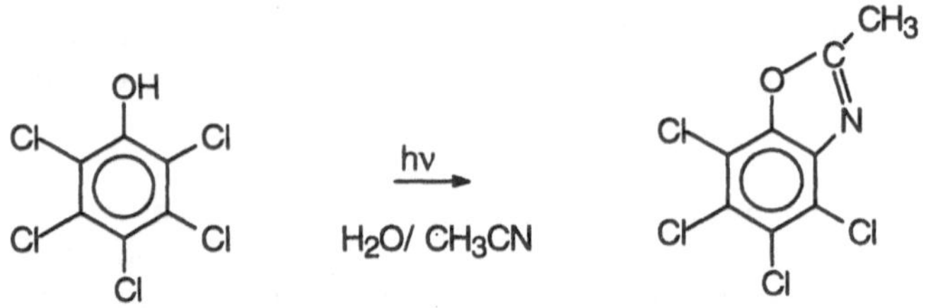

Scheme 6

Quantitative structure-activity relationships (QSARs) between photochemical reactivity and structural parameters were tentatively developed in order to predict the fate of halogenoaromatics in the environment [46]. The reaction selected to test this approach was the photohydrolysis of 3-halogenophenols as well as of 1,3-di- and 1,3,5-trihalogenobenzenes, which has the advantage to be very selective and well described in the literature. Irradiations were performed in polychromatic light with lamps emitting in the range 250–350 nm with an emission maximum at 300 nm. In each cases, photohydrolysis was the main transformation process, and hydroxylated products were formed. With substrates bearing two different halogen atoms, the halogen having the lowest carbon-halogen bond strength was released first. Pseudo-first order rate constants were measured for all compounds. Quantum yields were evaluated as a function of the wavelength using 3-chlorophenol as a reference. They were found to be independent of the wavelength as it is for 3-chlorophenol. For 3-halogenophenols, the following results were obtained: $\phi = 0.155$, 0.151, 0.096, and 0.090 with X = F, Br, I, and Cl, respectively. A good correlation ($r^2 = 0.94$) was obtained based on the energy of the C-X bond and the sum of all steric factors. This methodology can be used to predict the rates of photolysis of organic compounds in natural conditions, provided the reaction mechanism is known.

3.3.3
Substituted Halogenophenols

The mechanism of the photolysis of 4-chlororesorcinol shows features of the behavior of both 2-chlorophenol and 4-chlorophenol [47]. The transient intermediates detected by laser flash photolysis could be assigned by comparison with those obtained from these two compounds. Both a ketene and a carbene are observed upon excitation with a 10 ns pulse. The ketene adds a water molecule and yields the fulvene diol. The carbene reacts with oxygen with formation of a benzoquinone-O-oxide or abstracts a hydrogen atom from an alcohol giving rise to the 3-hydroxyphenoxyl radical.

In basic solution, photocontraction of the ring is the major process leading to two cyclopentadiene acids. In acidic and deoxygenated medium, 1,3,4-trihydroxybenzene and biphenyl derivatives are formed. In oxygen-saturated solution, hydroxybenzoquinone is the main photoproduct. Finally, resorcinol is found as the main product in the presence of 2-propanol. The complete reaction mechanism is outlined in Scheme 7.

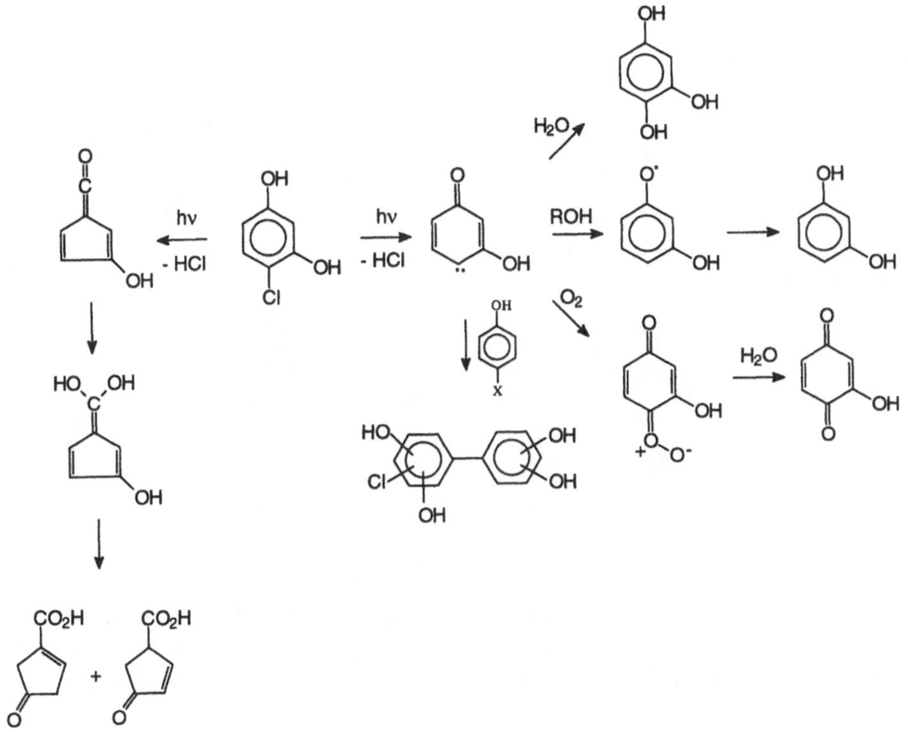

Scheme 7

The photoconversion of chlorohydroquinone into hydroquinone, chloro-benzoquinone and benzoquinone was explained by a radical mechanism [48] Scheme 8. It was proposed, but not proved, that the first step is the homolytic scission of the C–Cl bond the chlorine atom released being assumed to oxidize chlorohydroquinone into chlorosemiquinone radical. Absorption spectrum obtained by microsecond flash photolysis can be attributed to a mixture of semiquinone and chlorosemiquinone anions. Hydroquinone and chlorobenzo-quinone probably result from disproportionation of these two radicals. The formation of benzoquinone is enhanced in dilute solution. It was assumed to arise from oxidation of the primary radical by chlorine atoms.

The photodegradation mechanism of the herbicide bromoxynil (3,5-di-bromo-4-hydroxybenzonitrile) was first studied by Kochany et al [49]. They reported that 3-bromo-4-hydroxybenzonitrile and 4-hydroxybenzonitrile were the main photoproducts. More recently, it was shown that 3-bromo-4,5-dihy-

in dilute solution

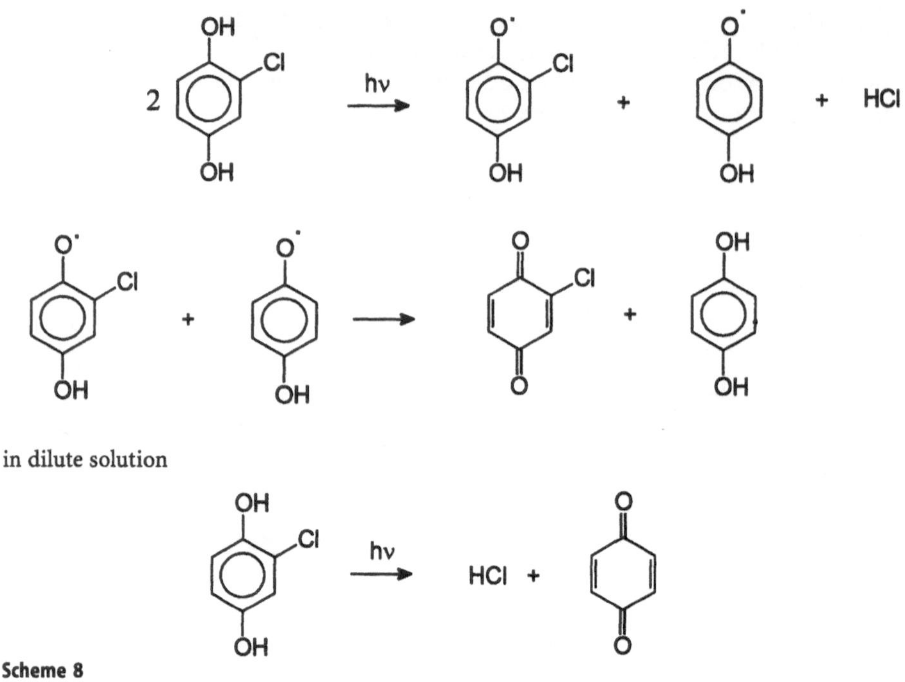

Scheme 8

droxybenzonitrile accounted for 65% of the photochemical conversion [50]. This discrepancy might be attributed to the presence of small amounts of organic matter in the former experiments. This remains to be confirmed.

3.3.4
Photosensitized Dehalogenations

As mentioned above, phenol and the dihydroxybenzenes are potentially efficient photosensitizers. Several studies have been carried out to investigate this possibility. Phenol was found to sensitize the transformation of halogenophenols [51, 52]. Formation of resorcinol was observed when a mixture phenol and 3-chlorophenol was irradiated at 254 nm. In the experimental conditions used, the light intensity absorbed by phenol was twice as high as that absorbed by 3-chlorophenol. The rate of 3-chlorophenol transformation was six times as high as that expected from direct excitation and the disappearance of phenol was negligible. From the Stern–Volmer plot, it was deduced that the intersystem crossing quantum yield for phenol is about 0.5, in reasonable agreement with the value obtained by the photophysical study described earlier [4].

Hydroquinone also has the properties required for a sensitizer. The intersystem crossing yield is high ($\phi = 0.39$ [5]). The lifetime of the triplet states is long enough to expect efficient energy transfer processes ($\tau = 1.3\,\mu s$) and the energy level of the triplet state is about 311 kJ mol^{-1} [53]. Finally, hydroquinone can be selectively excited since it absorbs at wavelengths longer than 300 nm. Upon irradiation of deoxygenated solutions containing hydroquinone and 3-chloro- or 3-bromophenol, resorcinol is the main product and 2,5,3'-trihydroxybiphenyl is formed as a minor product. No reaction was observed with 3-fluorophenol [54]. It was concluded that the energy transfer process leading to resorcinol formation takes place with 3-chloro- and 3-bromophenol but not with 3-fluorophenol for energetical reasons. The formation of 2,5,3'-trihydroxybiphenyl indicates that photosensitization by energy transfer is not the only pathway. This product was assumed to result from a reaction of 3-chloro- and 3-bromophenol with the triplet state of hydroquinone.

The transformation of 2-halogenophenols (bromo, chloro, fluoro) into cyclopentadiene carboxylic acids can be sensitized by hydroquinone, but as for 3-halogenophenols the formation of trihydroxybiphenyl was also observed [54]. It was deduced that the energy levels of the triplet states of 2-halogenophenols are lower than 311 kJ mol^{-1}. Moreover, it can be concluded that the ring contraction reaction seems to be possible from the triplet state in this case.

3.4
Nitro- and Nitrosophenols

The absorption spectra of the nitrophenols show a much greater overlap with the solar spectrum than those of other phenolic compounds. The molecular forms have an absorption band exhibiting a maximum in the 330–350 nm range. Moreover, as the pK$_a$ of the ground states molecules is quite low (7.3 for 2-nitrophenol, 8.0 for 3-nitrophenol and 7.0 for 4-nitrophenol), a significant proportion of the molecules are dissociated in neutral solutions and absorb visible light. The three isomers were found to be quite photostable [55–58]. Upon excitation at 365 nm, values of quantum yields of phototransformation are in the range $10^{-6} - 2 \times 10^{-5}$. Irradiation at 253.7 nm leads to a much more efficient photodecomposition reaction. With 2-nitrophenol, the photochemical process is about 160 times as efficient at 253.7 nm as at 365 nm [57].

Many photoproducts were detected. Photohydrolysis occurs with the three isomers and nitrite ions are released. This process, which is neither affected by oxygen nor by alcohol, was attributed to the heterolytic scission of the C–N bond (Scheme 9).

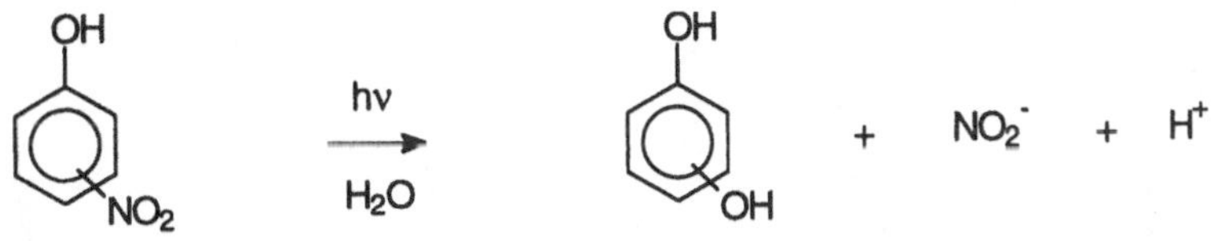

Scheme 9

With *meta* and *para* derivatives, photohydrolysis is the main primary process [55, 56]. The irradiation of the *ortho* derivative additionally leads to the formation of cyclopentadienic acids (Scheme 10). This reaction is a minor pathway in acidic medium but becomes the major reaction in basic solution. This ring contraction resembles that observed with 2-halogenophenols.

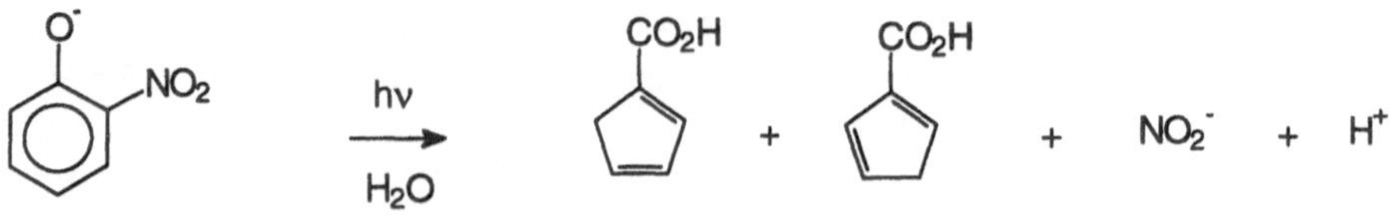

Scheme 10

Formation of nitrosophenol was observed with 2- and 4-nitrophenols in acidic solutions [55, 57]. A N-O bond scission step analogous to that observed in the photolysis of nitrite ions was proposed. The resulting radical was assumed to be reduced into nitrosophenol (Scheme 11).

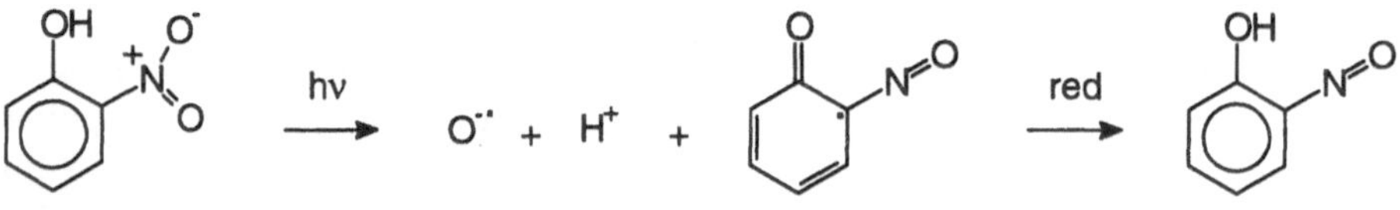

Scheme 11

Dihydroxynitrobenzenes are produced in a secondary stage involving the excitation of nitrite ions which is known to yield hydroxyl radicals.

4-nitrosophenol is photochemically converted into benzoquinone [59]. In acidic medium (pH = 1) the quantum yield of 4-nitrosophenol transformation is wavelength-dependent: $\phi = 4.2 \times 10^{-3}$ at 365 nm and 7.0×10^{-4} at 313 nm. In basic medium (pH = 9) 4-nitrosophenol is photostable ($\phi < 10^{-6}$). The intermediate formation of an oxaziridine was proposed, but not proved, to explain the conversion into benzoquinone. The wavelength effect is attributed to the tautomeric equilibrium between nitrosophenol and quinone-mono-oxime, the former being much more photostable than the latter (Scheme 12).

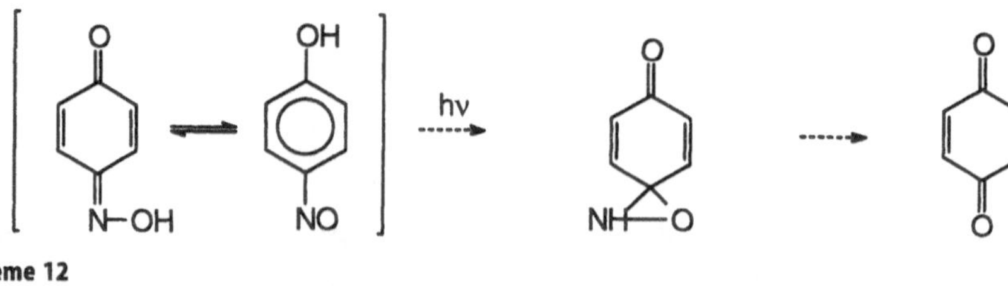

Scheme 12

3.5
Hydroxybenzonitriles

Upon irradiation of 2-hydroxybenzonitrile in neutral medium, benzoxazole is produced, which hydrolyses into 2-hydroxyformanilide in acid solutions [60], as shown in Scheme 13.

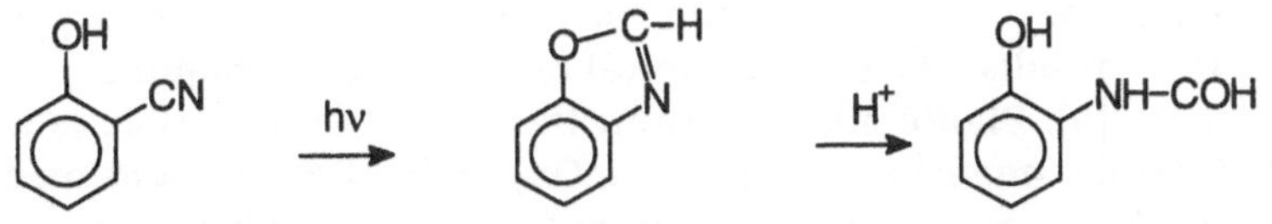

Scheme 13

Thus the phototransformation results in an exchange of the carbon and nitrogen atoms of the cyano group. Attempts were made to trap intermediates of this reaction by using low-temperature techniques. The formation of the isonitrile was suggested but could not be proved. On the other hand, 2-hydroxybenzoisonitrile was detected by IR spectroscopy at low temperature [61] as an intermediate in the photoconversion of indoxazene into benzoxazole (Scheme 14).

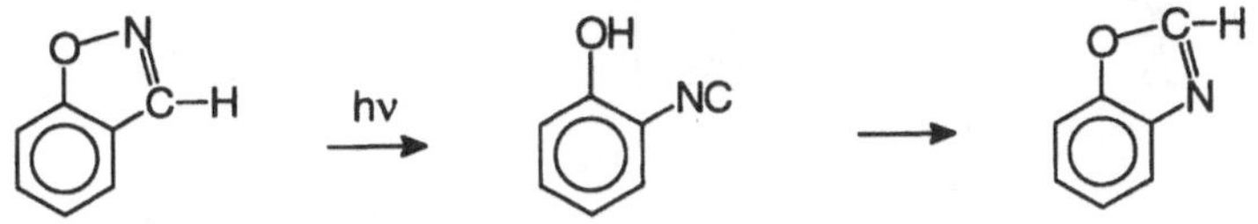

Scheme 14

The photolysis of 4-hydroxybenzonitrile was recently studied by means of transient absorption and product analysis techniques [62]. 4-hydroxybenzoisonitrile was found to be the only photoproduct in neutral deoxygenated solutions. Kinetic measurements revealed that it is a secondary product formed after two photochemical steps. The quantum yield of the reaction is equal to 0.0032 at pH = 2.0, 0.009 at pH = 5.6, and 0.0071 at pH = 9.4. Transient absorption spectroscopy measurements enabled us to conclude that the triplet excited state is involved in the formation of the isonitrile. The photounstable intermediate is probably an azirine. The proposed reaction sequence is shown in Scheme 15.

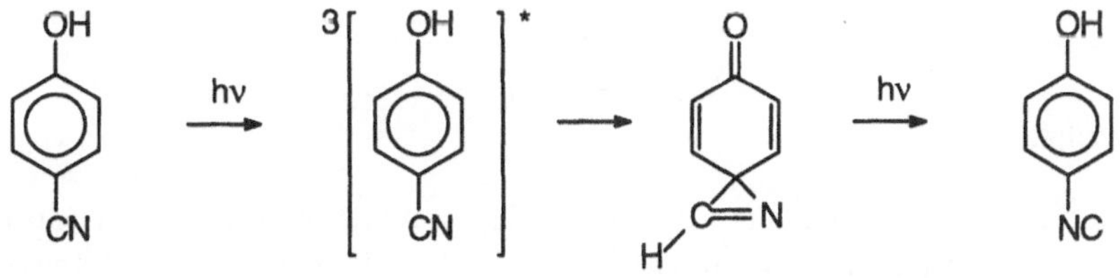

Scheme 15

3.6
Dihydroxybenzenes

The complex mixtures obtained upon irradiation of dihydroxybenzenes in water were tentatively analyzed many years ago [63, 64]. Polyhydroxybiphenyls are formed upon irradiation of resorcinol and catechol. Benzoquinone, hydroxybenzoquinone and 2,5,2′,5′-tetrahydroxybiphenyl were identified in irradiated solutions of hydroquinone. Radical mechanisms were suggested to interpret these results. The photochemical reactivity of hydroquinone was later reinvestigated [5]. Photoionization from the singlet excited state is a minor pathway if triplet quenchers are present. Oxygen and nitrate ions oxidize the triplet with formation of p-benzoquinone and hydroxybenzoquinone. The formation of 2,5,2′,5′-tetrahydroxybiphenyl results mainly from the reaction of hydroquinone in the excited triplet state with p-benzoquinone. The formation of an exciplex was assumed. Scheme 16 outlines the reactions starting from the hydroquinone triplet state.

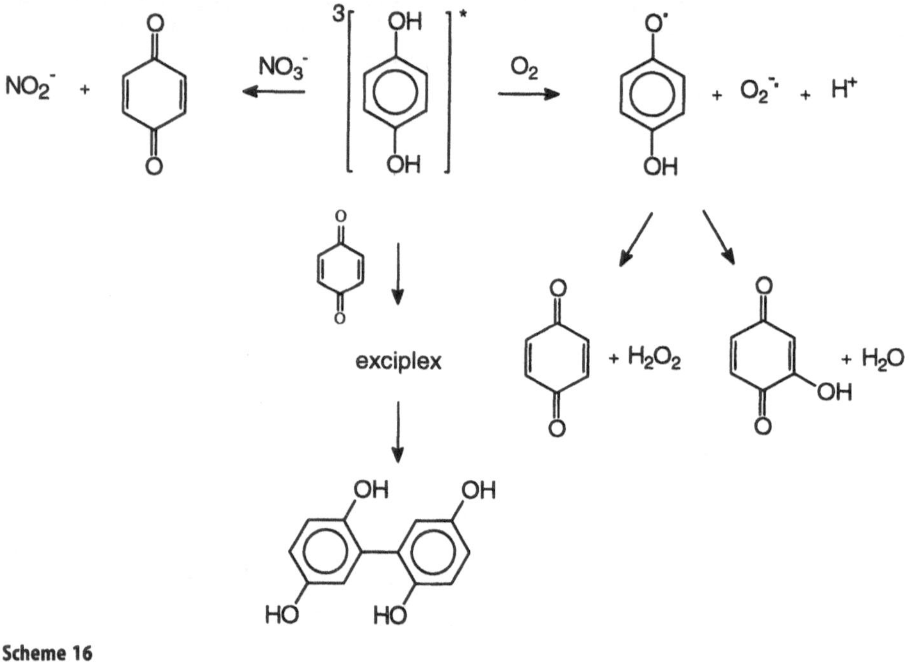

Scheme 16

3.7
o-Phenylphenol

When o-phenylphenol is irradiated at 254 in neutral or basic air-saturated solutions, three photoproducts are isolated: phenylhydroquinone, phenylbenzoquinone, and 2-dihydroxydibenzofuran. The first two are primary photoproducts but the third one is formed from the phototransformation of phenyl-benzoquinone [65]. Two isomeric acids resulting from the scission of the phenolic ring are produced in acidic solution along with the three other photoproducts [66]. The proposed reaction mechanism is depicted in Scheme 17.

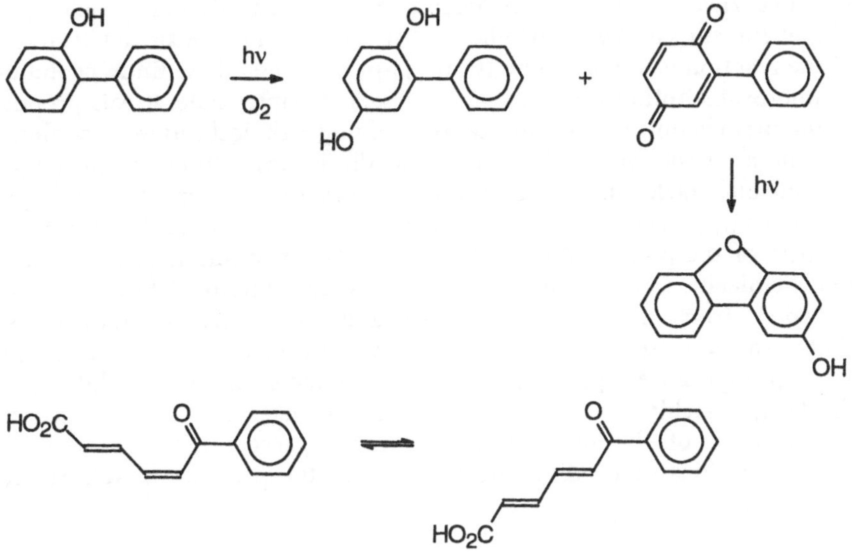

Scheme 17

3.8
Methoxy Derivatives

Aqueous anisole and *p*-dimethoxybenzene are able to eject electrons from singlet excited states. Radical cations are produced in this reaction which have been detected by transient absorption spectroscopy [67]. Several publications were devoted to the photochemistry of 4-chloroanisole. Anisole, 4-methoxy-phenol, and 4-chlorophenol are formed in neutral deoxygenated solution upon irradiation of 4-chloroanisole at 253.7 nm. Anisole is not formed in solutions saturated with N_2O or oxygen [68]. Several transient species were detected by microsecond flash photolysis [69]. The results were explained by assuming that the primary processes occurring after excitation are ionization with formation of the radical cation and homolytic cleavage of the C–Cl and O–Me bonds.

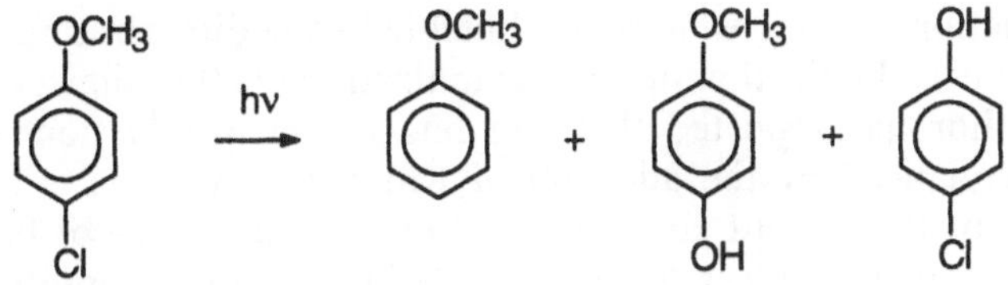

Scheme 18

The methoxy group was found to exert an activating and *ortho/para* directing influence in light-induced nucleophilic substitution reactions [70]. Irradiation of 4-fluoro- and 4-chloroanisole in *tert*-butanol/water (1:3) mixtures in the presence of the CN⁻ results in a clean and efficient substitution of the halogen by CN⁻. In the absence of CN⁻, photohydrolysis occurs. With the bromo and iodo derivatives the reaction yields a mixture of anisole and 4-cyanoanisole and the efficiency is lower. The photosubstitution can be sensitized by acetone and the reaction is quenched by trans-1,3-pentadiene. These findings indicate that photosubstitution proceeds via the excited triplet state. It was proposed that the first chemical step is ionization yielding a radical cation. According to calculations of the charge distribution in the radical cations of anisole and fluoroanisole, nucleophilic substitution occurs in the positions of highest positive charge. The proposed mechanism was finally confirmed by the similarity of the product distribution when the substitution was carried out photochemically or anodically. More recently, the intermediate formation of solvated electrons in the substitution mechanism was ruled out on the basis of a time-resolved study [71]. It was suggested that dissociation into radical ions occurs through an exciplex formation. The attack of the nucleophile on the radical cation yields a neutral radical which rearranges into final products.

Substitution of the methoxy group was observed upon irradiation of aqueous solutions of 1,2-dimethoxybenzene in the presence of sulfuric acid [72] (Scheme 19).

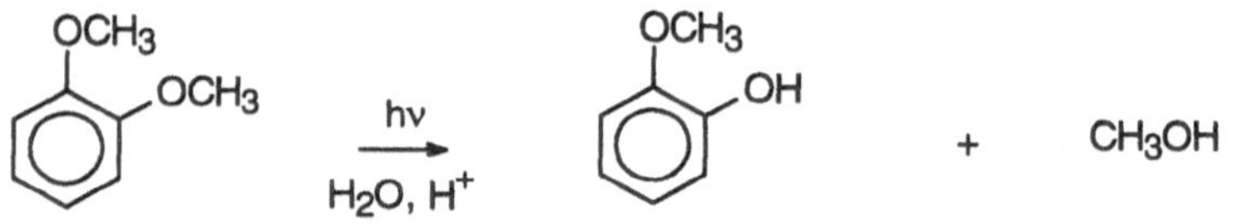

Scheme 19

3.9
Tyrosine

Tyrosine is an aromatic amino acid which plays an essential role in the photochemistry of proteins [2]. In spite of this, the analysis of the products of steady-state irradiation of aqueous tyrosine has been performed only recently. In the absence of oxygen, bityrosine I and 2-amino-4-ethenyl-hex-4-enic acid II are formed upon irradiation at 254 nm [73] (Scheme 20).

The formation of bityrosine is attributable to the dimerization of tyrosinyl radicals produced by the monophotonic ionization via the singlet excited state. Another author has reported that the rate of the production of tyrosinyl radicals is enhanced by the addition of KBr when tyrosine is irradiated at 235 nm, i.e., in the second absorption band. This influence of bromide ions points to a heavy-atom effect. It was suggested that in these conditions the O-H bond is cleaved via an upper excited triplet state [74].

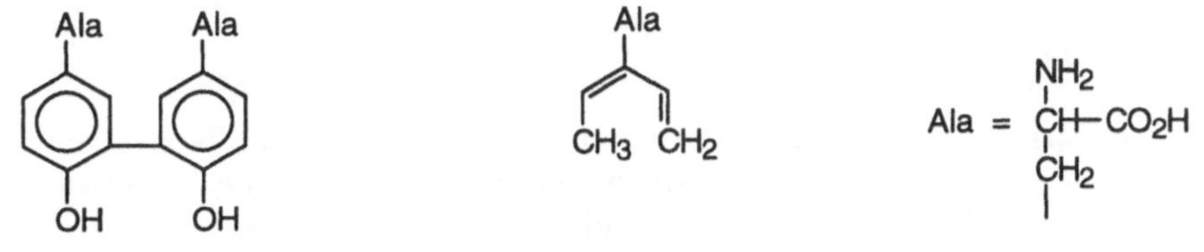

Scheme 20

The formation of product II is not influenced by oxygen and thus must proceed from the singlet state. The sequence shown in Scheme 21 was proposed to explain this transformation [73].

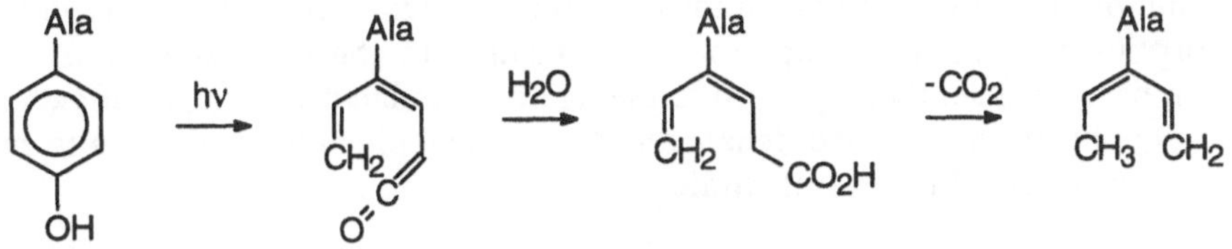

Scheme 21

In the presence of oxygen, the quantum yield of tyrosine photolysis is higher (0.097 instead 0.0091) and other photoproducts (Scheme 22) are formed [73].

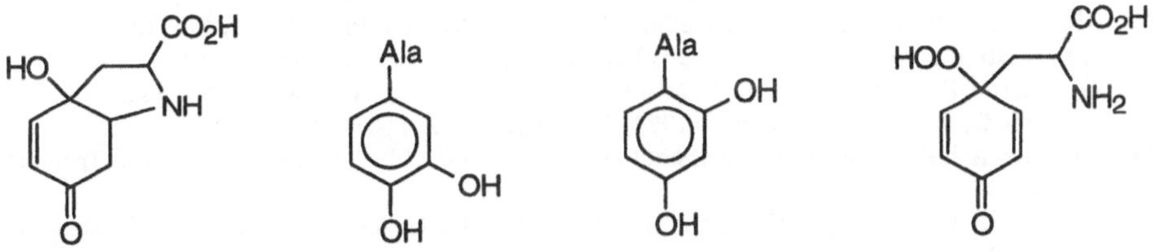

Scheme 22

The quenching of triplet tyrosine by oxygen is likely to lead to the formation of tyrosinyl radicals and hydroperoxyl radicals/superoxide ions which react together yielding the peroxide VI and subsequently the product III. Singlet oxygen is probably produced in the quenching reaction, too. This species should be able to abstract a hydrogen atom from tyrosine with formation of tyrosinyl radicals and hydroperoxyl radicals which are precursors of I and III. The formation of an endoperoxide via a non-radical route could also occur, the rearrangement of the intermediate giving rise to products IV and V [73].

4
Conclusions

Studies carried out by means of product analysis and time-resolved techniques have clarified the mechanisms of the direct photolysis of a range of substituted phenols, including halogen-, nitro-, and cyano-substituted derivatives. Among these, monohalosubstituted phenols, and in particular 4-chlorophenol, have been found to be photochemically quite active. The quantum yield of 4-chlorophenol dehalogenation in neutral aqueous solution is as high as 0.68. Other derivatives, in particular nitrophenols, are rather photostable. Quantum yields generally get lower upon polysubstitution. In many cases the triplet states of the photoexcited phenols appear to play an essential role in the photolysis mechanism. Triplet quenching by oxygen involving electron or energy transfer contributes to the higher photodegradation yields observed in oxygenated as compared to oxygen-free solutions. Of particular environmental relevance is the photosensitizing ability of some compounds, such as hydroquinone, whose absorption spectrum overlaps to some extent with the solar spectrum, and which have sufficiently long-lived triplet states populated with high efficiency. The possibility to sensitize dehalogenation of halophenols by this route has been demonstrated in several studies.

References

1. Creed D (1984) Photochem Photobiol 39:563
2. Ross JBA, Laws WR, Rousslang KW, Wyssbrod HR (1992) Top Fluoresc Spectrosc 3:1
3. Monti S, Köhler G, Grabner G (1993) J Phys Chem 97:13011
4. Köhler G, Grabner G J Phys Chem (submitted for publication)
5. Boule P, Rossi A, Pilichowski JF, Grabner G (1992) New J Chem 16:1053
6. Mártire DO, Braslavsky SE, Garcia NA (1991) J Photochem Photobiol A 61:113
7. Jin F, Leitich J, von Sonntag C (1993) J Chem Soc Perkin Trans I 1993:1861
8. Grabner G, Köhler G, Marconi G, Monti S, Venuti E (1990) J Phys Chem 94:3609
9. Grabner G, Richard C, Köhler G (1994) J Am Chem Soc 116:11470
10. Armstrong DR, Cameron C, Nonhebel DC, Perkins PG (1983) J Chem Soc Perkin Trans II 1983:563
11. Armstrong DR, Cameron C, Nonhebel DC, Perkins PG (1983) J Chem Soc Perkin Trans II 1983:569
12. Armstrong DR, Cameron C, Nonhebel DC, Perkins PG (1983) J Chem Soc Perkin Trans II 1983:575
13. Armstrong DR, Cameron C, Nonhebel DC, Perkins PG (1983) J Chem Soc Perkin Trans II 1983:581
14. Armstrong DR, Cameron C, Nonhebel DC, Perkins PG (1983) J Chem Soc Perkin Trans II 1983:589
15. Ye M, Schuler RH (1989) J Phys Chem 93:1898
16. Jonsson M, Lind J, Reitberger T, Eriksen TE, Merenyi G (1993) J Phys Chem 97:8229
17. Terzian R, Serpone N, Fox MA (1995) J Photochem Photobiol A 90:125
18. Boule P, Guyon C, Tissot A, Lemaire J (1987) ACS Symp Ser 327:10
19. Guyon C, Boule P, Lemaire J (1982) Tetrahedron Lett 23:1581
20. Boule P, Guyon C, Lemaire J (1984) Toxicol Environ Chem 7:97
21. Richard C, Grabner G (1994) Poster presented at the XVth IUPAC Symposium on Photochemistry, Prague 1994; manuscript submitted for publication.

22. Urwyler B, Wirz J (1990) Angew Chem Int Ed Engl 29:790
23. Meier H, Zeller KP (1975) Angew Chem 87:52; Vleggaar JJM, Huizer AH, Kraakman PA, Nijssen WPM, Visser RJ, Varma CAGO (1994) J Am Chem Soc 116:11754
24. Vacek G, Galbraith JM, Yamaguchi Y, Schaefer III HF, Nobes HR, Scott AP, Radom L (1994) J Phys Chem 98:8660
25. Boule P, Guyon C, Lemaire J (1982) Chemosphere 11:1179
26. Tissot A, Boule P, Lemaire J (1984) Chemosphere 13:381
27. Omura K, Matsuura T (1971) Tetrahedron 27:3101
28. Lipczynska-Kochany E, Bolton JR (1991) J Photochem Photobiol A 58:315
29. Lipczynska-Kochany E, Kochany J, Bolton JR (1991) J Photochem Photobiol A 62:229
30. Oudjehani K, Boule P (1992) J Photochem Photobiol A 68:363
31. Durand APY, Brown RG (1995) Chemosphere 31:3595
32. Durand APY, Brattan D, Brown RG (1992) Chemosphere 25:783
33. Ouardaoui A, Steren CA, van Willigen H, Yang C (1995) J Am Chem Soc 117:6803
34. Durand APY, Brown RG, Worrall D, Wilkinson F (1996) J Photochem Photobiol A 96:35
35. Thomas VR, Schreiner AF, Breemen RV, Xie T, Chen CL, Gratzl JS (1995) Holzforschung 49:139
36. Thomas VR, Schreiner AF, Xie T, Chen CL, Gratzl JS (1995) J Photochem Photobiol A 90:183
37. Lipczynska-Kochany E, Kochany J (1993) J Photochem Photobiol A 73:23
38. Ononye AI, Bolton JR (1986) J Phys Chem 23:6270
39. Boule P, Guyon C, Lemaire J (1984) Chemosphere 13:603
40. Boule P, Tissot A, Lemaire J (1985) Chemosphere 14:1789
41. Wong A, Crosby D (1981) J Agric Food Chem 29:125
42. Vollmuth S, Zajc A, Niessner R (1994) Environ Sci Technol 28:1145
43. Lamparski LL, Shehl RH, Johnson RL (1980) Environ Sci Technol 14:196
44. Choudhry GG, van der Wielen FWM, Webster GRB, Hutzinger O (1985) Can J Chem 63:469
45. Choudhry GG, Graham NJ, Webster GRB (1987) Can J Chem 65:223
46. Peijnenburg WJGM, de Beer KGM, de Haan MWA, den Hollander HA, Stegeman MHL, Verboom H (1992) Environ Sci Technol 26:2116
47. Krajnik P, Richard C, Grabner G (1995) Poster presented at the Gordon Research Conference on Physical Organic Chemistry, Plymouth (USA), July 1995, manuscript submitted for publication
48. Rossi A, Tournebize A, Boule P (1995) J Photochem Photobiol A85:213
49. Kochany J, Choudhry GG, Webster GRB (1991) Environ Sci Res 42:259
50. Machado F, Collin L, Boule P (1995) Pestic Sci 45:107
51. Boule P, Guyon C, Lemaire J (1984) Toxicol Environ Chem 7:97
52. David-Oudjehani K, Boule P (1993) New J Chem 17:567
53. Vesley GF (1971) J Phys Chem 75:1775
54. Oudjehani K, Boule P (1995) New J Chem 19:199
55. Alif A, Boule P, Lemaire J (1987) Chemosphere 16:2213
56. Alif A, Boule P, Lemaire J (1990) J Photochem Photobiol A 50:331
57. Alif A, Pilichowski JF, Boule P (1991) J Photochem Photobiol A 59:209
58. Lipczynska-Kochany E (1992) Water Pollut Res J Can 27:97
59 Pilichowski JF, Boule P, Billard JF (1995) Can J Chem 73:2143
60. Ferris JP, Antonucci FR, Trimmer RW (1973) J Am Chem Soc 95:919
61. Ferris JP, Antonucci FR (1974) J Am Chem Soc 96:2014
62. Scavarda F, Bonnichon F, Richard C, Grabner G (1997) New J Chem (in press)
63. Joschek HI, Miller SI (1966) J Am Chem Soc 88:3273
64. Perbet G, Filiol C, Boule P, Lemaire J (1979) J Chim Phys 76:89
65. Seffar A, Dauphin G, Boule P (1987) Chemosphere 16:1205
66. Sarakha M, Dauphin G, Boule P (1989) Chemosphere 18:1391
67. Grabner G, Rauscher W, Zechner J, Getoff N (1980) Chem Commun 1980:222
68. Abd El-Hameed FSM, Krajnik P, Getoff N (1993) Z Naturforsch 48a:799

69. Abd El-Hameed FSM, Krajnik P, Getoff N (1993) Z Naturforsch 48a:515
70. den Heijer J, Shadid OB, Cornelisse J, Havinga E (1977) Tetrahedron 33:779
71. Lemmetyinen H, Konijnenberg J, Cornelisse J, Varma CAGO (1985) J Photochem 30:315
72. Pollard R, Zhang G, Wan P (1993) J Org Chem 58:2605
73. Jin F, Leitich J, von Sonntag C (1995) J Photochem Photobiol A 92:147
74. Shimizu O (1984) Photochem Photobiol 39:507

9 Phototransformation of Pesticides in Aqueous Solution

P. Méallier

Laboratoire de Photochimie Industrielle – L.A.C.E. – U.M.R. 5634
Université Claude Bernard Lyon I, 43 boulevard du 11 novembre 1918, F-69622 Villeurbanne
Cedex. *E-mail: lace@univ-lyon1.fr*

The role of the light on the transformation of pesticides in water depends on many parameters. Transformation are categorized as being direct or indirect photodegradation. We investigated the influence of the spectroscopic properties of pesticides and their effects on the nature of the photochemical reactions with the oxygen species, adjuvants of formulation, humic acids, and water. Chemical reactions, especially elimination, substitution and hydrolysis, are generally accelerated by light, while other specific reactions such as photo-Fries rearrangement are initiated by it. With organo-halogenated pesticides singlet or triplet states are involved in the scission of the carbon-chlorine bond, while the triplet state is often the first step for the reaction of the other pesticides. In this paper some reactions are presented to illustrate these two types of mechanisms.

Keywords: pesticides, phototransformation, aqueous solution, formulation, humic substances.

Contents

The Handbook of Environmental Chemistry Vol. 2 Part L
Environmental Photochemistry (ed. by P. Boule)
© Springer-Verlag Berlin Heidelberg 1999

1
Introduction

Over the last few years, owing to the extensive use of these compounds, the amount of pesticides in aqueous media has increased. Their elimination from natural waters is due to biological and chemical degradation. Although photo-degradation is one of the main pathways in abiotic degradation, this research field has been studied only for the last thirty years.

The first contributions on the photodegradation of pesticides in water were published by D.G. Crosby [1], J.R. Plimmer [2], R.G. Zepp [3], in the U.S.A. Since 1981 this theme has become a main topic of research in European countries.

In 1986, the O.E.C.D. published a first draft guideline on the "photochemical-oxidation degradation in the atmosphere" followed by a second draft test guideline on the "phototransformation of chemicals in water" in 1987.

Many laboratories have undertaken studies about the extent and pathways of photodegradation for the main groups of pesticides including carbamates [4–6], phenylureas [7–8], organophosphorus [9], triazines [10–11], nitroanilines [12] and amides [13].

2
Photochemical Reactions

2.1
Direct Photodegradation

Direct photochemical reactions are initiated by the absorbance of visible or ultraviolet light by the substrate.

$$\text{Pesticide} + h\nu \rightarrow \text{Pesticide}^*$$

$$\text{Pesticide}^* \rightarrow \text{Photoproducts}$$

$$\text{Pesticide}^* + X \rightarrow \text{Photoproducts}$$

Where Pesticide* = pesticide in excited state, and X = solvent, pesticide or other molecules.

Most pesticides absorb light in the ultraviolet region between 250 and 300 nm. Therefore, ultraviolet sources are commonly used for the study of their photochemical behavior.

The absorption of UV light by pesticides is limited by five parameters:

1. The transparency of the natural water.

The average absorbance A of several streams in the USA has been reported [14, 15]. For a depth of 1 m, A is determined as being:

$$A = 4 \text{ at } 500 \text{ nm}$$

$$A = 12 \text{ at } 300 \text{ nm}$$

Table 1. Absorption maxima (λ: nm) and molecular extinction coefficients (ε: 1 mol^{-1}cm^{-1})

Pesticides	λ_1	ε_1	λ_2	ε_2
Chlorpropham	236	10250	206	24100
Simazine	264	735	207	30460
Diuron	247	16200	211	20500
Fenuron	236	14350	201	24200
Propyzamide	284	800	206	5540
Propanil	258	16670	208	25170
Linuron	245	16000	210	21000
Phenmedipham	271–279	2600–2400	234	30500
Carbetamide	272–280	890–450	235	17080
Isoproturon	275	800	238	17600

Table 2. Water solubility of some pesticides at 20 °C

Pesticide	Water Solubility (mg l^{-1}, pH = 7)	Pesticide	Water solubility (mg l^{-1}, pH = 7)
Chorphropham	89	Propanil	130
Linuron	81	Propyzamide	15
Phenmedipham	4.7	Fenuron	3.85
Carbetamide	3.5	Diuron	42
Isoproturon	65	Simazine	6.2

In pure sea water, the penetration depth is much higher:

   A = 1 at 5 m for UVB

   A = 1 at 20 m for green light.

2. The molecular extinction coefficient.

Some values are given in Table 1.

3. The water solubility.

Some values are given in Table 2.

4. The solar spectrum.

Considering the solar spectrum, it appears that direct photolysis is not always the main factor which induces the transformation of pesticides. The absorption field of pesticides is often located at wavelengths shorter than 300 nm and the solar radiations are only efficient at wavelengths higher than 290 nm (Table 3).

5. The pH value of the water.

The variation of the pH value causes a shift of the absorption maximum of the solution but the most important effect is observed on the hydrolysis constant [15–17]. Phenmedipham is given as an example in Table 4.

Table 3. Yearly average, mid-day sunlight intensities at sea level for latitude 40–50 degrees North [14]

Wavelength range nm	Sunlight intensities $I_0(\lambda)$ photons cm^{-2} s^{-1} 10 nm^{-1}
290–300	3.6×10^{11}
300–310	3.1×10^{13}
310–320	1.9×10^{14}
320–330	4.0×10^{14}
330–340	5.2×10^{14}
340–350	6.0×10^{14}
350–360	6.2×10^{14}
360–370	6.8×10^{14}
370–380	7.4×10^{14}
380–390	8.8×10^{14}
390–400	1.2×10^{15}

Table 4. K_{obs}: hydrolysis constant of phenmedipham at 25 °C

pH	8.19	9.39	11.8	12.8	13.8
$10^3 \times K_{obs} s^{-1}$	0.28	3.99	1.21	7.91	25.7

2.2
Indirect Photodegradation

In indirect photolysis, a sensitizer or a radical initiator present in the solution is capable of absorbing light in the first step of transfering its energy to the reactant or producing a reactive specie in a second step.

The nature of the reactive state of the molecule can be characterized as singlet state, triplet state or radical.

These different possibilities are illustrated by the following general mechanism:

$$Y + h\nu \longrightarrow Y^*$$

$$Y^* + Pesticide \longrightarrow Pesticide^* + Y$$

$$Y^* + Pesticide \longrightarrow Pesticide^{\bullet} + Y^{\bullet}$$

$$Pesticide^* \longrightarrow Photoproducts$$

$$Pesticide^* + X \longrightarrow Photoproducts$$

$$Pesticide^{\bullet} + X \longrightarrow Photoproducts$$

Where Pesticide* = pesticide in the excited state, Y=sensitizer or radical initiator, and X=all substances present in the solution.

Table 5. Concentrations of Reactive Species in Waters

Species	Concentration mol l^{-1}
Singlet molecular oxygen	$10^{-14} - 10^{-13}$
Peroxialkyl (ROO$^{\bullet}$)	$10^{-11} - 10^{-10}$
Hydroxyl ($^{\bullet}$OH)	$10^{-19} - 10^{-18}$
Hydrated electron	$10^{-17} - 10^{-15}$
Hydroperoxyl ($O_2^{\bullet-}$)	$10^{-9} - 10^{-8}$

Table 6. Rate constants for reaction $^{\bullet}$OH and $^{\bullet}$O^{-} radicals with various compounds. (Rate constants are for pH $\approx$ 7, unless indicated by a number in parentheses after rate constant value) [21]

Substrate	$k/10^9 \ M^{-1} \ s^{-1}$	
	$^{\bullet}$OH	$^{\bullet}$O^{-}
H_2O_2	0.027	–
O_2	8	3.6 (11)
O_3	0.11	–
HCO_3^-	0.0085	–
CO_3^{2-}	0.39 (11)	–
CN^-	7.6	0.26 (14)
Fe^{2+}	0.43 (3)	3.8 (5)
CH_3OH	0.97	0.75 (14)
HCOH	~1 (1)	–
HCOOH	0.13 (1)	–
C_2H_5OH	1.9	1.2 (13)
CH_3COOH	0.016	–
CH_3COCH_3	0.11	–
CH_3CN	0.022	0.21 (14)
CH_2Cl_2	0.058 (~10)	–
$CHCl_3$	~0.005 (5.7)	–
$CHCl=CCl_2$	4.2	–
$ClCH_2COOH$	0.043 (1)	–
diethyl ether	3.6	0.95 (13)
benzene	7.8	–
toluene	3.0 (3)	2.1 (13)
phenol	6.6	–
benzophenone	8.8	–
benzaldehyde	4.4 (9)	–
benzoic acid	4.3	–
benzoate	5.9	~0.04 (14)
chlorobenzene	5.5	–
benzoquinone	1.2	–
anisole	5.4	–
nitrobenzene	3.9	–
benzonitrile	4.4	0.07 (14)

When the reactant is oxygen, a specific mechanism is involved and produce singlet oxygen and superoxide anion. This two excited forms oxygen can react with substrate or produce hydrogen peroxide.

$$A \xrightarrow{h\upsilon} A^*$$

$$A^* + D \longrightarrow D^{\bullet +} + A^{\bullet -}$$

$$A^{\bullet -} + {}^3O_2 \longrightarrow O_2^{\bullet -} + A^{\bullet}$$

$$O_2^{-\bullet} + HO_2^{\bullet} + H^+ \longrightarrow O_2 + H_2O_2$$

Where A = electron acceptor, and D = electron donor.

An estimation of the average concentrations of the reactive species in waters was published in 1990 [18]. They are given in Table 5. The rate constants for reactions of hydroxyl radicals and their anionic form with various compounds are gathered together in Table 6 [19].

Chemicals used as sensitizers or radicals initiators include methylene blue, rose Bengal, neutral red, toluidine blue, riboflavin, humic acids, chlorophyll, acetone, nitrate and nitrite anions and iron salts.

Nitrates and nitrites are present in neutral waters and absorb light between 290 and 400 nm. They can generate hydroxyl radicals.

$$NO_2^- + H_2O \rightarrow NO + {}^{\bullet}OH + OH^-$$

$$NO_3^- + H_2O \rightarrow NO_2 + {}^{\bullet}OH + OH^-$$

Consequently nitrates and nitrites are a potential source of hydroxyl radicals.

The same result is obtained with iron species which absorb light and give rise to the formation of $^{\bullet}OH$. The simultaneous reoxidation of iron (II) into iron (III) by oxygen confers a catalytic aspect to this process.

In laboratory studies, acetone or Riboflavine is very often used to accelerate the photodegradation of pesticides. In this case, acetone acts as a sensitizer or as a radical initiator, Table 7 [20].

With acetone and acetophenone a triplet-triplet energy transfer was observed for the wavelengths higher than 290 nm and a radical initiation for the

Table 7. First order initial kinetic rates for the photodegradation of hexachlorocyclopentadiene

First order kinetic rates (h^{-1})

pH	Direct photolysis	Riboflavin	
		5 mg l^{-1}	10 mg l^{-1}
5	0.37	0.81	0.46
7.4	0.33	0.90	0.37
9.2	0.47	0.95	0.35

Table 8. Triplet state energies of sensitizers [23]

Sensitizer	Triplet state energy kJ M^{-1}	Sensitizer	Triplet state energy kJ M^{-1}
Acetone	330	Rose bengal	186
Benzophenone	289	Phenazine	184
Biphenyl	275	Eosin	180
Naphtalene	254	Anthracene	178
Benzyl	224	Crystal violet	163
Pyrene	203	Naphtacene	122

Table 9. Photodegradation of carbetamide in water [24]

	Water	Water + acetone 1% (v/v)
Quantum Yield	3.7×10^{-2}	9.3×10^{-2}

Table 10. Photolysis data for the degradation of bromoxynil in water, buffers and hydrogeno-carbonated solutions [26]

Solution	First order rate constant [10^{-3} s^{-1}]	Half-life [min]
Buffer pH = 8.3	1.04	11.5
5 mM NaHCO$_3$	1.02	12.5
10 mM NaHCO$_3$	0.96	14
50 mM NaHCO$_3$	0.89	15
Buffer pH = 11.6	1.08	11
5 mM NaHCO$_3$	0.88	15
10 mM NaHCO$_3$	0.54	23
50 mM NaHCO$_3$	0.42	29.5

wavelengths shorter than 290 nm [20]. Table 8 lists the triplet state energies of compounds which are very often used as sensitizing agents. The influence of acetone on the photodegradation of carbetamide appears in Table 9.

The radical mechanism is limited by the presence of radical scavengers such as carbonate ions. Carbonates and hydrogenocarbonates are major components of natural waters and they play a very important role in the photosynthesis and the regulation of pH in the aquatic environment [23].

Carbonates and hydrogenocarbonates have an inhibiting effect on the photolysis of pesticides. Bromoxynil is given as an example in Table 10. This effect is due to the quenching of hydroxyl radicals and hydrated electrons. Rate constants are given in Table 11.

When the fate of a commercial pesticide is studied in the environment, in river- or in lake water, for example, the presence of humic acids and the influence of the adjuvants of formulation have to be considered. Because of their ability to act as radical sources [25], as quenchers of reactions [26] or as anti-UV (Table 12), these substances can modify the kinetic of phototransformation.

Table 11. Rate constants of reaction of hydroxyl radicals and hydrated electrons with carbonate or hydrogenocarbonate ions according to Buxton et al. [24]

	$k\ M^{-1}\ s^{-1}$
$^{\bullet}OH + CO_3^{2-} \rightarrow OH^- + CO_3^{\bullet-}$	3.9×10^8
$^{\bullet}OH + HCO_3^- \rightarrow CO_3^{\bullet-} + H_2O$	8.5×10^6
$e^- + CO_3^{2-} \rightarrow \dots$	3.9×10^5

Table 12. Initial rates of 2,4,6-trimethylphenol [TMP] consumption as a function of the concentration of humic acids (m_{HA}) [39]

[TMP] (M)	m_{HA} (g l^{-1})	$I_{a(HA)}$ (Einstein)	$-d[TMP]/dt$ (Ms^{-1})
1.0×10^{-3}	0.029	6.8×10^{-7}	3.9×10^{-9}
1.0×10^{-3}	0.1	1.3×10^{-6}	5.7×10^{-9}
1.0×10^{-3}	0.10	4.8×10^{-5}	1.7×10^{-7}
1.0×10^{-3}	0.28	5.3×10^{-5}	1.0×10^{-7}

Humic materials are natural products which are present in soils, sediments and aquatic systems and result from living matter. Their absorption range extends from 200 to 600 nm and depends on their molecular weight [27]. Humic sustances have a complex photochemical reactivity. They act as precursors for the production of reactives intermediates such as $^1O_2$ [28–30], e_{aq}^- [31–33], $O_2^{\bullet-}$, H_2O_2 [34, 35]. The excited state which is involved in these reactions is designated as the triplet state [36].

The role of humic acids in the photoreactions does not depend on the chemist who prepares the formulation but the choice of the adjuvants may be determined according to the triplet state of the pesticide and to its absorption range [17–22]. Carbetamide is given as an example in Table 13.

Table 13. Photodegradation quantum yield of carbetamide as a function of adjuvants

Adjuvants	Quantum Yield
none	3.7×10^{-2}
S 25	2.7×10^{-2}
S 3 D 323	2.9×10^{-2}
NP 17 OE	2.9×10^{-2}
NP 10 OE	3.1×10^{-2}
Ricin	3.4×10^{-2}
S 60	3.5×10^{-2}

Where S 25 = Ethoxylated tristyrylphenol (25 OE), S 3 D 33 = Phosphate acid of ethoxylated tristyrylphenol (16 OE), NP 17 OE = Nonylphenol (17 OE), NP 10 OE = Nonylphenol (10 OE), Ricin = Castor oil, and S 60 = Dodecylbenzene calcium sulfonate.

An energy transfer from the adjuvant to the pesticide increases the degradation rate. In the opposite case, the rate decreases. A judicious formulation permits us to control the pesticide life-time in the environment.

2.3
Quantum Yield

In direct photodegradation, each quantum absorbed activates exactly one molecule in the primary excitation process. The primary process may be defined as the process involving the actual initiation of the chemical reaction or as the process involved in the formation of an excited state due to the direct absorption of a quantum of light. Secondary processes involve reactions or dissipation of energy, as for example when the molecule returns to the ground state.

An important parameter to characterize the photostability of a molecule is the phototransformation quantum yield. This parameter ϕ_r can be defined as the ratio between the number of moles which are transformed and the number of einsteins (mole of photons) which are absorbed. For solutions of low absorbance and with monochromatic light sources it can be calculated from the following relationship [38]:

$$\phi_{r(\lambda)} = \frac{R}{I_a(\lambda)}$$

Where R = rate of disappearance of the irradiated compound in $mol\,l^{-1}s^{-1}$; $I_{a(\lambda)}$ = the absorbed light intensity of the compound in solution at the monochromatic wavelength λ, in einsteins s^{-1}; and $\phi_{r(\lambda)}$ = the photodegradation quantum yield

With polychromatic light sources, the calculation is more complex and it is necessary to split the excitation range into 10 or 5 nm wave-band intervals. The total of absorbed light is the summation of absolute values of incident light in each spectral band.

$$\phi_r = \frac{R}{10^3 \sum_{\lambda_2}^{\lambda_1} I_a(\lambda).\Delta\lambda}$$

Where R = rate of disappearance of the compound, $\sum_{\lambda_2}^{\lambda_1} I_a(\lambda).\Delta\lambda$ is the computed total absorbed light intensity in the spectral range ($\lambda_1 \rightarrow \lambda_2$), and 10^3 = constant for converting $mol\,l^{-1}\,s^{-1}$ into $mol\,cm^{-3}\,s^{-1}$.

To calculate the quantum yield of transformation in these conditions it is necessary to know:

- the emission spectrum of the source,
- the absorption spectrum of the chosen actinometer*,
- the photodegradation kinetic of this actinometer,
- the absorption spectrum of the studied substance,
- the photodegradation kinetic of this substance.

---

* A chemical actinometer is a photochemical system used for measuring light intensity. Its quantum yield is supposed to be known.

Another parameter characterizing the fate of a compound is the lifetime τ.

$$\tau = \frac{1}{k}$$

Where k is the pseudo-first order rate constant of direct photolysis in s^{-1}: $c = c_o e^{-kt}$. At $t = \tau$ the concentration is divided by e.

Sometimes the half-life $\tau_{1/2}$ is preferred to the life-time τ. The half-life is defined as the time which corresponds to half conversion and $\tau_{1/2}$ can be easily deduced from τ:

$$\tau_{1/2} = \tau \times \ln 2 = 0.693 \, \tau$$

Franck and Klöpper have calculated the intensity of the solar spectrum in Central Europe [39] and Zepp and Cline in North America [14]. From these data it is possible to evaluate the life-time (or half-life) of a chemical compound irradiated by sunlight.

$$\tau = 1/\phi_r . k_a$$

$$k_a = \sum_{\lambda_2}^{\lambda_1} 2.303.10^3.I_o(\lambda).\varepsilon(\lambda).\Delta\lambda$$

Where τ = life-time (in s), k_a = the pseudo first order constant for direct photolysis (in einstein $mol^{-1} s^{-1}$), ϕ_r = the quantum yield in polychromatic light between λ_1 and λ_2 (in mole einstein^{-1}), $I_{o(\lambda)}$ = the solar intensity at the wavelength λ (einstein $cm^{-2} s^{-1} nm^{-1}$), $\Delta\lambda$ = the wavelength interval, and $\varepsilon(\lambda)$ = the molar absorption coefficient at the wavelength λ.

3
Examples of Typical Reactions

In the phototransformation of pesticides in aqueous solutions, we have to take into account the dual action of the light: light induces specific reactions such as isomerization, elimination and increases the rate of pure chemical reactions as oxidation and hydrolysis. Oxidation resulting from direct excitation may be different from reactions induced by chromophores present in the solution.

These reactions are observed with all pesticides but some of them are more frequent than others depending on the organic groups and atoms of the molecule. Only the first step of the phototransformation of a pollutant in water involves light, the second step sometimes results from oxidation and often from hydrolysis or biotransformation.

3.1
Oxidation

In photooxidation, oxygen acts simultaneously as a radical initiator, producing free peroxyl radicals, and as an effective radical scavenger (oxygen quenches the excited triplet states). Whereas in the phototransformation of fluridone [40], the presence of oxygen in the solution decreases the direct photolysis rate,

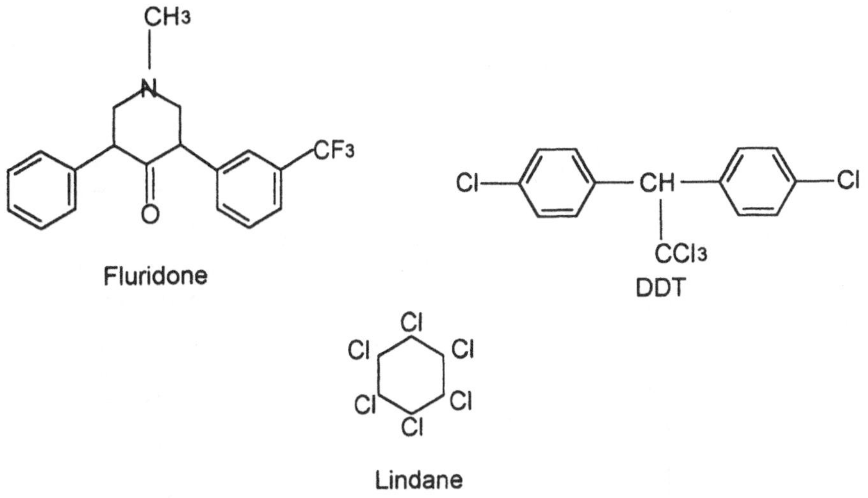

CH3

Fluridone

DDT

Lindane

Scheme 1

in a solution containing a photosensitizer such as rose Bengal or benzophenone, oxygen promotes the transformation of D. D. T. or lindane [41].

Sometimes, the difference between direct photolysis and sensitized photoreaction lies in the nature of the photoproducts [42].

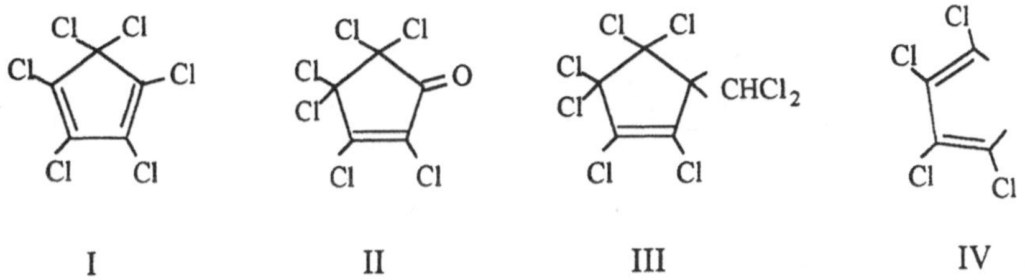

I   II   III   IV

Scheme 2

With hexachlorocyclopentadiene (I) in aqueous/organic mixtures with riboflavin as sensitizer, the major products are different in direct photolysis and in indirect photolysis. In the first case the main products are hexachlorocyclopentadione (II) and dichloromethylhexachlorocyclopentadiene (III), while in the second experiment the major product is tetrachlorobutadiene (IV).

The photodegradation of triazines [11] has been studied with various sensitizers. Photoproducts are the same, but the rate constant depends on the sensitizer.

The presence of oxygen in the irradiation mixture is essential for the photodecomposition reaction. The nature of the sensitizer is also important: Riboflavin (RF) and flavin mononucleotide (FMN) are more efficient than rose Bengal (RB) and methylene blue (MB) (Scheme 3).

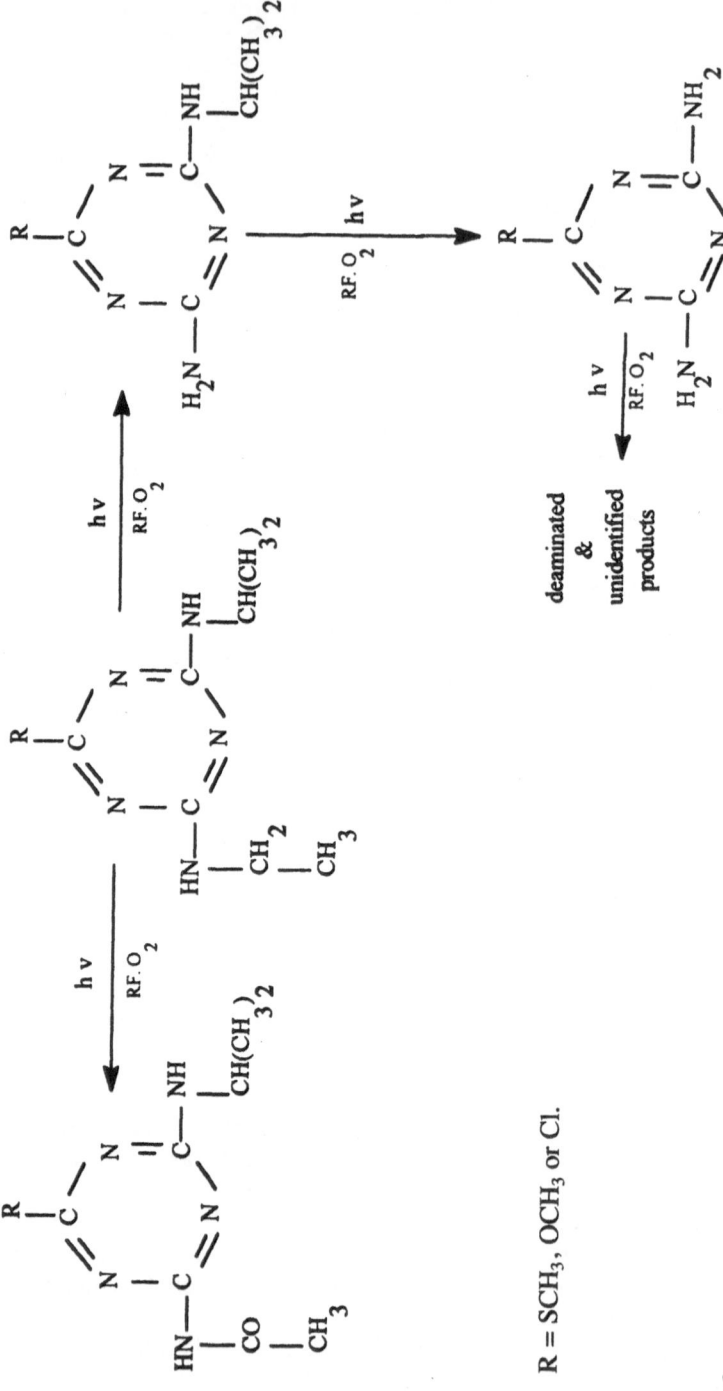

R = SCH₃, OCH₃ or Cl.

Scheme 3

3.2
Hydrolysis

Pure oxidation is very often the first step of the phototransformation of pesticides in water. However hydrolysis of the initial compound or hydrolysis of the photoproducts is the main reaction [43]. The transformation of chlor-propham does not depend on the presence of oxygen and can be explained by a "photohydrolysis" of this pesticide [44].

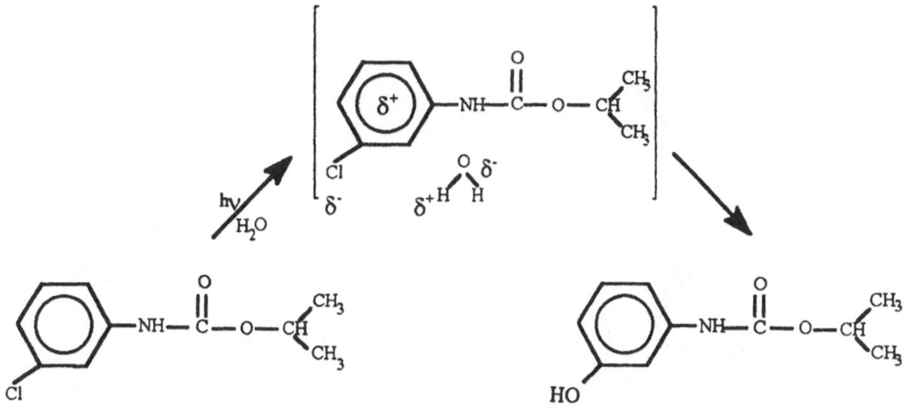

Scheme 4

The same result was obtained with diuron but a wavelength effect was reported, reaction in *para* being favored at long wavelengths [45].

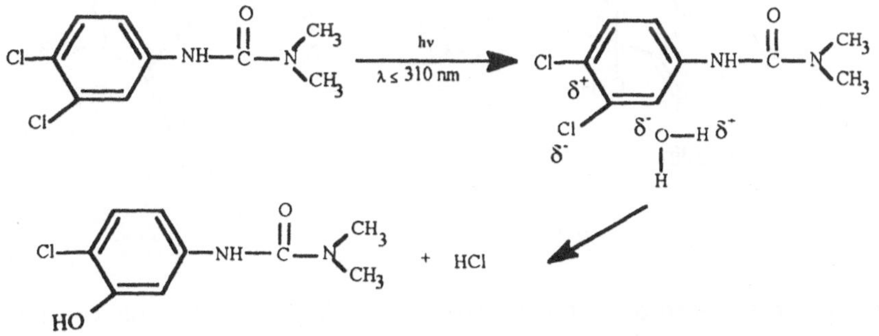

Scheme 5

3.3
Elimination – Reduction

Photo-elimination takes place predominantly with halogenated compounds. The C–X bond scission is mainly observed with urea, carbamates, ketones and aldehydes containing H atoms in the γ position. No general law can be applied for the halogen loss position and for the degradation of ketones, urea and carbamates [46].

With fungicides such as iprodione, procymidone and vinclozolin in iso-propanol, the substitution of a chlorine atom by hydrogen is the dominant reaction. In water this reaction plays a minor role [47].

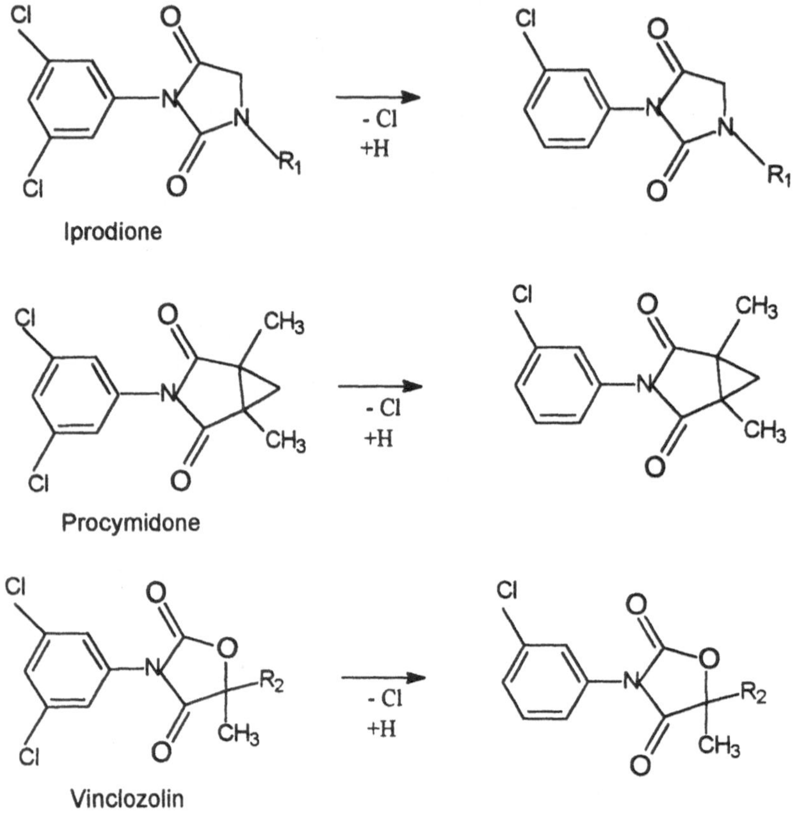

Scheme 6

In agreement with these results, the photoreduction of organic compounds in water rarely occurs in the absence of a reductant. This reaction was never-theless noted with bromoxynil, a herbicide which presents a strong electron withdrawing substituent on the aromatic ring [48, 49]. However the main pho-toproducts result from photohydrolysis [50](Scheme 7):

With isoproturon, the reactive state which initiates the radical mechanism is a triplet state. Two reactions may be expected by scission of the N–C bonds. The major compounds are the mono (II) and the di-demethylated (III) products [51–52].

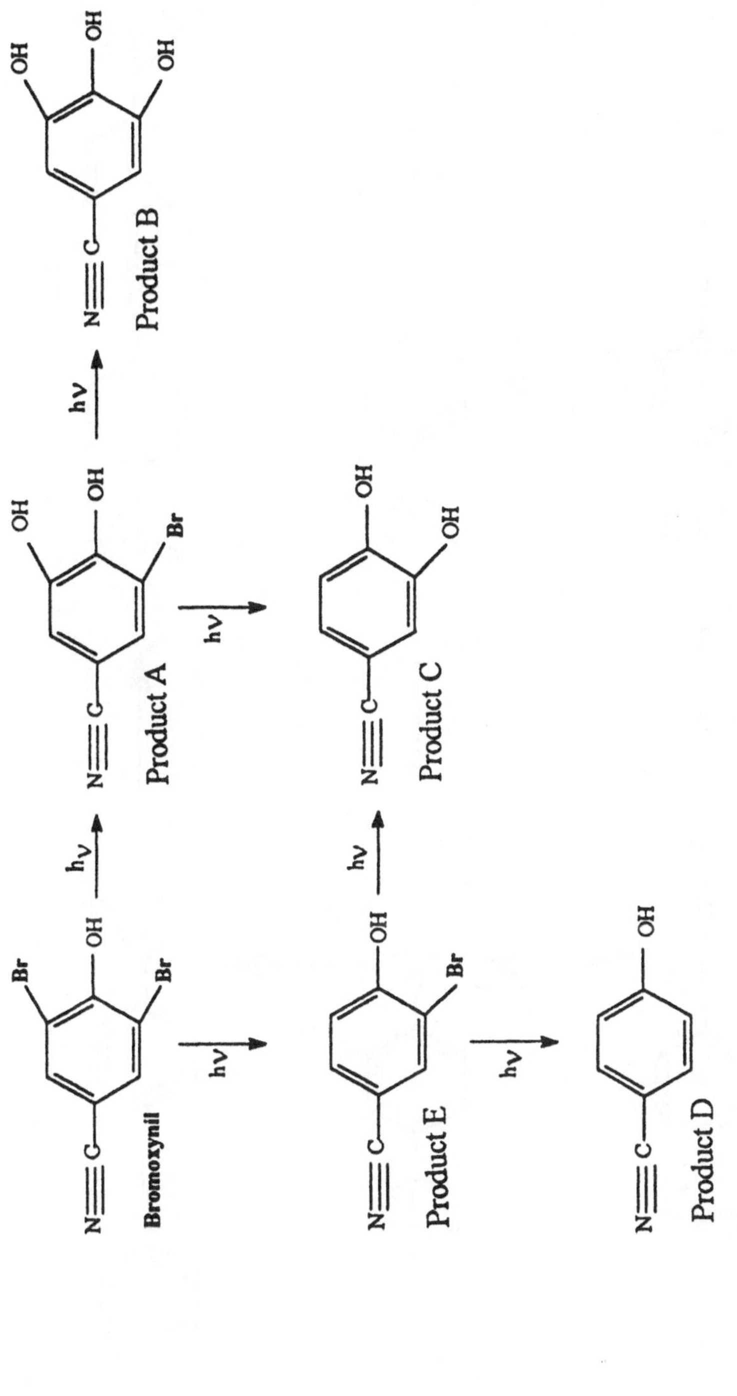
Bromoxynil
Product A
Product B
Product C
Product D
Product E
hν

Scheme 7

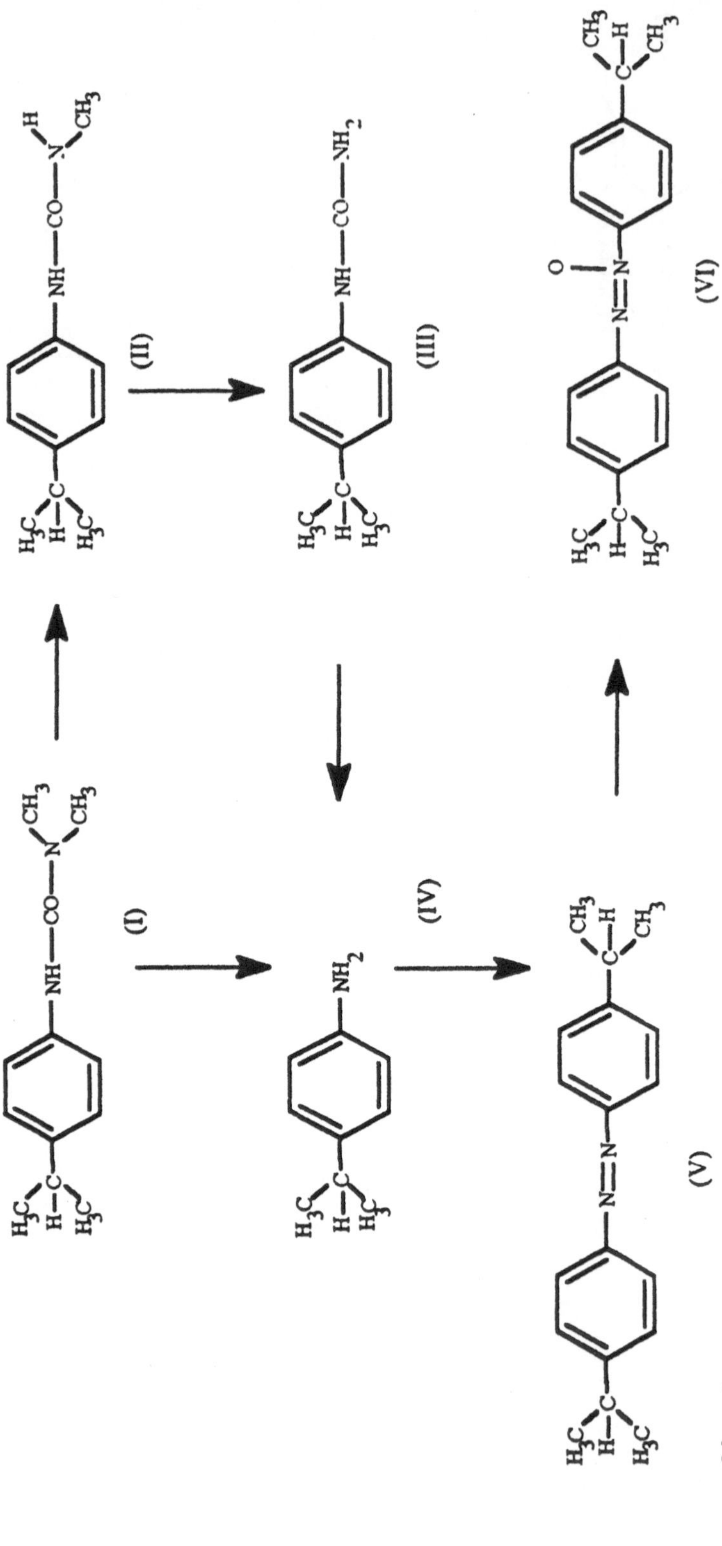

Scheme 8

In this photochemical decomposition two photoproducts (V and VI) result from the recombination of radicals.

3.4
Photoisomerization

When a chemical compound is photolyzed in a solvent causing a strong cage effect, a photo-Fries rearrangement is predominant. This first step is frequently observed with urea and carbamates. It is strongly suggested that the photo-Fries rearrangement of aryl esters is initiated in the lowest excited singlet state [53, 54].

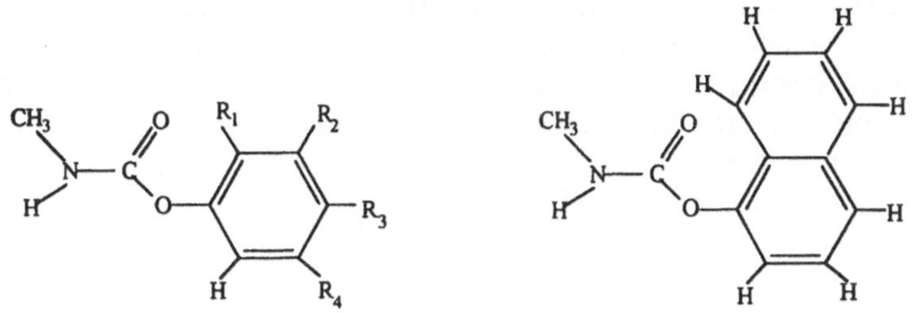

Sevin

	R_1	R_2	R_3	R_4
Phenmec	H	H	H	H
Metacrate	H	CH$_3$	H	H
Landrin	H	CH$_3$	CH$_3$	CH$_3$
Zectran	H	CH$_3$	N(CH$_3$)$_2$	CH$_3$

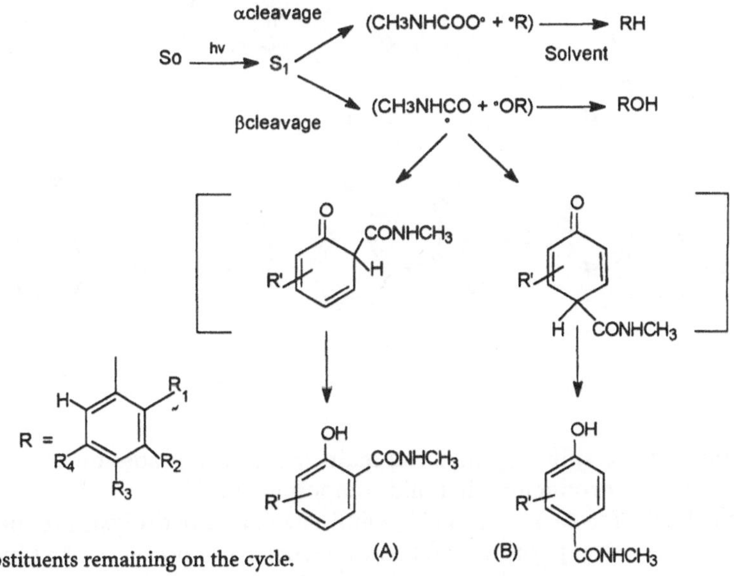

R' = substituents remaining on the cycle.

Scheme 9   Proposed reaction mechanism

A typical result of photo-rearrangement of O-aryl N-methyl carbamate was proposed by T. Katagi in a theoretical study on Phenmec, Metacrate, Landrin, Zectran and Sevin [55].

2-Hydroxy-N-methylbenzamide (A) and 4-hydroxy-N-methylbenzamide (B) derivatives are typical photo-Fries products which are assumed to be formed by radical combination between RO• and CH₃NHCO•.

3.5
More Complex Reactions in Water

In natural water, only a few pesticides give specific reactions. In a large majority of cases, the phototransformation results from several mechanisms: oxidation, elimination, reduction, isomerization and photohydrolysis. The types of reaction involved sometimes depend on the wavelength which is absorbed and on the other components which are present in water.

The phototransformation of the systemic insecticide propaphos is a typical example of oxidation and hydrolysis.

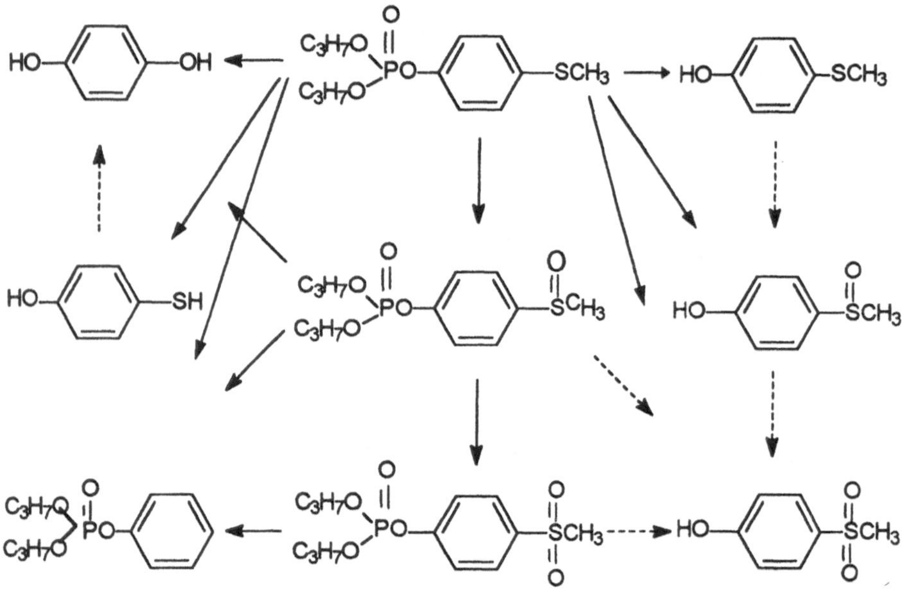

Scheme 10

Oxidation on the sulfur atom is the main route of degradation. The second route is hydrolysis producing phenolic compounds [56].

With the herbicide carbetamid [57] elimination and hydrolysis are the main pathways for the direct phototransformation whereas in presence of a sensitizer the following products are obtained (Scheme 11).

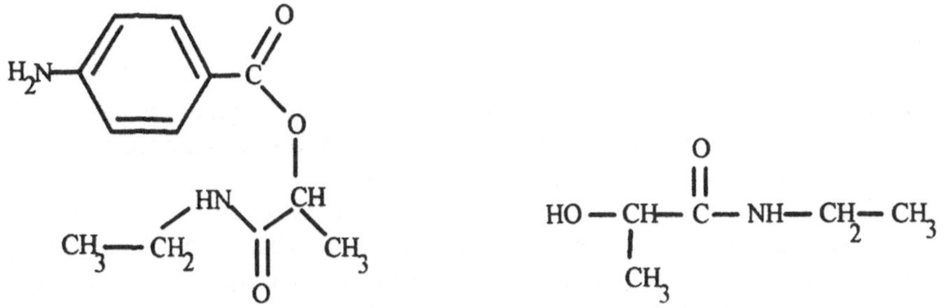

N-ethyl-lactamide-4-aminobenzoate ester   N-ethyl lactamide

Scheme 11

- With acetophenone:
  These photoproducts result from photoelimination and photo-Fries rearrangement.
- With hydrogen peroxide or titanium dioxide:
  Photo-Fries rearrangements, photo-elimination and photohydrolysis are involved in the formation of photoproducts.

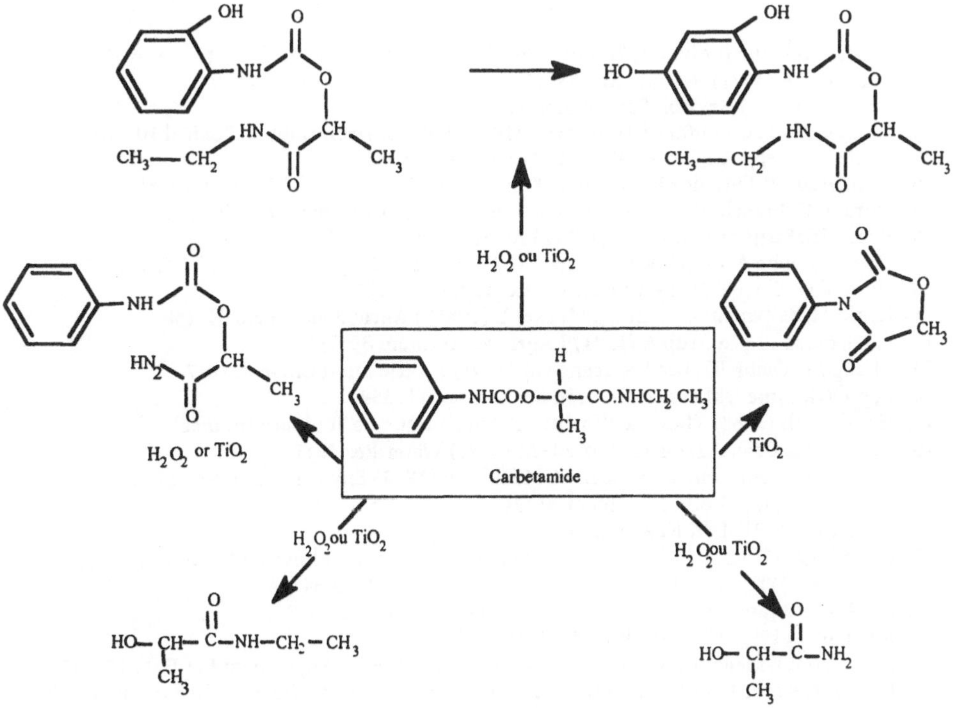

Scheme 12

4
Conclusion

Biodegradation is the most likely pathway for the elimination of chemicals from the environment. Direct and indirect photoreaction are also significant routes for the degradation of these substances in water.

The influence of sunlight depends on the transparency of the water but also on the presence of various compounds which are able to generate reactive chemical species. The photons absorbed do not induce just specific reactions such as elimination and rearrangement but also cause the hydrolysis of the initial molecule or photoproducts.

Free radicals are often involved in photolysis. The kinetics are enhanced by organic radicals, $^{\bullet}OH$, $HO_2^{\bullet}$ and reduced by carbonates and hydrogeno-carbonates.

The quantum yield of phototransformation is an important parameter for assessing the fate of a chemical. It is recommended by the members of the O.E.C.D. to calculate the quantum yield in solar light for all compounds which have a molar extinction coefficient $\varepsilon > 10$ at a wavelength $\lambda \geq 295$ nm. From this parameter, the absorption spectrum of the compound and sunlight spectrum, it is possible to estimate the photochemical lifetime of various pollutants in surface waters.

References

1. Crosby DG (1966) Abstr. 152nd meeting Am Chem Soc, New York, section A, p 32
2. Plimmer JR (1971) Residue Rev 33:47
3. Zepp RG (1975) Environ Sci Technol, 11:359
4. Macheterre L, Choudhry GG, Webster GRB (1988) Environ Contam Toxicol 103:61
5. Wolfe NL, Zepp RG, Paris DF (1978) Water Res 12:565
6. Samanidou V, Fytianos K, Pfister G, Bahadir M (1988) Sci Total Environ 76:85
7. Durand G, Barceló D, Albaiges J, Mansour M (1990) Chromatographia 29:120
8. Rosen JD, Strusz RF, Still CC, (1969) J Agric Food Chem 17:206
9. Chuckwudebe A, March RB, Othman M, Fukuto TR (1989) J Agric Food Chem 37:539
10. Durand G, Barceló D (1990) J Chromatographia 502:275
11. Reitjo M, Saltzman S, Acher AJ, Muzkat L (1983) J Agric Food Chem 31:138
12. Pal S, Moza PN, Kettrup A (1991) J Agric Food Chem 39:797
13. Chang LL, Giang BY, Lee KS, Tseng CK (1991) J Agric Food Chem 39:617
14. Zepp RG, Cline DM (1977) Environ Sci Technol 11:359
15. Hamida NB (1982) Thèse de l'Institut Polytechnique de Toulouse (France)
16. Nubbe ME, Dean Adams V, Moore WM (1995) Water Res 29:1287
17. Emmelin C, Guittonneau S, Brun H, Méallier P (1993) Environ Chem Rev 14:283
18. Klöpffer W (1991) EPA Newsletter 41:24
19. Bolton JR (1991) EPA Newsletter 43:40
20. Choudhry GG, Roof AAM, Hutzinger O (1979) Toxicol and Environ Chem Rev 2:259
21. Tanaka FS, Wien RG, Hoffer BL (1986) J Agric Food Chem 34:547
22. Méallier P, Mamouni A, Mansour M (1990) Chemosphere 20:267
23. Kochany J (1992) Chemosphere 24:1119
24. Buxton GV, Greenstock CL, Helman WP, Ross AB (1988) J Phys Chem Ref Data 17:513
25. Hazen JL, Krebs PJ (1989) 2nd International Symposium on adjuvants for agrochemicals, Blackburg USA 16:195

26. Bridges DC, Falb LN, Smith AE (1989) 2nd International Symposium on adjuvants for agrochemicals, Blackburg USA 18:215
27. Guilford J, Guilherme LI (1996) New J Chem 20:221
28. Haag WR, Hoigné J (1986) Environ Sci Technol 20:341
29. Aguer JP, Richard C (1993) Toxicol Environ Chem 39:217
30. Frimmel F, Bauer H, Putzien J, Murasecco P, Braun AM (1987) Environ Sci Technol 21:241
31. Zepp RG, Braun AM, Hoigné J, Leenheer JA (1987) Environ Sci Technol 21:485
32. Power JF, Sharma DK, Cooper CH, Bonneau R, Joussot-Dubien J (1986) Photochem Photobiol 44:11
33. Aguer JP, Richard C (1994) J Photochem Photobiol A: Chem 84:69
34. Baxter RM, Carey JH (1983) Nature 306:575
35. Cooper WJ, Zika RG (1980) Science 207:886
36. Canonica S, Jans U, Stemmerler K, Hoigné J (1995) Environ Sci Technol 29:1822
37. Aguer JP, Richard C (1996) J Photochem Photobiol A: Chem 93:193
38. Lemaire J, Klais O, Leahey J, Merz W, Philp J, Wilmes R, Wolff CJM (1985) Chemosphere 14:1
39. Frank R, Klöpffer W (1988) Chemosphere 17:985
40. Saunders DG, Mosier JW (1983) J Agric Food Chem 31:237
41. Prakash S, Tandon GS, Seth TD, Joshi PC (1994) Biochem and Biophys Res Comm 199:1284
42. Nubbe ME, Dean AV, Moore WM (1995) Water Res 29:1287
43. Kochany J, Choudhry GG, Webster GRB (1990) Pestic Sci 28:69
44. Faure V (1996) Thèse de l'Université de Clermont-Ferrand (France)
45. Jirkovsky J, Faure V, Boule P Pestic Sci (1997) Pestic Sci 50:42
46. Kotzias D, Korte F (1981) Ecotoxicol and Environ Safety 5:503
47. Schwack W, Bourgeois B (1989) Z Lebensm Unters Forsch 188:346
48. Kochany J, Choudhry GG, Webster GRB (1990) Arch Environ Contam Toxicol 19:325
49. Guittonneau S, Momege S, Schafmeier A, Viac PO, Méallier P (1995) Revue des Sciences de l'eau 8:201
50. Machado F, Collin L, Boule P (1995) Pestic Sci 45:107
51. De Saint Laumer C, Emmelin C, Méallier P (1996) Fresenius (in press)
52. Kulshrestha G, Mukerje SK (1986) Pestic Sci 17:489
53. Kalmus CE, Heracles DM (1974) J Am Chem Soc 96:449
54. Herweh JE, Hoyle CE (1980) J Org Chem 45:2195
55. Katagi T (1991) J Pestic Sci 16:57
56. Koshioka M, Kanazawa J, Murai T (1986) J Pestic Sci 11:557
57. Mamouni A (1989) Thèse de l'Université de Lyon (France)

10 Photodegradation of Lipidic Compounds During the Senescence of Phytoplankton

J.-F. Rontani

Laboratoire d' Océanographie et de Biogéochimie (LOB) – UMR 6535, Centre d' Océanologie de Marseille – OSU, Faculté des Sciences de Luminy – case 901, 13288 Marseille, France.
E-mail: rontani@com.univ-mrs.fr

This paper reviews the rates, pattern and possible mechanisms of light-induced degradation of lipidic components in detritus derived from phytoplankton. After a brief discussion of the main photooxidation reactions which can occur in the euphotic zone of the oceans, emphasis is given to the photodegradation of lipids during the senescence of phytoplankton. Production and quenching of excited states of chlorophyll and toxic oxygen species (singlet oxygen, superoxide ion, hydroxyl radical and hydrogen peroxide) in healthy and senescent cells are discussed and the photooxidation of the main lipidic cell components (chlorophylls, carotenoids, fatty acids, sterols, alkenones and hydrocarbons) is examined. Areas requiring further work are indicated.

Keywords: Photodegradation, lipids, phytoplankton, senescene, heterogenas, photooxidative processes.

Contents

The Handbook of Environmental Chemistry Vol. 2 Part L
Environmental Photochemistry (ed. by P. Boule)
© Springer-Verlag Berlin Heidelberg 1999

1
Introduction

It is generally considered that photooxidation reactions play an important role in the degradation of organic matter in the upper layers of the oceans [1–3]. The present paper reviews a particular aspect of these processes: *the heterogeneous photooxidative reactions associated with phytodetritus*. Though most of the organic components of phytoplankton are susceptible to being photodegraded during senescence, there has been very little research in this area. This is easily understandable in the light of our still fragmentary knowledge of the photochemical processes involved.

These relatively unknown processes could represent a significant source of low molecular weight organic compounds in near-surface waters [4–6], and may also play a role in the formation of marine humic acids [7].

It is hoped that this short review will stimulate research in this very interesting area and that the questions raised will be answered in the near future.

2
Photooxidation of Organic Compounds in the Marine Environment

Excellent contributions to this field have been made by Zafiriou [1], Zafiriou et al. [2], Zepp and Cline [8], Zepp [9] and Crosby [10]. The main photooxidative reactions which can occur in the marine environment will be briefly described in this section.

2.1
Direct Photooxidative Reactions

These processes involve light-absorbing entities called chromophores, which undergo oxidative change as a direct consequence of absorbing photons [2]. Comparatively few natural molecules of known structure can react in this way (carbonyl compounds, riboflavin, tryptophane, thiamine, vitamin B12, chlorophylls, polyunsaturated fatty acids) [2]. In contrast, these reactions must play an important role in the photochemistry of coloured complex organic molecules of unknown structure such as humic and fulvic acids [1, 2].

2.2
Indirect Photooxidative Reactions

Indirect processes are common in natural waters and are especially important because they can alter molecules that resist photolysis, such as transparent species or chromophores whose reactive states are inefficiently populated by absorption [2]. These reactions involve substances called "photosensitizers". These compounds have two systems of electronically excited states, the singlet (^{1}sens) and the triplet (^{3}sens) [11]. Most of the photosensitized oxidations occur by way of the triplet sensitizer [12] (Fig. 1). Many dyes, pigments and aromatic hydrocarbons are effective sensitizers [11].

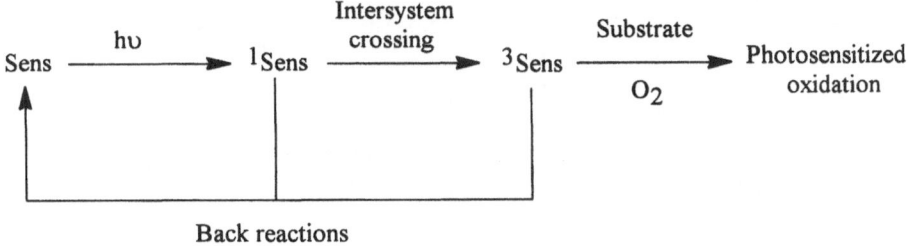

Fig. 1. Mechanism for photosensitized oxidation

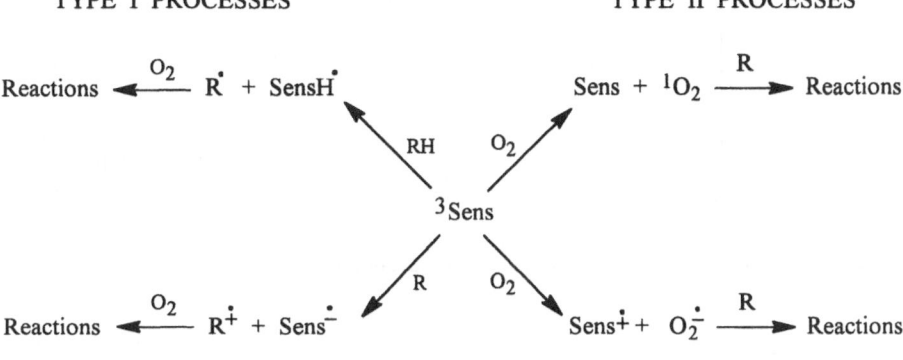

Fig. 2. Reactions of the triplet sensitizer

The sensitizer triplet can react:

– either with another molecule directly, to give radicals after hydrogen atom or electron transfer (type I processes) [12, 13] (Fig. 2),
– or with oxygen (Fig. 2). In this case, the excitation is transferred from the sensitizer to the oxygen, to produce an electronically excited singlet state of oxygen ($^1O_2$). Less efficient electron transfer from sensitizer to oxygen can occur, affording superoxide ion $(O_2^{\bullet-})$[11, 14].

Superoxide ion can also be produced by reaction of reduced sensitizers (sens$^\bullet$ or sens$^{\bullet-}$) with oxygen [15].

3
Photooxidation of Organic Compounds in Senescent Phytoplanktonic Cells

3.1
Production of Photochemically Excited States of Chlorophyll and Toxic Oxygen Species in Chloroplasts

The organisation of the pigments into the lipo-protein membranes of the chloroplasts consists of a series of pigment-protein complexes which cooperate

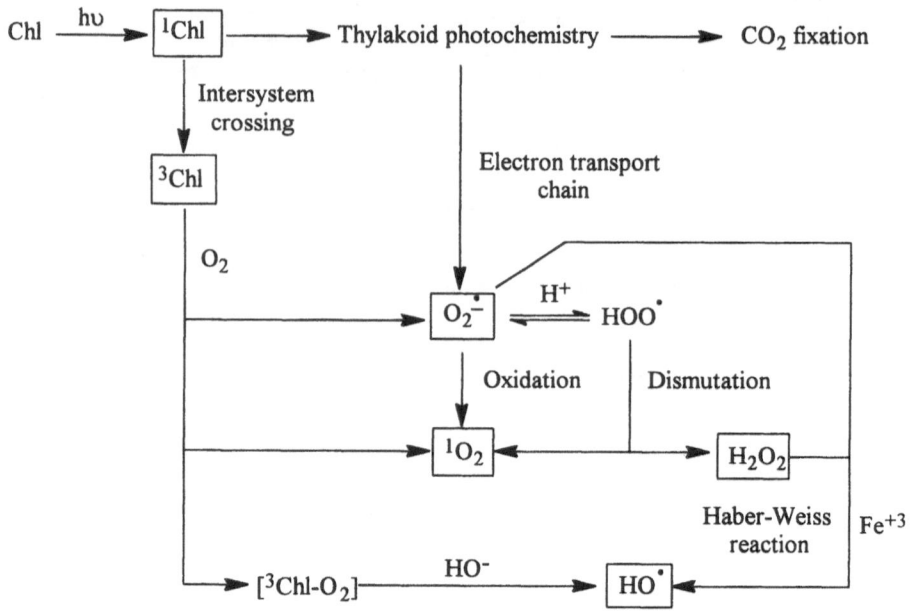

Fig. 3. Scheme summarising toxic oxygen species production in chloroplasts

together to promote electron flow [16]. When a chlorophyll molecule absorbs a quantum of light energy, an excited singlet state (^{1}Chl) is formed. If the excitation energy is not dissipated in the photochemical reactions of photosynthesis [11], intersystem crossing may occur to form the longer live triplet state (^{3}Chl) [17]. ^{3}Chl is not only potentially damaging itself in type I reactions [17] but it may also generate toxic oxygen species by reaction with ground state oxygen ($^3O_2$). Three pathways (Fig. 3) emerge as very probable for the reaction of ^{3}Chl with molecular oxygen: (i) formation of an active complex between ^{3}Chl and $^3O_2$ which can then oxidize HO^- to the extremely powerful oxidizing agent $HO^\bullet$ [18, 19], (ii) energy transfer from ^{3}Chl to $^3O_2$ leading to the production of very reactive $^1O_2$ [20], and (iii) electron transfer from ^{3}Chl to $^3O_2$ resulting in the formation of $O_2^{\bullet-}$ [21].

Since CO_2 is the ultimate sink for electrons generated in light reactions in chloroplasts, a loss in Calvin cycle efficiency may favour O_2 receiving the electrons, leading to formation of $O_2^{\bullet-}$ [22]. Superoxide ion may subsequently be oxidized to $^1O_2$ by strong oxidants accumulated at the donor side of photosystem II (PSII) [23], or disproportionate to H_2O_2 and $^1O_2$ in the presence of a proton donor (Eqs. (1) and (2)). The experiments of Foote [11] suggest the yield of $^1O_2$ in this last process is a few tenths of one percent.

$$O_2^{\bullet-} + H^+ \rightleftharpoons HOO^\bullet \tag{1}$$

$$2\,HOO^\bullet \longrightarrow H_2O_2 + {}^1O_2 \tag{2}$$

In the presence of Fe^{+3} superoxide ion intereacts with hydrogen peroxide to give hydroxyl radicals (Eq. (3)). This reaction, known as the Haber-Weiss reaction, may be of considerable importance during plant senescence [24]:

$$O_2^{\bullet -} + H_2O_2 \xrightarrow{Fe^{+3}} HO^{\bullet} + HO^- + O_2 \tag{3}$$

3.2
Photoprotection in Healthy Cells

In view of their susceptibility to oxidative damage, one would expect to find many antioxidant protective mechanisms in chloroplasts.

Carotenoids (Car) quench ^3Chl and $^1O_2$ by energy transfer mechanisms (Eqs. (4) and (5)) at very high rates [11] (Table 1). They have a dual role: preventing $^1O_2$ formation and helping to remove any that does manage to form [29]:

$$Car + {}^3Chl \longrightarrow Chl + {}^3Car \xrightarrow{radiationless} Car \tag{4}$$

$$Car + {}^1O_2 \longrightarrow {}^3O_2 + {}^3Car \xrightarrow{radiationless} Car \tag{5}$$

Tocopherols can remove $^1O_2, O_2^{\bullet -}, HOO^{\bullet}$ and $HO^{\bullet}$ in membranous environments by acting as sacrificial chemical scavengers, these processes resulting in the irreversible oxidation of the tocopherol molecule [25] (Fig. 4).

Table 1. Antioxidant properties of chloroplast components

	k (mol^{-1} s^{-1})	References
Quenching of ^3Chl by β-carotene	2×10^9	11
Quenching of $^1O_2$ by β-carotene	3×10^{10}	11
Quenching of $^1O_2$ by α-carotene	2.5×10^8	25
Scavenging of $O_2^{\bullet -}$ by ascorbic acid	2.7×10^5	26
Scavenging of $HO^{\bullet}$ by ascorbic acid	7×10^9	27
Scavenging of $O_2^{\bullet -}$ by SOD	2×10^9	28

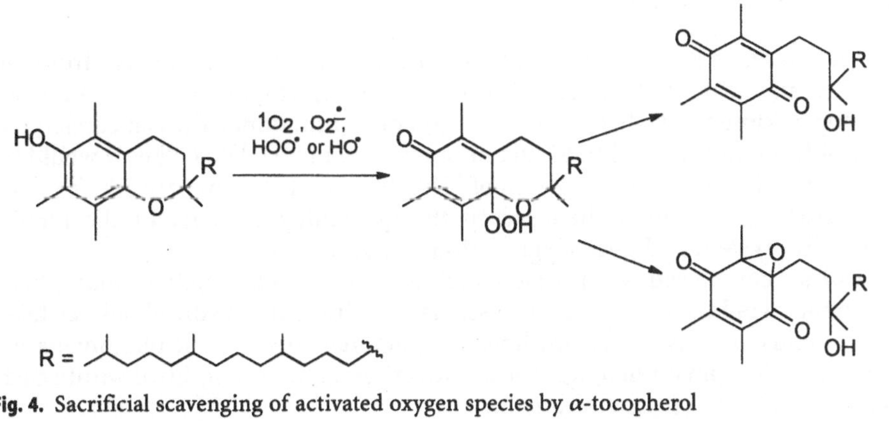

Fig. 4. Sacrificial scavenging of activated oxygen species by α-tocopherol

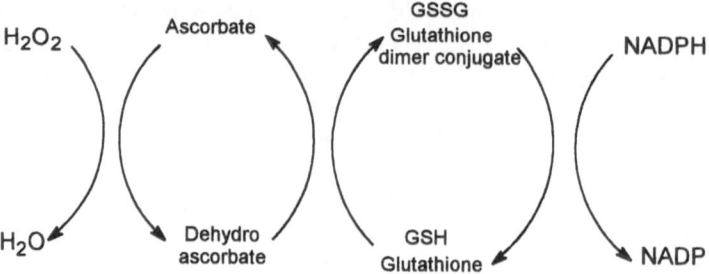

Fig. 5. The "ascorbate-glutathione cycle" in chloroplasts. Reprinted with kind permission from Elsevier Science Ireland Ltd.

In common with carotenoids, tocopherols and ascorbic acid are also efficient physical deactivators (i.e. "quenchers") of $^1O_2$ [28] (Table 1).

Superoxide dismutase enzyme (SOD) (Eq. (6)) and ascorbic acid may also scavenge $O_2^{\bullet-}$. The rate constant for reaction of ascorbic acid with $O_2^{\bullet-}$ is much smaller than that for reaction of $O_2^{\bullet-}$ with SOD (Table 1), but the molar concentration of ascorbic acid in chloroplast is much greater than that of SOD [28]:

$$2O_2^{\bullet-} + 2H^+ \xrightarrow{\text{SOD}} H_2O_2 + O_2 \tag{6}$$

Catalase activity by virtue of its lowering H_2O_2 levels provides less substrate for $HO^{\bullet}$ in the Haber-Weiss reaction [24]:

$$H_2O_2 + H_2O_2 \xrightarrow{\text{catalase}} 2H_2O + O_2 \tag{7}$$

It has been proposed [30] that chloroplasts use a cycle of reactions (Fig. 5) involving ascorbate and glutathione for the removal of H_2O_2.

Antioxidants such as tocopherols and ascorbic acid act as a second line of defence in minimizing the damage wrought by $HO^{\bullet}$ produced in spite of the actions of scavengers of $O_2^{\bullet-}$ and H_2O_2 [31].

3.3
The Photodynamic Effect

As we can see in Sect. 3.1, in healthy cells the primary route for energy from the ^{1}Chl is the fast photochemical reactions of photosynthesis [11]. In dead phytoplanktonic cells this pathway would not be functional; thus an accelerated rate of formation of ^{3}Chl (and that of pheopigments triplets (^{3}Pheo)) would be expected [5]. The rate of formation of ^{3}Chl, ^{3}Pheo and toxic oxygen species ($^1O_2$, $O_2^{\bullet-}$, H_2O_2, $HO^{\bullet}$) might then exceed the quenching capacity of the photoprotective system and photodegradation can occur [32].

Unless complexed with water-soluble proteins, chlorophylls and pheopigments would tend to remain associated with other hydrophobic cellular compounds, such as membrane lipids, in phytodetritus [5]. The photooxidative effect of chlorophyll/pheopigment sensitization might be amplified within such a hydrophobic micro-environment.

Moreover, the lifetime of $^1O_2$ produced from sensitizers in a lipid-rich hydrophobic microenvironment could be longer, and its potential diffusive distance greater, than if produced by sensitizers in aqueous solution [33].

3.4
Photodegradation of Lipidic Cell Components

3.4.1
Chlorophylls

It is generally considered that photodegradation processes play a major role in the photosynthetic pigment decrease observed during senescence [34]. Light-dependent degradation of chlorophylls showed a good fit to first-order kinetics in detrital chlorophytes and diatoms [5, 35]. Table 2 lists first-order rate constants (k_1) for chlorophyll degradation with respect to cumulative light exposure calculated for killed cell experiments.

The breakdown of chlorophyll has so far been studied almost exclusively with respect to the porphyrin moiety of the molecule [36]. It is surprising that the phytol moiety has been neglected, because this unsaturated chain (considered to be the major source of acyclic isoprenoids with twenty or fewer carbon atoms in the biosphere [37]) is also sensitive to photochemical processes. In previous work, we showed that in dead phytoplanktonic cells the chlorophyll phytyl chain is photodegraded only five to eight times slower than the tetrapyrrolic structure [35]. The photodegradation of this isoprenoid chain is a second-order process, which fits the equation $-d[\text{phytol}]/dD = k_2 [\text{phytol}]^2$ (where D = light dose) (Table 2).

Different mechanisms were proposed in order to explain the light-induced degradation of chlorophylls. In vitro, the photooxidation of various bacterio-chlorophylls containing a *meso*-substituted methyl group at position 20 is believed to involve $^1O_2$ addition to the C_1–C_{20} double bond and subsequent ring cleavage [38, 39] (Fig. 6).

In the absence of this methyl group at position 20, the attack of $^1O_2$ focuses on the C_4–C_5 double bond [40].

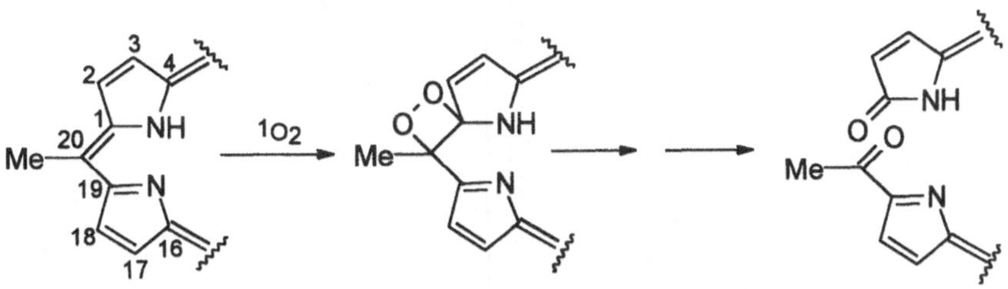

Fig. 6. Mechanism of ring opening at the C_1–C_{20} double bond of bacterio-chlorophyll c. Reprinted with kind permission from American Society for Biochemistry and Molecular Biology

Table 2. Calculated rate constants (k_1 and k_2) for light-dependent aerobic degradation of chlorophylls (Chl) in killed phytoplanktonic cell experiments

	Dunaliella tertiolecta			Skeletonema costatum		Phaeodactylum tricornutum			References
	k_1 (m² Ein⁻¹)	k_2 (μmol⁻¹ Ein⁻¹ m²)	Regression (r^2)	k_1 (m² Ein⁻¹)	Regression (r^2)	k_1 (m² Ein⁻¹)	k_2 (μmol⁻¹ Ein⁻¹ m²)	Regression (r^2)	
Chl-a	0.19		0.97	0.29	0.99	0.56		–	5, 35
Chl-b	0.12		0.92						5
Chl-c				0.32	0.99				5
Phytylchain		0.029	0.99				0.161	0.99	35

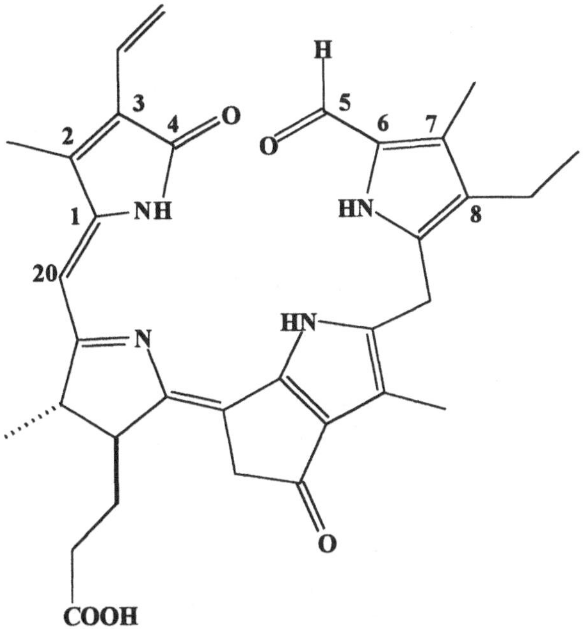

Fig. 7. Red bilin derivative detected in the algae *Chlorella protothecoides*

On the other hand, interaction of $O_2^{\cdot -}$ with chlorophyll is supposed to lead to the opening of the cyclopentanone ring V [41] in analogy to the well known mechanism of allomerization [42].

There is not a great deal known about the structure of the photoproducts of the chlorophyll tetrapyrrolic structure [43], probably due to the instability of these substances [32]. Recently, a linear tetrapyrrole structure formed by ring opening at the C_4-C_5 carbon bond of chlorophyll-a was identified in senescent *Chlorella protothecoides* [44, 45]. This compound (Fig. 7) and the chlorophyll-a catabolite detected in senescent Barley leaves [46] represent the first examples of natural products whose structures correspond to those of the in vitro photooxidation products of porphyrins and chlorins [40].

It was previously demonstrated in vitro that the chlorophyll phytyl chain can be photodegraded by $^1O_2$ [47] and oxy-free radicals [48]. The photosensitized oxidation (involving $^1O_2$) of the phytyl chain leads to the production of photo-products of type **a** and **b** quantifiable after alkaline hydrolysis respectively in the form of 6,10,14-trimethylpentadecan-2-one (1) and 3-methylidene-7,11,15-trimethylhexadecan-1,2-diol (2) (Fig. 8).

Oxy-free radicals add mainly to the phytol olefinic carbon 2; this is to be expected owing to the extra stabilization of the radical formed compared with that resulting from the attack of the carbon 3 [48]. Addition of HO• and subsequent oxidation of the radical formed yield the ketone (1) and 2,3-dihydroxy-phytanate (3) according to the mechanisms described in the Fig. 9.

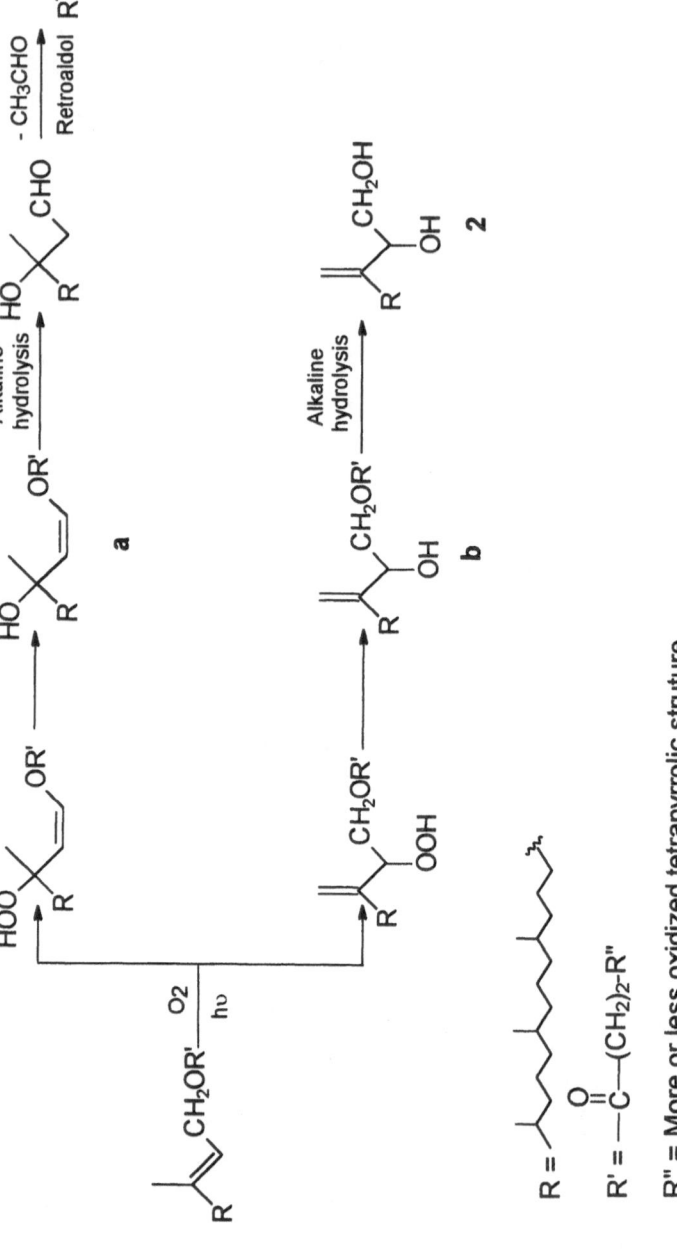

Fig. 8. Photodegradation of chlorophyll phytyl chain involving $^1O_2$ and reactions of photoproducts during alkaline hydrolysis. Reprinted with kind permission from Elsevier Science

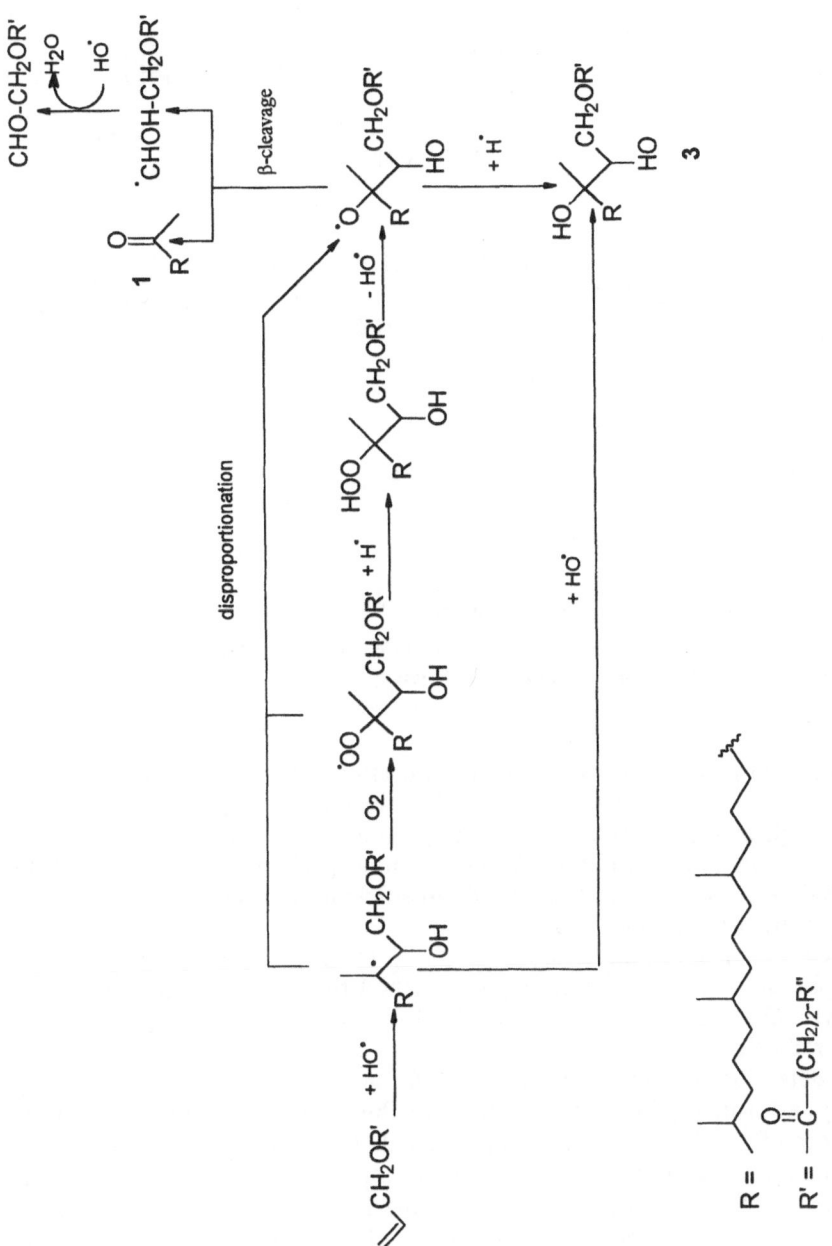

Fig. 9. Action of hydroxyl radicals upon the double bond of the chlorophyll phytyl chain. Reprinted with kind permission from Elsevier Science S.A.

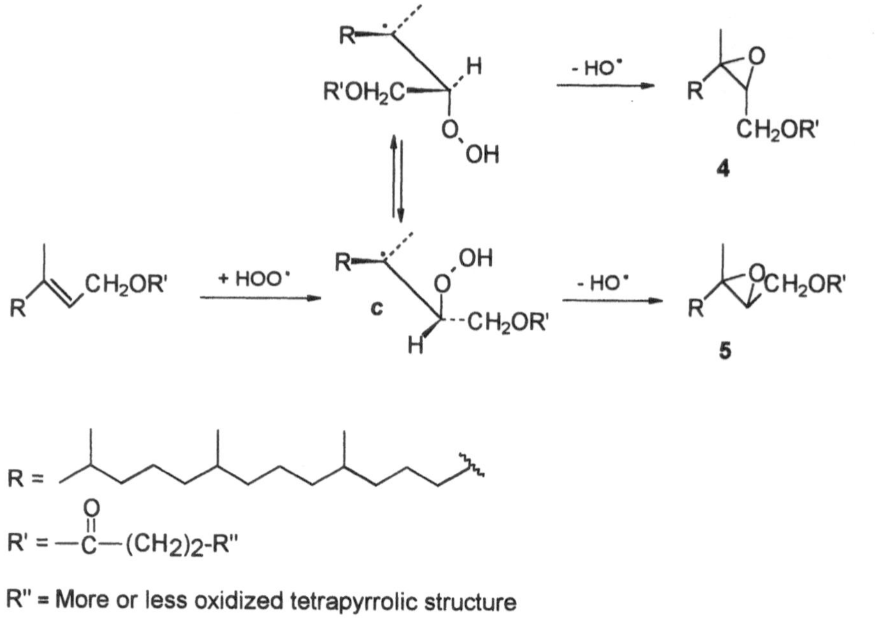

Fig. 10. Action of peroxyl radicals upon the double bond of the chlorophyll phytyl chain. Reprinted with kind permission from Elsevier Science S. A.

Reaction of peroxyl radicals on the unsaturation of the phytyl chain results in the formation of a mixture of *Z* and *E* epoxy-esters (**4** and **5**) (Fig. 10). The relatively high stereoselectivity observed (production of a mixture of *Z* and *E* epoxy-esters in a ratio of about 1.5 to 8.5) can be attributed to an insufficiently long lifetime of the radical **c** (Fig. 10) for an efficient rotation about the 2,3 bond take place before cyclization [48].

Analysis of isoprenoid photoproducts of chlorophyll after irradiation of killed phytoplanktonic cells clearly established that the photodegradation of the phytyl chain in phytodetritus involved mainly $^1O_2$ and, to a small degree, peroxy radicals [35].

The detection of high quantities of 6,10,14-trimethylpentadecan-2-one (**1**) [50, 51] and 3-methylidene-7,11,15-trimethylhexadecan-1,2-diol (**2**) [52] after hydrolysis of marine sediments confirmed that a non-negligible part of the phytoplanktonic chlorophyll phytyl chain must be photodegraded during senescence in the euphotic zone of the oceans. Due to its stability in sediments [49] diol **2** could be proposed as a specific marker for the photodegradation of chlorophylls with a phytol ester group in the marine environment [49].

3.4.2
Carotenoids

In phytodetritus, pigments (i.e. chlorophylls, pheopigments and carotenoids) apparently remain in a close, molecular-scale association at relatively high

Table 1. Calculated first-order rate constants (k_1) for light-dependent degradation of carotenoids in aerated killed cell experiments (from Nelson [5]

Carotenoid	Dunaliella tertiolecta		Skeletonema costatum	
	k_1 (m² Ein⁻¹)	Regression (r^2)	k_1 (m² Ein⁻¹)	Regression (r^2)
Lutein	0.24	0.95		
Neoxanthin	0.53	0.99		
Violaxanthin	0.37	0.98		
β-Carotene	0.37	0.96	0.43	0.96
Fucoxanthin			0.25	0.99
Diadinoxanthin			0.51	0.97

localized concentrations, even though the structure of the thylakoid membrane has been disrupted [5]. Thus, the sensitized photooxidation of carotenoids would be enhanced.

In a previous paper, Nelson [5] studied the rates of light-dependent degradation of pigments (and notably of carotenoids) in detritus derived from phytoplankton. The photodegradation of carotenoids showed a good fit to first-order kinetics in detrital chlorophyte and diatom cells (Table 3).

Using D_2O (which prolongs $^1O_2$ lifetime) and radical scavengers (gallate and (+) catechin), Takahama [53] demonstrated that carotenoid photobleaching in isolated chloroplasts involves $^1O_2$ and radical chain reactions. The photosensitized oxidation (involving $^1O_2$) of carotenoids has been studied in vitro [54, 55]. Loliolide (**6**), *iso*-loliolide (**7**) and dihydroactinidiolide (**8**) were identified as major photoproducts (Fig. 11) depending on the functionality of carotenoids at C_3.

Recently, we gave evidence for the presence of loliolide and *iso*-loliolide after irradiation of killed phytoplanktonic cells [56]. Such compounds have also been detected in situ in suspended particles [57] and in sediments [57, 58]; unfortunately, the sedimentary loliolides can be diagenetic products of carotenoids apparently resulting from microbially mediated reactions under anaerobic conditions [57]. Thus, these compounds cannot constitute unequivocal indicators of photooxidative processes [5].

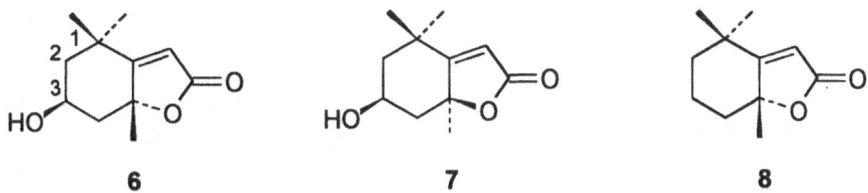

Fig. 11. Structure of the main carotenoid photoproducts

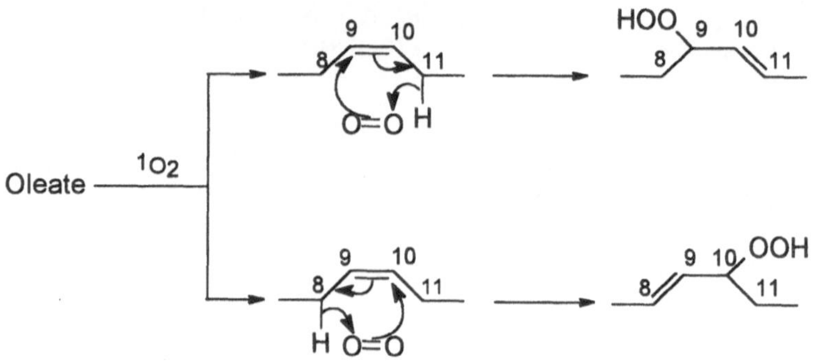

Fig. 12. Mechanism of photosensitized oxidation of oleate

3.4.3
Fatty Acids

Since damaging effects of $^1O_2$ are primarily concerned with the oxidation of membrane lipids [59], chloroplast membrane components are particularly susceptible to type II photooxidation [60, 61]. This is the case for unsaturated fatty acids, which generally predominate in algal lipids, particularly in the photosynthetic membranes [62].

Photosensitized oxidation of unsaturated fatty acids involves a direct reaction of $^1O_2$ with the carbon-carbon double bond by a concerted "ene" addition [63], and leads to the formation of hydroperoxides at each unsaturated carbon [64]. Thus, oleate produces two isomeric hydroperoxides (**9** and **10**) (Fig. 12) with allylic *trans* double bond. These hydroperoxides may subsequently play a part in initiating the normal free radical autoxidation of unsaturated fatty acids [65].

Several secondary oxidation products can also be formed in highly photooxidized unsaturated fatty acids [64]. Particular attention has been paid to the cyclic peroxides formed during these processes [66, 67] (Fig. 13). The mechanisms suggested for the formation of these compounds involve 1,3-cyclization of *homo*-allylic hydroperoxide isomers [68].

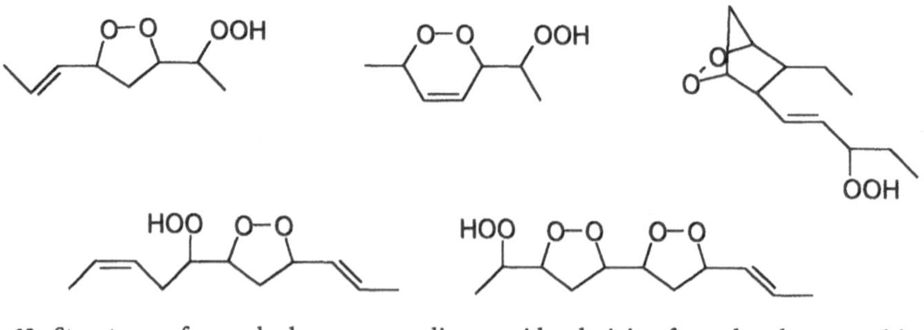

Fig. 13. Structures of some hydroperoxy cyclic peroxides deriving from the photo-sensitized oxidation of fatty acids

ω-Oxocarboxylic and α,ω-dicarboxylic acids have been identified in marine aerosols [4, 6] and marine sediments [69]. The predominence of C_9 species suggests that the photosensitized oxidation of biogenic unsaturated fatty acids could be a likely source for these ω-oxoacids. Phytoplankton derived unsaturated fatty acids, such as oleic and linoleic acids, do generally contain double bonds at the C_9 position [4] and Hock cleavage [63] of the 9- and 10-hydroperoxides resulting from photosensitized oxidation of such fatty acids produces predominantly the ω-oxononanoic acid (Fig. 14).

It is of interest to note that C_6–C_{10} aldehydes [70] and C_7–C_{11} ω-oxoacids (with C_9 predominant) [4] were detected in seawater samples from the same Peru upwelling region.

3.4.4
Sterols

As important unsaturated components of biological membranes, sterols are highly susceptible to photooxidative degradation during senescence of phytoplankton. Type I [71] and type II photooxidation [72, 73] of these compounds have been extensively studied in vitro (Fig. 15).

Though Gagosian and Nigritelli [74] have calculated that, at most, 0.3% of sterols produced in the Sargasso Sea is deposed on the ocean floor, in the literature there are practically no studies dealing with the photodegradation of these compounds in senescent phytoplanktonic cells.

We recently observed that the major sterol of *Emiliania huxleyi* (24α-methylcholesta-5,22(E)-dien-3β-ol) is quickly degraded during irradiation of sterile killed cells of this coccolithophorid [75]. This very interesting result strongly suggests that the photochemical processes associated with phytodetritus must play an important role in the degradation of these compounds in the euphotic layer of the oceans. This very important aspect should be investigated without delay.

3.4.5
Alkenones

Emiliania huxleyi and members of the class *Prymnesiophyceae* are now the recognized sources of long-chain (nC_{37}–nC_{39}) alken-2 and -3 ones (alkenones) in sediments [76, 77]. The ratio of di- to di- plus tri-unsaturated C_{37} alkenones (denoted $U_{37}^{k\prime}$) has been shown to have a linear response with temperature in culture and water column studies [78, 79]. These results have led to the use of C_{37} alkenones as paleotemperature indicators [80].

In recent work, we studied the photooxidation of these compounds in order to determine if photochemical processes could appreciably modify $U_{37}^{k\prime}$ ratios during algal senescence [75]. In solution, type II photochemical processes degraded alkenones more slowly than fatty acids with the same degree of unsaturation. This difference of reactivity was attributed to (i) the *trans* geometry of the alkenone double bonds [81], which is known to be less reactive toward singlet oxygen than the *cis* geometry of the fatty acid double bonds [82] and (ii)

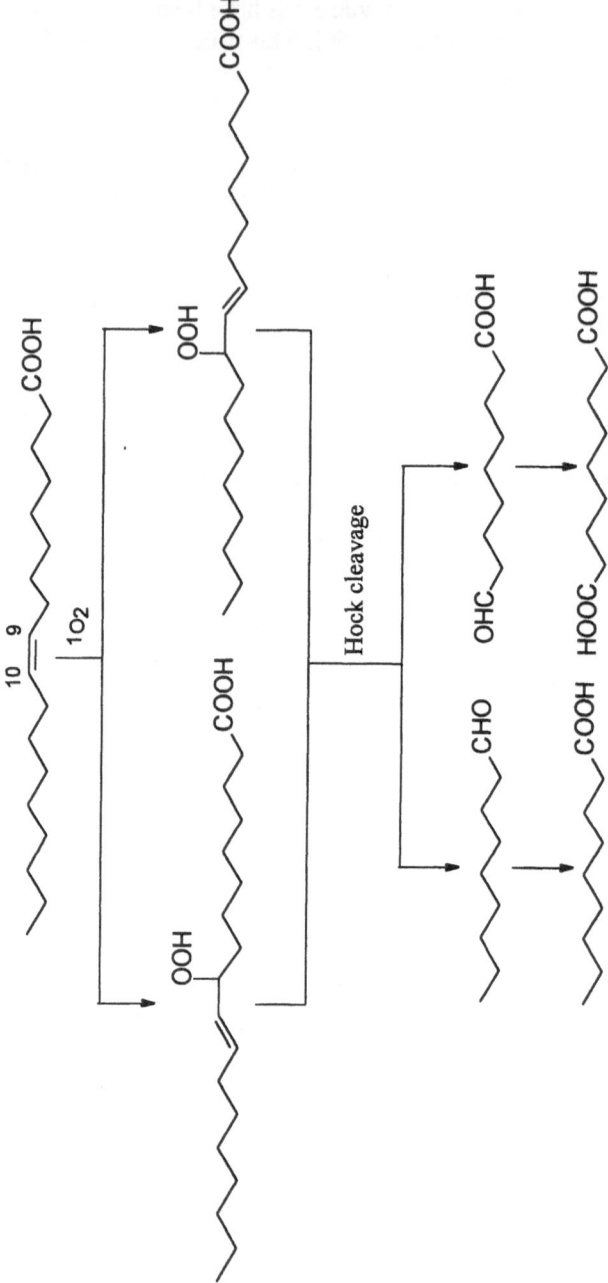

Fig. 14. Proposed production of ω-oxononanoic acid by photosensitized oxidation of oleic acid and subsequent Hock cleavage of the hydroperoxides formed

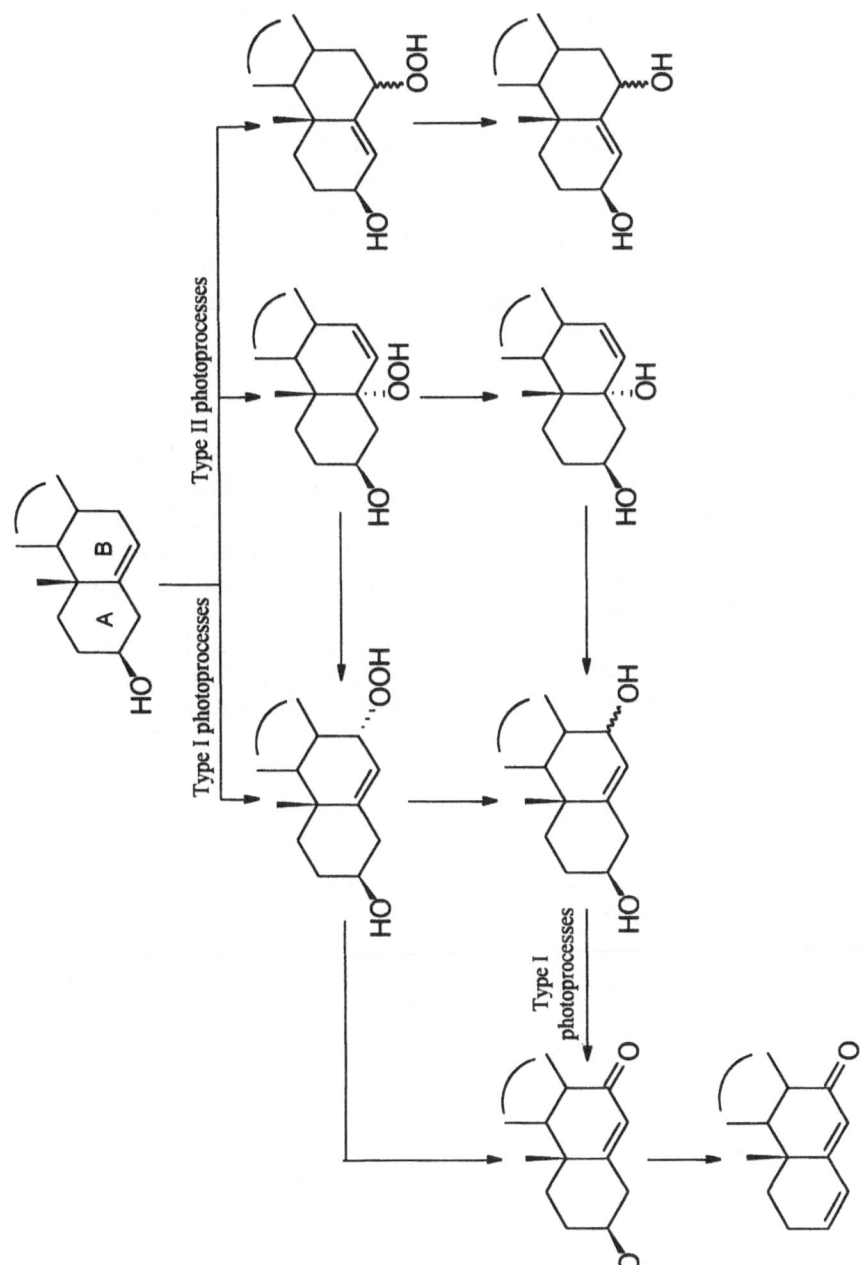

Fig. 15. Sterol photoproducts generated by $^1O_2$ and free radical pathways (only the major photoproducts are shown)

the separation of the double bonds by five saturated carbons in the alkenone structure. The photodegradation rate of these unsaturated ketones increased logically with their degree of unsaturation.

Though selective, photochemical degradation of alkenones is not fast enough in killed cells of *Emiliania huxleyi* to induce significant modifications of the $U^k_{37}{}'$ ratio before the photodestruction of the photosensitizing substances. This poor photo-reactivity was attributed not only to the structure of alkenones but also to a localization of these compounds elsewhere than in cell membranes [75]. Indeed, recently Conte and Eglinton [83] failed to detect alkenones in the membranes of *Emiliania huxleyi* after cell fractionation.

3.4.6
Hydrocarbons

Marine algae contain normal paraffins, often with nC_{15} and nC_{17} which tend to dominate the other n-alkanes [84]. They are also rich in n-alkenes (particularly in $nC_{21}:6$ [85]) and squalene (a highly unsaturated isoprenoid hydrocarbon known to be the biosynthetic precursor of steroids). Recently, relatively large quantities of several highly branched isoprenoid hydrocarbons were detected in diatomaceous algae [86–88].

Numerous authors have used hydrogen atom abstractor sensitizers during in vitro studies of the photooxidation of alkanes (Table 4). In the presence of such sensitizers saturated hydrocarbons can be strongly photooxidized; the nature of

Table 4. Photosensitized oxidation of linear and branched alkanes

Substrate	Sensitizer	Photoproducts	References
Hexadecane	1-Naphthol	n-Alkanols	89
Hexadecane	Xanthone	n-Alkanols	90
Pentadecane	Anthraquinone	n-Alkanones n-Alkenes	91
Hexadecane	Anthraquinone	n-Alkanones n-Alkanols 2,5-Diketones γ-Lactones n-Alkenes Aldehydes n-Alkanes Carboxylic acids	92, 93
Pristane	Anthraquinone	Isoprenoid ketones Isoprenoid alkanes Tertiary alcohols	94
3,6-Dimethyloctane	Anthraquinone	3,6-Dimethyloctan-3-ol 3,6-Dimethyloctan-3,6-diol	95

Table 5. Rate constants for reactions of $^1O_2$ with isolated acyclic olefins

Substrate	k (mol^{-1} s^{-1})	References
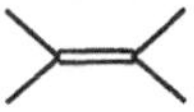 (tetramethylethylene)	2.2×10^7	82
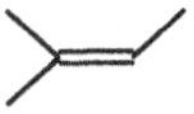 (trimethylethylene)	2.3×10^6	96
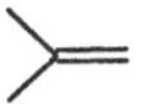 (isobutylene)	6.2×10^4	82
 (2-pentene)	4.8×10^4	82
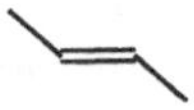 (2-methyl-1-butene)	7.2×10^3	82
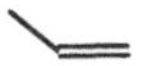 (1-butene)	4.6×10^2	98

the photoproducts formed depends on the sensitizer and the experimental conditions employed (Table 4).

The type II photooxidation of isolated acyclic olefins has been intensively studied for the past 30 years (Table 5).

These reactions may be categorized into two classes outlined in Fig 16. The first of these involves a (2 + 2) cycloaddition to sterically hindered olefins [99, 100]; the resulting dioxetane readily cleaves, thermally or photochemically, into two carbonyl-containing fragments [63]. The second type is the photooxygenation of olefins that contain at least one allylic hydrogen to yield allylic

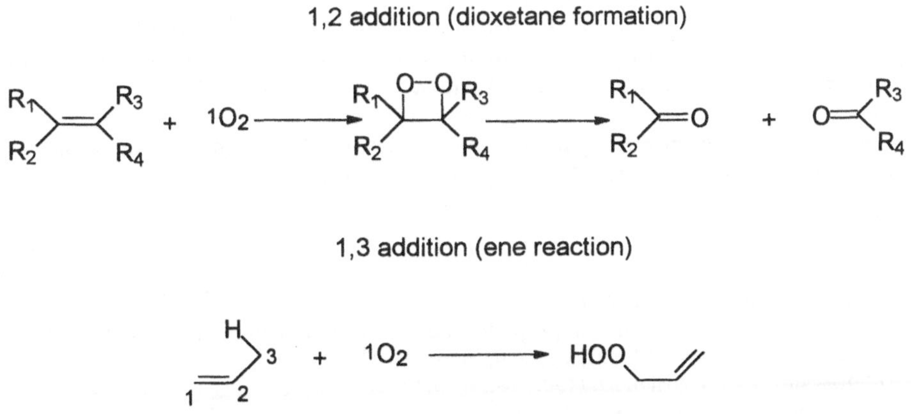

1,2 addition (dioxetane formation)

1,3 addition (ene reaction)

Fig. 16. Reaction of $^1O_2$ with isolated acyclic double bonds. Reprinted with kind permission from American Chemical Society

hydroperoxides in which the double bond has shifted to a position adjacent to the original double bond [63].

Despite these interesting results obtained in solution, the photooxidation of hydrocarbon components of phytoplankton during senescence has not been strongly investigated. Some recent observations [101] seem to indicate that these compounds are relatively protected from photochemical damage in senescing cells of phytoplankton; however, these results require confirmation and the actual extent of this phenomenon remains to be determined.

4
Conclusions

The available results reviewed in the present paper suggest that the hetero-genous photooxidative reactions associated with phytodetritus must play a major role in the degradation of organic matter in the euphotic layer of the oceans. More work is definitely required in order to elucidate the mechanisms of these very interesting processes.

Though it is very difficult to assess the relative quantitative importance of toxic oxygen forms within phytodetritus, $^1O_2$ seems to play a key role in the photodegradation of most of the lipidic cell components (e.g. chlorophylls, carotenoids, fatty acids). This is in good agreement with the well known longer lifetime of $^1O_2$ in hydrophobic microenvironments [33]. The action of this transient yet pernicious molecule during senescence of phytoplankton requires particular attention.

The localization of organic compounds within the structural organization of phytodetritus can strongly affect their photodegradation rates. If membrane lipids (pigments, sterols, fatty acids) appear to be quickly photodegraded during senescence of phytoplankton, this is not the case of other lipids such as alkenones and hydrocarbons localized elsewhere than in cellular membranes.

It was previously proposed that the photooxidative cross-linking of un-saturated fatty acids could play a role in the formation of humic acids [7]. The hydrophobic microenvironment of phytodetritus, which provides high localized concentrations of unsaturated lipids and visible light absorbing photosensitizers [5], could constitute an ideal site for such reactions. This very important area merits more detailed study.

Acknowledgements. Financial support over many years by the Groupement de Recherche HYCAR 1123 (CNRS/Elf Aquitaine) is gratefully acknowledged.

References

1. Zafiriou OC (1977) Mar Chem 5:497
2. Zafiriou OC, Joussot-Dubien J, Zepp RG, Zika RG (1984) Environ Sci Technol 18:358A
3. Dister B, Zafiriou OC (1993) J Geophys Res 98:2341
4. Kawamura K, Gagosian RB (1987) Nature 325:330
5. Nelson JR (1993) J Mar Res 51:155
6. Stephanou EG, Stratigakis N (1993) Environ Sci Technol 27:1403
7. Harvey GR, Boran DA, Chesal LA, Tokar JM (1983) Mar Chem 12:119

8. Zepp RG, Cline DM (1977) Environ Sci Technol 11:359
9. Zepp RG (1978) Environ Sci Technol 12:327
10. Crosby DG (1972) Adv Chem 111:173
11. Foote CS (1976) Photosensitized oxidation and singlet oxygen: consequences in biological systems. In: Pryor WA (ed) Free radical in biology 2. Academic Press, New York, p 95
12. Gollnick K (1968) Adv Photochem 6:1
13. Schenck GO, Koch E (1960) Z Electrochem 64:170
14. Kasche V, Lindquist L (1965) Photochem Photobiol 4:923
15. Kepka A, Grossweiner LI (1972) Photochem Photobiol 14:621
16. Rabinowitch HD, Fridovich I (1983) Photochem Photobiol 37:679
17. Knox JP, Dodge AD (1985) Phytochem 24:889
18. Harbour JR, Bolton JR (1978) Photochem Photobiol 28:231
19. Chauvet JP, Villain F, Viovy R (1981) Photochem Photobiol 34:557
20. Foote CS, Chang YC, Denny RW (1970) J Am Chem Soc 92:5216
21. Connolly JS, Gorman DS, Seely GR (1973) Ann NY Acad Sci 206:649
22. Biswal B (1995) J Photochem Photobiol B: Biol 30:3
23. Biswal UC, Biswal B (1990) Biol Edn 7:56
24. Leshem YY (1988) Free radicals in chemistry and biology 5:39
25. Fryer MJ (1992) Plant Cell Environ 15:381
26. Nishikimi M (1975) Biochem Biophys Res Commun 63:463
27. Anbar M, Neta P (1967) Int J Appl Radiat Isot 18:495
28. Halliwell B (1987) Chem Phys Lipid 44:327
29. Krinsky NI, Deneke SM (1982) J Natl Cancer Inst 69:205
30. Foyer CH, Halliwell B (1976) Planta 133:21
31. Borg DC (1976) Applications of electron spin resonance in biology. In: Pryor WA (ed) Free radicals in biology 1. Academic Press, New York, p 69
32. Merzlyak MN, Hendry GAF (1994) Proceeding of the Royal Society of Edinburgh 102B:459
33. Suwa K, Kimura T, Schaap AT (1977) Biochem Biophys Res Commun 75:785
34. Maunders MJ, Brown SB (1983) Planta 158:309
35. Rontani J-F, Beker B, Raphel D, Baillet G (1995) J Photochem Photobiol A:Chem 85:137
36. Peisker C, Düggelin T, Rentsch D, Matile P (1989) J Plant Physiol 135:428
37. Volkman JK, Maxwell JR (1986) Acyclic isoprenoids as biological markers. In: Johns RB (ed) Biological markers in the sedimentary record. Elsevier, Amsterdam, p 1
38. Brown SB, Smith KM, Bisset GMF, Troxler RF (1980) J Biol Chem 255:8063
39. Troxler RF, Smith KM, Brown SB (1980) Tetrahedron Lett 21:491
40. Fuhrhop JH, Wasser PKV, Subramanian J, Schrader U (1974) Liebigs Ann Chem 1450
41. Merzlyak MN, Kuprianova NS, Kovrizhnikh VA, Afanas'ev IB (1985) J Inorg Biochem 24:239
42. Seely GR (1966) The structure and chemistry of functional groups. In: Vernon LP, Seely GR (eds) The chlorophylls. Academic Press, New York London, p 67
43. Hendry GAF, Houghton JD, Brown SB (1987) New Phytol 107:255
44. Engel N, Jenny TA, Mooser V, Gossauer A (1991) FEBS Lett 293:131
45. Iturraspe J, Engel N, Gossauer A (1994) Phytochem 35:1387
46. Kräutler B, Jaun B, Bortlik B, Schellenberg K, Matile P (1991) Angew Chem 103:1354
47. Rontani J-F, Baillet G, Aubert C (1991) J Photochem Photobiol A: Chem 59:369
48. Rontani J-F, Aubert C (1994) J Photochem Photobiol A: Chem 79:167
49. Rontani J-F, Raphel D, Cuny P (1996) Org Geochem 24:825
50. Ten Haven HL, Baas M, De Leeuw JW, Schenck PA (1987) Mar Geol 75:137
51. Rontani J-F, Giral PJ-P, Baillet G, Raphel D (1992) Org Geochem 18:139
52. Rontani J-F, Grossi V, Faure R, Aubert C (1994) Org Geochem 21:135
53. Takahama U (1983) Plant Cell Physiol 24:495
54. Iseo S, Hyeon SB, Sakan T (1969) Tetrahedron Lett 4:279
55. Iseo S, Hyeon SB, Katsumura S, Sakan T (1972) Tetrahedron Lett 25:2517

56. Rontani J-F (1996) personal communication
57. Repeta DJ (1989) Geochim Cosmochim Acta 53:699
58. Klok J, Baas M, Cox HC, De Leeuw JW, Schenck PA (1984) Tetrahedron Lett 25:5577
59. Percival MP, Dodge AD (1983) Plant Sci Lett 29:255
60. Heath RL, Packer L (1968) Arch Biochem Biophys 125:189
61. Heath RL, Packer L (1968) Arch Biochem Biophys 125:850
62. Wood BJB (1974) Fatty acids and saponifiable lipids. In: Stewart WD (ed) Algal physiology and biochemistry. University of California Press, Berkeley, p 236
63. Frimer AA (1979) Chem Rev 79:359
64. Frankel EN (1984) J Am Oil Chem Soc 61:1908
65. Frankel EN, Neff WE, Selke E, Weisleder D (1982) Lipids 17:11
66. Frankel EN, Neff WE, Rohwedder WK, Khambay BPS, Garwood RF, Weedon BCL (1977) Lipids 12:1055
67. O'Connor DE, Mihelich ED, Coleman MC (1981) J Am Chem Soc 103:223
68. Pryor WA, Stanley JP, Blair E (1976) Lipids 11:370
69 Stephanou EG (1992) Naturwissenschaften 79:128
70. Gschwend P, Zafiriou OC, Gagosian RB (1980) Limnol Oceanogr 25:1044
71. Rontani J-F, Raphel D, Aubert C (1993) J Photochem Photobiol A: Chem 72:189
72. Nickon A, Bagli JF (1961) J Am Chem Soc 83:1498
73. Kulig MJ, Smith LL (1973) J Org Chem 38:3639
74. Gagosian RB, Nigritelli GE (1979) Limnol Oceanogr 24:838
75. Rontani J-F, Cuny P, Grossi U, Beker B (1997) Org Geochem 26:503
76. Volkman JK, Eglinton G, Corner EDS, Sargent JR (1980) Advances in Organic Geochemistry 1979, p 219
77. Marlowe IT, Brassell SC, Eglinton G, Green JC (1984) Org Geochem 6:135
78. Prahl FG, Wakeham SG (1987) Nature 330:367
79. Prahl FG, De Lange GJ, Lyle M, Sparrow MA (1989) Nature 341:434
80. Brassell SC, Eglinton G, Marlowe IT, Sarnthein M (1986) Nature 320:129
81. Rechka JA, Maxwell JR (1988) Advances in Organic Geochemistry 1987, p 727
82. Hurst JR, Wilson SL, Schuster GB (1985) Tetrahedron 41:2191
83. Conte MH, Eglinton G (1993) Deep Sea Res 40:1935
84. Volkman JK, Holdsworth DG, Neill GP, Bavor HJ (1992) Sci Total Environ 112:203
85. Saliot A (1981) Natural hydrocarbons in seawater. In: Duursma EK, Dawson R (eds) Marine organic chemistry. Elsevier, Amsterdam Oxford New York, p 327
86. Summons RE, Barrow RA, Capon RJ, Hope JM, Stranger C (1992) Aust J Chem 46:907
87. Volkman JK, Barrett SM, Dunstan GA (1994) Org Geochem 21:407
88. Belt ST, Cooke DA, Robert J-M, Rowland S (1996) Tetrahedron Lett 37:4755
89. Klein AE, Pilpel N (1973) J Chem Soc Faraday Trans 69:1729
90. Gesser HD, Wildmann TA, Tewari YB (1977) Environ Sci Technol 6:605
91. Ehrhardt M, Petrick G (1985) Mar Chem 16:227
92. El Anba-Lurot F, Guiliano M, Doumenq P, Bertrand J-C, Mille G (1996) Intern J Environ Anal Chem 63:289
93. Rontani J-F (1991) Intern J Environ Anal Chem 45:1
94. Rontani J-F, Giusti G (1987) J Photochem Photobiol A: Chem 40:107
95. Rontani J-F, Giusti G (1988) Tetrahedron Lett 29:1923
96. Monroe BM (1978) J Phys Chem 82:15
97. Koch E (1968) Tetrahedron 24:6295
98. Kopecky KR, Reich HJ (1965) Can J Chem 43:2265
99. Bartlett PD, Ho MS (1974) J Am Chem Soc 96:627
100. Takeshita H, Hatsui T, Jinnai O (1976) Chem Lett 1059
101. Rontani J-F (1996) Personnal communication

11 Photocatalytic Detoxification of Polluted Waters

Detlef Bahnemann

Institute for Solar Energy Research (ISFH, Institut für Solarenergieforschung GmbH Hameln/Emmerthal), Sokelantstrasse 5, D-30165 Hannover, Germany.
E-mail: isfh.Bahnemann@oln.comlink.apc.org

During the past 15 years research and development in the area of photocatalysis have been tremendous. The present review describes the basic principles of photocatalysis, focusing in particular on important mechanistic and kinetic aspects. The properties and requirements for efficient photocatalysts are discussed in detail. A few representative model compounds have been selected to illustrate the major reaction pathways in photocatalytic degradation processes. While tetrachloromethane has been chosen as a typical example for a reductive pathway, the degradation of 4-chlorophenol and naphthalene are described in detail with special emphasis on possible and often undesired intermediates. The couple benzoquinone/hydroquinone is presented to elucidate an electron-shuttle mechanism which results in an undesirable short-circuiting of the photocatalyst. Crucial reaction parameters such as pH, temperature, solute concentration and light intensity, are given together with current theoretical models to explain their effects on the overall process efficiency. Various solar reactors for photocatalytic water treatment are described in detail including the comparison of their overall performance. Finally, several examples for the treatment of real wastewater are presented together with some initial economic considerations.

Keywords: Photocatalysis, Solar Detoxification, Reaction Mechanisms, Novel Photocatalysis, Kinetic Models

Contents

The Handbook of Environmental Chemistry Vol. 2 Part L
Environmental Photochemistry (ed. by P. Boule)
© Springer-Verlag Berlin Heidelberg 1999

List of Symbols and Abbreviations (in the sequence as they appear in the text)

PTR	Parabolic Trough Reactor
TFFBR	Thin Film Fixed Bed Reactor
CPCR	Compound Parabolic Collecting Reactor
DSSR	Double Skin Sheet Reactor
TiO_2	Titanium Dioxide
ZnO	Zinc Oxide
Fe_2O_3	Iron (III) Oxide
CdS	Cadmium Sulfide
ZnS	Zinc Sulfide
NHE	Normal Hydrogen Electrode
VB	Valence Band
CB	Conduction Band
E_g	Bandgap Energy
PCBs	Polychlorinated Biphenyls
$^{\bullet}OH$	Hydroxyl Radical
h^+_{VB}	Valence Band Hole
e^-_{CB}	Conduction Band Electron
pH_{zpc}	pH-Value at the Zero Point of Charge
HBA	4-Hydroxybenzylalcohol
HQ	Hydroquinone
DHBA	Dihydroxybenzylalcohol
HBZ	4-Hydroxybenzaldehyde
iPrOH	Isopropanol
d [nm]	Particle Diameter

e^-_{aq}	Hydrated Electron
$E\ [V(\textit{vs. NHE})]$	One-electron Redox Potential
h^+_{tr}	Deeply trapped Hole
h^+_{tr*}	Shallowly trapped Hole
e^-_{tr}	Trapped Electron
DCA^-	Dichloroacetate
$[\equiv TiOH_{tot}]$	Total Concentration of Surface Hydroxyl Groups
γ	Absorption Coefficient
$K_{DCA^-}\ [l\ mol^{-1}]$	Equilibrium Constant for DCA^- Adsorption
A_{500}	Absorption of Transient at 500 nm
SCN^-	Thiocyanate Ion
κ	Transmission Coefficient
$\nu\ [s^{-1}]$	Frequency of Exited State Molecular Vibration
$E_V\ [eV]$	Flat Band Potential
$E_{F,redox}\ [eV(\textit{vs. NHE})]$	Redox Potential
λ	Reorientation Energy
ζ	Photonic Efficiency
4-CP	4-Chlorophenol
Φ	Quantum Yield
$Ti^+\text{-}{}^\bullet OH_s$	Surface-Trapped Hole
L-H	Langmuir-Hinshelwood
TOC [ppm]	Total Organic Carbon Content
HHQ	Hydroxyhydroquinone
BQ	Benzoquinone
HQ	Hydroquinone
HBQ	Hydroxybenzoquinone
4-CC	4-Chlorocatechol
HPR	Hydroxy Phenyl Radical
HPLC	High Performance Liquid Chromatography
FTIR	Fourier Transform Infrared Spectroscopy
CDHB	5-Chloro-2,4'-Dihydroxybiphenyl
THB	2,5,4'-Trihydroxybiphenyl
HPBQ	4-Hydroxyphenylbenzoquinone
DHBP	2,4'-Dihydroxy- and 4,4'-Dihydroxybiphenyl
PAHs	Polynuclear Aromatic Hydrocarbons
$>TiOH^\bullet$	Surface-bound Hydroxyl Radical
I_a	Absorbed Light Intensity
$A\ [l\ mol^{-1}\ s^{-1}]$	Preexponential Factor
$E_a\ [J\ mol^{-1}]$	Activation Energy
$R\ [J\ mol^{-1}\ K^{-1}]$	Gas Constant
$T\ [\ K]$	Temperature
$k\ [l\ mol^{-1}\ s^{-1}]$	Second Order Rate Constant
Red.	Substrate (Pollutant) Molecule
PSA	Plataforma Solar de Almeria
NREL	National Renewable Energy Laboratory (USA)
AM	Air Mass
PLEXIGLAS®	Polyacrylmetamethacrylate (Brandname)

UV-A	Portion of Solar Spectrum below 400 nm
ε^{hv}	Dimensionless Photoreactor Efficiency
c^o [mol l^{-1}]	Initial Substrate Concentration
$v^\bullet$ [l s^{-1}]	Flow Rate
I [mol photons l^{-1} s^{-1}]	Light Intensity
$\Phi_{reactor}$	Photochemical Reaction Yield
Δc [mol l^{-1}]	Amount of Degraded Substrate Concentration
S_{TOC}	Mass Area Ratio
A_r	Surface Element of the Reactor
R_{eff}	Effective Degradation Rate for a Specific Reactor
$E_{uv,tot}$	Accumulated UV radiation
η [µg TOC J^{-1}]	Efficiency
LLNL	Lawrence Livermoore National Laboratory
TCE	Trichloroethylene
BTEX	Benzene, Toluene, Ethylbenzene, and Xylenes
COD	Chemical Oxygen Demand
BOD$_5$	Biological Oxygen Demand (5 days incubation at 37 °C)
ECD	Electron Capture Detector
A	Illuminated Reactor Area
$n^\bullet_{in}$	Incoming Molar Flow
λ_{ex} [nm]	Laser Excitation Wavelength

1
Introduction

Persistent organic chemicals are present as pollutants in wastewater effluent from industrial manufacturers and normal households, and in landfill leachates. They can be found in groundwater wells and surface waters. In all cases they have to be removed to protect our water resources or to achieve drinking water quality. Therefore, many processes have been proposed over the years and are currently being employed to destroy these toxins. The so-called photocatalytic detoxification has been discussed as an alternative method to clean-up polluted water in the scientific literature since 1976 [1]. Lately, considerable public attention has been focused on this possibility of combining heterogeneous catalysis with solar technologies to achieve the mineralization of toxins present in water [2, 3]. Compilations of substances which can be mineralized using photocatalysis are now available [4–6]. Several reviews have recently been published discussing the underlying reaction mechanisms and illustrating examples of successful laboratory and field studies [7–15]. While the overall stoichiometry of most mineralizations appears to be understood, details of the complex reaction mechanism are still not known.

2
Mechanisms of Photocatalysis

Semiconductors (e.g., TiO$_2$, ZnO, Fe$_2$O$_3$, CdS, and ZnS) can act as sensitizers for light-induced redox processes due to their electronic structure which is characterized by a filled valence band and an empty conduction band [16]. Absorption of a photon with an energy greater than the bandgap energy leads to the formation of an electron/hole pair. In the absence of suitable scavengers, the stored energy is dissipated within a few nanoseconds by recombination [17]. If a suitable scavenger or surface defect state is available to trap the electron or hole, recombination is prevented and subsequent redox reactions may occur. The valence band holes are powerful oxidants (+1.0 to +3.5 V vs NHE depending on the semiconductor and pH), while the conduction band electrons are good reductants (+0.5 to –1.5 V vs NHE) [18]. Most organic photodegradation reactions utilize the oxidizing power of the holes either directly or indirectly; however, to prevent a buildup of charge one must also provide a reducible species to react with the electrons. In bulk semiconductor electrodes only one species, either the hole or electron, is available for reaction due to band bending [19]. However, in very small semiconductor particle suspensions both species are present on the surface. Therefore careful consideration of both the oxidative and the reductive paths is required.

Figure 1 shows a cartoon which is frequently used to illustrate photocatalytic processes. It consists of a superposition of the energy bands of semiconducting TiO_2 (valence band VB, conduction band CB) and the geometrical image of a spherical particle. Absorption of a photon with an energy hv greater than or equal to the bandgap energy E_g generally leads to the formation of an electron/ hole pair in the semiconductor particle. Subsequently these charge carriers either recombine and dissipate the input energy as heat, get trapped in metastable surface states, or react with electron donors and acceptors adsorbed on the surface or bound within the electrical double layer.

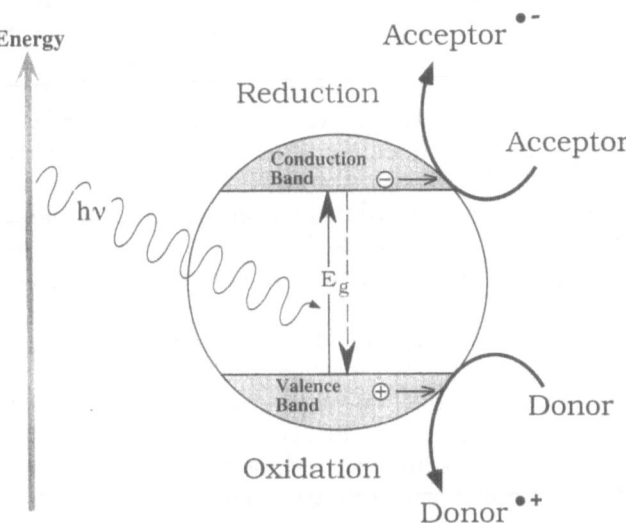

Fig. 1. Energetic principles of photocatalysis

The application of illuminated semiconductors for the remediation of contaminants has been used successfully for a wide variety of compounds [20–27] such as alkanes, aliphatic alcohols, aliphatic carboxylic acids, alkenes, phenols, aromatic carboxylic acids, dyes PCBs, simple aromatics, halogenated alkanes and alkenes, surfactants, and pesticides, as well as for the reductive deposition of heavy metals (e.g., Pt^{4+}, Au^{4+}, Rh^{3+}) from aqueous solution to surfaces [28–31]. In many cases, complete mineralization of organic compounds has been reported.

A general stoichiometry for the heterogeneously-photocatalyzed oxidation of a generic chlorinated hydrocarbon to complete mineralization can be written as follows:

$$C_xH_yCl_z + \left(x + \frac{y-z}{4}\right) O_2 \xrightarrow{hv, TiO_2} xCO_2 + zH^+ + zCl^- + \left(\frac{y-z}{4}\right) H_2O \qquad (1)$$

The exact nature of the main oxidizing species formed on the surface of the semiconductor particles following the absorption of a photon is yet unclear. A significant body of literature exists that the initial oxidation of a pollutant molecule may either occur by indirect oxidation via a surface-bound hydroxyl radical (i.e., a trapped hole at the particle surface) or directly via the valence-band hole before it is trapped either within the particle or at the particle surface.

In support of hydroxyl radical as the principal reactive oxidant in photo-activated TiO_2 is the observation that intermediates detected during the photocatalytic degradation of halogenated aromatic compounds are typically hydroxylated structures [20, 32–36]. These intermediates are consistent with those found when similar aromatics are reacted with a known source of hydroxyl radicals. In addition, ESR studies have verified the existence of hydroxyl and hydroperoxyl radicals in aqueous solutions of illuminated TiO_2 [37–39]. Mao et al. [40] have found that the rate of oxidation of chlorinated ethanes correlates with the C-H bond strengths of the organics, which indicates that H atom abstraction by $^•OH$ is an important factor in the rate-determining step for oxidation. The strong correlation between degradation rates and concentration of the organic pollutant adsorbed to the surface [21, 41–47] also implies that the hydroxyl radicals are adsorbed at the surface. On the other hand, Mao et al. [40] have observed that trichloroacetic acid and oxalic acid are oxidized primarily by valence band holes on TiO_2 via a photo-Kolbe process. It should be noted that these compounds also have no hydrogen atoms available for abstraction by $^•OH$.

Another approach to clarify the nature of the oxidizing species in photocatalytic systems is to use probe molecules the $^•OH$-adducts of which are well known and can be analyzed qualitatively and quantitatively [48, 49]. Often there are, however, two possible routes to arrive at the same reaction products. Valence band holes can form hydroxyl radicals upon trapping at surface hydroxyl groups of a metal oxide semiconductor:

$$h^+_{VB} + OH_S^- \rightarrow {}^•OH_S \qquad (2)$$

The reaction of a probe molecule P will then result in the observed product distribution

$$\cdot OH_S + P \rightarrow P^{\cdot}\text{-}OH \rightarrow products \tag{3}$$

Alternatively, P can be oxidized directly by valence band holes:

$$P + h^+_{VB} \rightarrow P^{+\cdot} \tag{4}$$

followed by the hydrolysis of the intermediate radical cation $P^{+\cdot}$:

$$P^{+\cdot} + H_2O \rightarrow P^{\cdot}\text{-}OH + H^+_{aq} \tag{5}$$

thereby leading to the same products.

Fox et al. have attempted to identify the hydroxy adducts ($P^{\cdot}$-OH) directly by their characteristic transient absorption spectrum using a laser flash photolysis technique combined with diffuse reflectance measurements [50, 51]. Employing different probe molecules, such as Pt^{2+}, phenothiazine, or methylviologen, they were unable to detect any transient signals which could be explained by $\cdot OH$-adduct formation. However, these observations do not prove that hydroxyl radicals are not involved in redox processes on metal oxide particles since acid-base catalysis favored on the highly hydroxylated surfaces might result in the rapid depletion of the $\cdot OH$-adducts before they could be detected within the time-resolution of the kinetic apparatus.

An alternative attempt to answer the question of the intermediacy of $\cdot OH$ radicals in photocatalytic systems is to employ probe molecules which will clearly undergo different reaction pathways depending upon whether the initial oxidative attack is by a hydroxyl radical or by a hole, respectively. As will be shown in the following, the acetate molecule appears to fulfill these conditions. It has been established in detailed radiation chemical investigations that hydroxyl radicals attack acetate ions mainly at the methyl group [52]:

$$\cdot OH + CH_3COO^- \rightarrow \cdot CH_2COO^- + H_2O \tag{6}$$

In the presence of air the radicals thus formed react quickly with molecular oxygen, leading to a rather complex product distribution with glycolate and glyoxalate being some of the major products [53]. On the other hand, the direct one-electron oxidation of acetate, e.g., on Pt electrodes, results in the well-known Kolbe decarboxylation with the formation of methyl radicals:

$$h^+_{VB} + CH_3COO^- \rightarrow CH_3COO^{\cdot} \rightarrow \cdot CH_3 + CO_2 \tag{7}$$

A considerably different product distribution results when these methyl radicals react with O_2 [54], i.e., both, glycolate and glyoxalate, are no longer formed. Acetate is indeed readily degraded when oxygen-containing suspensions of TiO_2 and acetate are illuminated with near-UV light [55]. While the resulting degradation rates depend strongly on the pH-value of the solution, the main products in alkaline solution (pH 10.6) are glycolate and formate. Glyoxalate is also detected, however, with a maximum yield being reached at relatively short illumination times. Glycolate and glyoxalate are also both readily oxidized in the presence of TiO_2 under bandgap illumination [55]. While formate is the main ionic product of the photocatalytic oxidation of glycolate, glyoxalate is detected as a primary oxidation product over the whole pH range studied. This can be taken as further indication of the intermediacy of hydroxyl radicals since a decarboxylation of glycolate should only result in C1-products such as methanol or formate.

The formation of glycolate and glyoxylate can be taken as strong evidence for the oxidation of acetate via hydroxyl radicals. The relative importance of this reaction path seems to be higher with increasing pH. In alkaline solution the surface of the photocatalyst TiO_2 is negatively charged (pH of zero point of charge for TiO_2: $pH_{zpc} = 6.0 - 6.4$ [56]) and the resulting electrostatic repulsion should hinder the adsorption of the negatively charged carboxyl group of the acetate molecule, thus favoring an attack of surface-bound hydroxyl radicals onto the methyl group. On the other hand, negatively charged carboxyl groups are directed towards positively charged surface groups of the semiconductor particles at pH values below the pH_{zpc} and an attack leading to the subsequent decarboxylation of the acetate molecule will be favored.

On the other hand, Carraway et al. [22] have obtained experimental evidence for the direct hole oxidation of tightly-bound electron-donors such as formate, acetate, and glyoxylate at the semiconductor surface. In the case of the gem-diol form of glyoxylate, the photocatalytic oxidation appears to proceed via a direct hole transfer (i.e., electron transfer from the surface-bound substrate) to form formate as a primary intermediate as follows:

$$HCOCO_2^- + H_2O \leftrightarrow HC(OH)_2CO_2^- \tag{8}$$

$$HC(OH)_2CO_2^- + h^+_{VB} \rightarrow HC(OH)_2CO_2 \tag{9}$$

$$HC(OH)_2CO_2^\bullet \rightarrow HC(OH)_2^\bullet + CO_2 \tag{10}$$

$$HC(OH)_2^\bullet + h^+_{VB} \rightarrow HCO_2^- + 2H^+ \tag{11}$$

Grabner et al. [57] have used time-resolved absorption spectroscopy to demonstrate the formation of phenoxyl and $Cl_2^{\bullet-}$ radical anions in the photo-oxidation of phenol on TiO_2 colloids. The formation of $Cl_2^{\bullet-}$ is postulated to occur by the direct valence-band hole oxidation of Cl^-, which was introduced into the solution as HCl to adjust the pH. These investigators used laser pulse energy-dependent measurements of transient formation to determine oxidation quantum yields and concentrations of surface-adsorbed Cl^-. Richard [58] has argued that both holes and hydroxyl radicals are involved in the photooxidation of 4-hydroxybenzylalcohol (HBA) on ZnO or TiO_2. Her results suggest positive holes and hydroxyl radicals have different regioselectivities in the photocatalytic transformation of HBA. Hydroquinone (HQ) is thought to result from the direct oxidation of HBA by h^+_{VB}, dihydroxybenzylalcohol (DHBA) from the reaction with $^\bullet$OH, while 4-hydroxybenzaldehyde (HBZ) is produced by both pathways. In the presence of isopropanol (iPrOH), used as $^\bullet$OH quencher, the formation of DHBA is completely inhibited and the formation of HBZ is partly reduced.

3
Kinetic Aspects of Photocatalytic Processes

In principle the same reactions occur at semiconductor particles and electrodes. Differences exist only insofar as at particles two reactions proceed simultaneously upon excitation, a reduction by electron transfer and an oxidation by hole transfer, whereas at electrodes only one process occurs if the electrode is polarized [19]. Accordingly, the state of particles corresponds to the open circuit condition of a semiconductor electrode, i.e., the charge carriers are not separated by any band bending and will consequently both have to react via the same semiconductor surface to avoid their undesired recombination. This makes the analysis of the mechanism for reactions at particles much more difficult, especially, since it is always the slowest process that determines the overall rate.

The primary processes occurring upon bandgap irradiation of extremely small titanium dioxide particles (diameter $d \approx 2.5$ nm) have recently been studied extensively, employing ultrafast laser flash photolysis equipment with picosecond [59] or even subpicosecond time-resolution [60, 61]. While quantitative discrepancies were apparent which could, for example, be explained by slightly different preparation techniques of the TiO_2 colloids, the authors agreed principally on the underlying qualitative concepts of photocatalysis. Following their generation within the short light pulse, both electron and hole are extremely rapidly trapped in surface states of the semiconductor particle. As has been shown earlier, the short wave-length absorption corresponds to the absorption of trapped holes and the long wavelength absorption to trapped electrons [62, 63]. Since the trapped electron exhibits a strong transient optical absorption around 650 nm while the trapped hole absorbs predominantly at shorter wavelengths, i.e., around 430 nm, a very broad, featureless transient absorption spectrum ranging from 400 to 800 nm is consequently observed immediately following the laser flash. The recombination kinetics of the trapped charge carriers have been studied in detail by both groups of authors [59–61].

Concerning the nature of the electron and hole trapping centers, several problems arise. Assuming that the absorptions of the trapped electrons around 650 nm (1.7 eV) and of holes at 430 nm (2.7 eV) correspond to transitions between classical surface states and the conduction and valence band, respectively, these surface states would have to be located near the middle of the bandgap. This assignment, however, cannot be correct since the reduction of O_2 would not be thermodynamically possible if the electrons originated from an energy state 1.7 eV below the conduction band of titanium dioxide. Accordingly, it has been proposed that the transient absorption around 650 nm is due to an excitation of a trapped electron within a surface molecule such as, for instance, a hydrated Ti(III)-molecule ($t_{2g} \rightarrow e_g$ transition) [64]. In the case of the trapped holes the corresponding optical transition may be correlated to an excitation within a peroxide formed at the surface [64].

Studying the decay kinetics of these transient absorption spectra in the presence of molecular oxygen, air, or molecular nitrogen, respectively, a rate constant $k_{12} = 7.6 \times 10^7 \, l \, mol^{-1} \, s^{-1}$ has been determined for the reaction

$$e^-_{tr} + O_2 \rightarrow O_2^{\bullet-} \tag{12}$$

of the trapped electrons with molecular oxygen [64]. This rate constant is considerably smaller than that reported for the reaction of hydrated electrons (e^-_{aq}) with molecular oxygen in homogenous aqueous solutions (i.e., $k = 1.9 \times 10^{10} \, l \, mol^{-1} \, s^{-1}$ [65]). However, this difference is readily explained by the much more negative one-electron redox potential of e^-_{aq} (i.e., $E = -2.87 \, V$ (vs NHE) [66]) compared with the value for the flat-band potential of a colloidal TiO_2 particle (i.e., $E = -0.12 - 0.059*(pH) \, V$ (vs NHE) [67]). Hence, while the hydrated electrons possess a considerable driving force for the reduction of molecular oxygen to form the superoxide radical $O_2^{\bullet-}$ ($E = -0.33 \, V$ (vs NHE) [68]), at pH 3 the conduction band electrons of the colloidal TiO_2 particles barely reach a one-electron potential negative enough to reduce O_2. It is in general problematic to use values of redox potentials derived in homogenous solutions to explain processes occurring at interfaces. However, it has for example be shown by Grätzel et al. that the one-electron redox potentials of various viologens determined in homogenous aqueous solution can satisfactorily be employed to determine, e.g., the flat-band potential of semiconductor particles yielding values which are in good agreement with those of the corresponding bulk electrodes [67]. Thus, values of redox potentials derived from homogenous studies have generally been used with the assumption that this treatment yields the best possible approximation of the interfacial situation.

The following mechanistic scheme has recently been deduced from the flash photolytic investigations of pure aqueous TiO_2 suspensions. The electron/hole pairs formed upon laser light absorption (reaction at Eq. 13) are trapped almost instantaneously (reactions at Eqs. 14 and 15):

$$TiO_2 + h\nu \rightarrow e^- + h^+ \tag{13}$$

$$e^- \rightarrow e^-_{tr} \tag{14}$$

$$h^+ \rightarrow h^+_{tr} \tag{15}$$

Since in the presence of molecular oxygen a certain amount of O_2 will be adsorbed at one of the active TiO_2 surface sites, electrons can also be trapped at these sites (reaction at Eq. 16). This process results in the relative increase in the concentration of h^+_{tr} as compared with that of e^-_{tr}:

$$O_{2,ads} + e^- \rightarrow O_{2,ads}^{\bullet-} \tag{16}$$

While most trapped charge carriers recombine quickly, i.e., within less than 200 ns (reaction at Eq. 17), a minority survives, possibly in different and/or deeper traps and will be available for reactions with substrates such as O_2 (reaction at Eq. 12):

$$e^-_{tr} + h^+_{tr} \rightarrow TiO_2 \tag{17}$$

Bahnemann et al. proposed two different trap sites for holes on the surface of the TiO_2 particles [64]. While holes which are trapped in energetically deep traps, h^+_{tr}, are characterized by their transient absorption around 450 nm, those initially residing in shallow traps, h^+_{tr*}, do not possess such spectral features. Following their generation all holes are rapidly trapped in either of these energy states (reactions at Eqs. 15a and 15b):

$$h^+ \rightarrow h^+_{tr} \tag{15a}$$

$$h^+ \Leftrightarrow h^+_{tr*} \tag{15b}$$

The holes trapped in shallow traps can be excited thermally into the valence band resulting in an equilibrium with the free holes as indicated in the reaction at Eq. (15b). Shallowly trapped holes, h^+_{tr*}, will therefore have a comparable reactivity as detrapped holes, h^+. Both types of trapped holes will recombine with the trapped electrons within the first 200 ns after their generation following the reaction at Eq. (17); only holes excited thermally from the shallow traps have the chance to migrate to the energetically more favored h^+_{tr} site (cf. reaction at Eq. 18):

$$h^+_{tr*} \rightarrow h^+_{tr} \tag{18}$$

Since the trapped electrons can alternatively also react with O_2 (reaction at Eq. 12) and a certain fraction of electrons is originally trapped at $O_{2,ads}$ sites (reaction at Eq. 16), a surplus of h^+_{tr} remains after the completion of these processes.

When the dynamics of the photocatalytic oxidation of the model compound dichloroacetate (DCA^-) were studied in detail it was concluded that h^+_{tr} do not react with DCA^-, although either free holes are directly transferred to (adsorbed) A^- molecules (reaction at Eq. 19) or shallowly trapped h^+_{tr*} are detrapped (cf. reaction at Eq. 15b) to react with dichloroacetate in the nanosecond timescale via reaction at Eq. (19) [64]:

$$h^+ + DCA^- \rightarrow DCA^• \tag{19}$$

A similar reactivity of h^+_{tr} had previously been reported by Bahnemann et al. who studied reactions in colloidal TiO_2/Pt suspensions with an average particle diameter of approximately 12 nm [62,63]. While the addition of ethanol as a hole scavenger resulted in a considerable increase of the rate of disappearance of the h^+_{tr}-absorption, the addition of citrate and acetate mainly led to a decrease of its initial absorption height. It was concluded that strongly adsorbed ionic species would primarily react with free holes while weakly adsorbed molecules will mainly react with long-lived h^+_{tr} in a diffusion-controlled process. It has recently been proposed that deeply trapped h^+_{tr} are chemically equivalent to surface-bound hydroxyl radicals while weakly trapped holes apparently possess an electrochemical potential close to that of free holes and should therefore be considered to be chemically similar to the latter. These shallow traps are most likely created by lattice imperfections of the semiconductor nanocrystals.

A kinetic model has been derived to analyze the time-resolved observations during the one-electron oxidation of DCA^- assuming that the direct charge transfer to dichloroacetate proposed in the reaction at Eq. (19) requires that the trapping molecules are adsorbed on the TiO_2 surface prior to the absorption of the photon [64]. Otherwise, this reaction would not be able to compete with the normal hole trapping reactions at Eqs. (15a) and (15b). The equilibrium concentration of adsorbed DCA^-, $[DCA^-_{ads}]$, can be calculated from the total concentration of surface hydroxyl groups, $[\equiv TiOH_{tot}]$, following the reaction at Eq. (20) [56]:

$$[DCA^-_{ads}] = \frac{K_{DCA}[DCA^-]}{1 + K_{DCA}[DCA^-]} [\equiv TiOH_{tot}] \tag{20}$$

A detailed FTIR study has shown that DCA^- forms a bidentate surface complex over almost the entire pH regime with $K'_{DCA^-} = K_{DCA^-}$ $[\equiv TiOH_{tot}] = 0.02$ measured at pH 2.5 [56]. Assuming that the absorption A of the transient measured at 500 nm is proportional to the concentration of trapped holes, i.e., $A_{500} = \gamma p_{tr}$ (with γ being the absorption coefficient of the trapped holes), and defining $p_{tr,0}$ as the concentration of h^+_{tr} at $[DCA^-] = 0$, the following kinetic equation has been derived to describe the direct hole trapping at adsorbed dichloroacetate molecules [64]:

$$\frac{1}{(A_{500})^{-1} - (A_{500,0})^{-1}} = \frac{\gamma k'_{tr}}{k[\equiv TiOH_{tot}]} + \frac{\gamma k'_{tr}}{k K_{DCA}[\equiv TiOH_{tot}]} \frac{1}{[DCA^-]} \tag{21}$$

(with $A_{500,0}$: absorption of transient at 500 nm in the absence of DCA^-)

Figure 2 shows that a plot of $(1/A_{500} - 1/A_{500,0})^{-1} = f(1/[DCA^-])$ indeed exhibits the linear relationship predicted by Eq. (21).

Division of the intercept value from Fig. 2 (3.8×10^{-4}) by the slope of the straight line (1.7×10^{-4} mol l^{-1}) yields, according to Eq. (21), $K_{DCA^-} = 2.2$ l mol^{-1}.

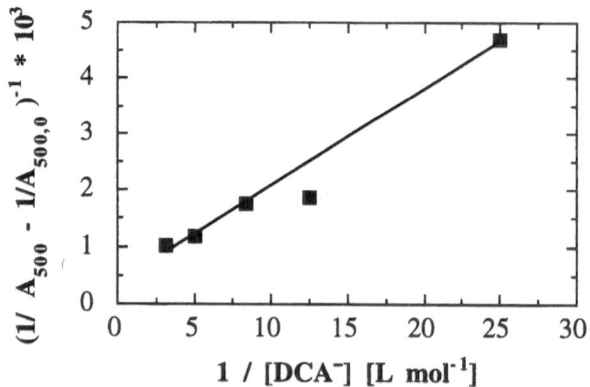

Fig. 2. Initial transient absorption observed upon laser excitation ($\lambda_{ex} = 355$ nm) at 500 nm in the presence of various DCA^- concentrations: pH 2.0; 1.0×10^{-4} mol l^{-1} colloidal TiO_2/Pt (1%) particles; absorbed photon concentration per pulse, 1.6×10^{-5} mol l^{-1}; air-saturated solution; calculated via Eq. (21) and plotted as a function of the reciprocal dichloroacetate concentration (adopted from [64])

Taking $[\equiv TiOH_{tot}] = 4 \times 10^{-3} \, mol \, l^{-1}$ [56], $K'_{DCA^-} = 0.01$ is readily calculated which agrees very well with the value determined from totally independent adsorption studies of DCA^- on TiO_2 (K'_{DCA^-} (pH 2.5) = 0.02 [56]) supporting the derived kinetic model. Hence, the direct one-electron oxidation of DCA^- immediately follows the hole transfer from the bulk to the TiO_2 surface while surface-bound hydroxyl radicals are apparently unreactive towards DCA^-.

A similar kinetic model has been used to analyze the one-electron oxidation of another model compound, namely thiocyanate ions (SCN^-), by holes (reaction at Eq. 22):

$$SCN^-_{ads} + h^+ \xrightarrow{\ k_{22}\ } SCN^\bullet_{ads} \tag{22}$$

with $k_{22}' = 6 \times 10^5 \, s^{-1}$ being calculated from the respective experimental data under the assumption that the electron/hole recombination occurs with a rate constant of $k_{rec} = 3 \times 10^7 \, s^{-1}$ as determined by Rothenberger et al. [17] ($k_{22}' = k_{22}$ SCN^-_{ads}) [64]. Since the hole transfer to an adsorbed molecule occurs, this process should be adiabatic. According to the theory by Marcus the electron transfer rate constant, k_{ET}, for an adiabatic process can be derived from Eq. (23) [69]:

$$k_{ET} = \kappa \, \nu \exp \left[- \frac{(E_v - E_{F,redox} - \lambda)^2}{4 \, kT\lambda} \right] \tag{23}$$

(with $\kappa \cong 1$: transmission coefficient; $\nu \cong 10^{13} \, s^{-1}$: frequency of molecular vibration in the exited state; $E_V \cong +3.0$ eV: flat band potential of holes in TiO_2; $E_{F,redox} = 1.33$ eV vs NHE: redox potential for $SCN^\bullet/SCN^-$ [66]; λ: reorientation energy).

Taking $k_{ET} \cong k_{22}' = 6 \times 10^5 \, s^{-1}$ a reorientation energy $\lambda = 0.64$ eV is calculated for an adsorbed thiocyanate anion. A slightly higher but still reasonable value for the orientation energy (i. e., $\lambda = 0.78$ eV) is computed when shorter electron/hole lifetimes as recently reported by Bowman et al. [60, 61] are used for this calculation.

4
Conventional and Novel Photocatalysts

Anatase, titanium dioxide (TiO_2), the material with the highest photocatalytic detoxification efficiency, is a wide bandgap semiconductor ($E_g \sim 3.2$ eV [70]). Thus, only light below 400 nm (i. e., 5 % of the solar energy reaching the surface of the earth) is absorbed and capable of forming electron-hole pairs which are a prerequisite for the photocatalytic process. In recent years, Degussa P25 TiO_2 has set the standard for photoreactivity in environmental applications. P25 is a non-porous 70:30 anatase:rutile mixture with a BET surface area of $55 \pm 15 \, m^2 \, g^{-1}$ and crystallite sizes of 30 nm in 0.1 μm diameter aggregates. The so-called photonic efficiency ζ is generally used to compare the activity of different photocatalysts. It is usually derived from the experimental data as the ratio of the photocatalytic degradation rate (in units $mol \, l^{-1} \, s^{-1}$) and the incident light intensity following a recently published definition [71]. In most cases, the degradation rate is calculated from the initial slope of the individual concentration vs time profiles.

Many researchers claim that rutile is a catalytically inactive [72–74] or much less active form [37, 43, 75, 76] of TiO_2, while others find that rutile has selective activity toward certain substrates. Highly annealed ($T \geq 800\,°C$) rutile appears to be photoinactive [72–74] in the case of 4-chlorophenol oxidation. However, Domènech [77] has shown that TiO_2 in the rutile form is a substantially better photocatalyst for the oxidation of CN^- than is the anatase form; on the other hand, he also showed that Degussa P25 was a better catalyst than rutile for the photoreduction of $HCrO_4^-$ [78, 79].

Tanaka et al. [80] have shown that the photocatalytic degradation of several compounds over different mineral phases and preparation methods of TiO_2 was dependent upon the calcination temperature for some samples and independent for others. They found that for trichloroethylene degradation in water the rate of photodegradation increased with calcination temperature up to 500 °C or in some cases up to 600–700 °C and then decreased above those temperatures. They also noted that commercial anatase forms (Degussa P25 and TP-2) were better for Cl_2CCClH degradation than commercially available rutile (Katayama, TP-3, and TM-1) and that specific surface area did not appear to be a determining factor. Tanaka et al. concluded that calcined synthesized anatase was better than P25 and that both of these types were better than 100% rutile. However, when hydrogen peroxide was added as an electron acceptor, rutile showed greater photocatalytic activity.

Martin et al. [74] report an increase in photodegradation rates of 4-chlorophenol as the anatase form of TiO_2 is calcined progressively from 100 to 400 °C (i.e., the particles calcined at 400 °C yield the highest photodegradation rates) and then a decrease in photodegradation rate was noted for samples calcined above 500 °C. For comparison, the photonic efficiency was found to be 0.23 for anatase (400 °C) and 0.03 for rutile (800 °C).

Recently, the photocatalytic activities of several TiO_2 materials have been compared by Lindner et al. [81] studying the degradation of an aliphatic and an aromatic model compound under optimized experimental conditions (e.g., $c(TiO_2)$, pH of the solution). For dichloroacetic acid (DCA) a new commercially available anatase catalyst, i.e., Sachtleben Hombikat UV100, yielded significantly higher photonic efficiencies ($\zeta = 15\%$) than almost all the other materials, especially in comparison to rutiles. It was noted that for an unprejudiced comparison of different photocatalysts an optimization of the catalyst concentration and the pH of the solution is generally important [11, 82]. The optimum catalyst concentration depends mainly on the catalyst, i.e., while for P25 a maximum activity was reached at $0.5\,g\,l^{-1}$, for Hombikat UV100 a significant rise of the photonic efficiency was observed even beyond $5\,g\,l^{-1}$ [81]. The optimum pH is often found to be specific for the pollutant; while DCA is most rapidly degraded in acidic solution (pH 3) [83], alkaline conditions (pH 11) result in highest yields for chloroform [21]. For 4-chlorophenol (4-CP), however, the observed discrepancies are usually relatively small and the degradation is generally slower ($\zeta < 1\%$) [81]; as will be shown in detail below, in this case the formation of intermediates has to be considered too. It is important to note at this point that meaningful results for this pH-dependence can only be generated when the pH is kept constant throughout the experiment. While the addition of buffers for this purpose has to be done with extreme care to avoid competitive reactions with the buffer's constituents, the so-called pH-stat technique appears to be superior in that it minimizes the influence on other experimental parameters [83]. Employing the latter method, constant pH is maintained throughout the entire duration of the experiment by the automatic addition of acid or base as hydroxyl ions or protons are formed, respectively. Under the condition that the reaction mixture contains an inert salt in a sufficiently high concentration, the ionic strength will also not change notably during this treatment.

In order to enhance interfacial charge-transfer reactions of TiO_2 bulk-phase and colloidal particles, the properties of the particles have been modified by selective surface treatments such as surface chelation [41, 84], surface derivatization [85], platinization [81, 86 – 87], and by selective doping of the crystalline matrix [74 – 75, 88 – 106]. For the model compounds introduced above it could be shown that while for DCA the addition of $CuNO_3$ (1 – 10 mmol l^{-1}) or a platinization of the catalyst lead to a significant raise in ζ, this had almost no effect on the degradation of 4-CP [81]. For this model compound ζ could only be accelerated by the addition of bromate (3 mmol l^{-1} $KBrO_3$); the photocatalytic nature of the process could be demonstrated. The addition of other oxidants had no beneficial effect, and in some cases even an inhibition was observed. The stability and long time activity of standard and platinized TiO_2 materials could be demonstrated.

$Fe(III)$ doping of TiO_2 has been shown to increase the photonic efficiency for the photoreduction of N_2 [91, 94, 95] and of methylviologen [90] and to inhibit electron-hole pair recombination [107 – 109], while in the case of phenol degradation, $Fe(III)$ doping was reported to have little effect on efficiency [94, 95]. Enhanced photoreactivity for water cleavage [110] and N_2 reduction [95] with $Cr(III)$-doped TiO_2 has been noted, although others [111, 112] have found the opposite effect with $Cr(III)$-doping. Negative effects of doping [98] have been noted for Mo and V in TiO_2, while Grätzel and Howe [89] note an inhibition of the electron/hole recombination with the same dopants. Karakitsou and Verykios [75] reported that doping TiO_2 with cations of higher valency than that of $T(IV)$ resulted in enhanced photoreactivity, while Mu et al. [112] noted that doping with trivalent and pentavalent cations was actually detrimental to the photoreactivity of TiO_2.

Colloidal Ti/Fe mixed oxide nanoparticles were recently synthesized in aqueous solution with the Fe^{3+} content varying from 0.05 up to 50 at% and particle diameters between 4 and 6 nm [106]. The absorption spectra of these particles could be shifted considerably into the visible part of the solar spectrum with increasing content of ferric ions. The photophysical properties of these newly synthesized particles were examined in detail. Steady state illumination experiments showed a reversible bleaching of the absorption spectrum in the presence of a hole scavenger indicating that a negative excess charge could be stored. The photocatalytic activity of the Ti/Fe mixed oxide particles was compared to that of pure colloidal TiO_2 by studying the photo-

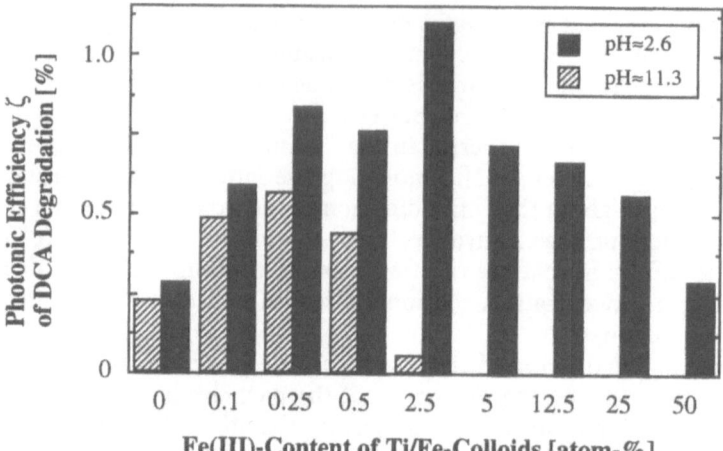

Fig. 3. Photonic efficiency ζ of the photocatalytic degradation of dichloroacetic acid with $0.5\,g\,l^{-1}$ Ti/Fe mixed oxide colloids of different iron content, $2.5 \times 10^{-3}\,mol\,l^{-1}$ DCA, O_2-sat. (adopted from [106])

catalytic degradation of the model compound dichloroacetic acid ($Cl_2HCCOOH$, DCA). The photonic efficiency, ζ, of the DCA degradation employing various mixed oxide colloids as photocatalysts is shown in Fig. 3 as a function of the respective iron content at pH ≈ 2.6 and pH ≈ 11.3.

The data presented in this figure show that at pH ≈ 2.6 the efficiency of the destruction of dichloroacetic acid obtained with the Ti/Fe-mixed oxide particles is always higher than for pure TiO_2 colloids. Also the yield increases strongly at low iron contents reaching a maximum at 2.5 at% Fe^{3+}. At this point the photonic efficiency is almost four times higher than for pure TiO_2 colloids. Therefore, it has been proposed by Bockelmann et al. [106] that in acidic solution small iron contents inhibit the recombination of photogenerated charge carriers. At pH ≈ 11.3 the photonic efficiencies of the destruction of dichloroacetic acid are generally smaller than at pH ≈ 2.3. This effect was explained by the pH-dependence of the adsorption of dichloroacetic acid on the TiO_2 surface. In acidic solution the metal oxide surface is positively charged whereas dichloroacetic acid is deprotonated ($pK_A = 1.48$) and therefore negatively charged and should thus be well adsorbed.

It is also obvious from the data shown in Fig. 3 that at pH ≈ 11.3 the efficiency of the detoxification of dichloroacetic acid increases only slightly up to 0.25 at% Fe^{3+} and subsequently decreases with increasing iron content. Already at a content of 5 at% iron there is no detectable detoxification in alkaline solutions. A possible explanation for this observation is that the recombination of charge carriers is the predominant process in alkaline solution. Assuming that the trapping of the photogenerated electrons at surface states forming Ti^{3+} or Fe^{2+} suppresses this recombination, it has been concluded that in alkaline solution the iron on the TiO_2 surface cannot be reduced to Fe^{2+} by the photogenerated

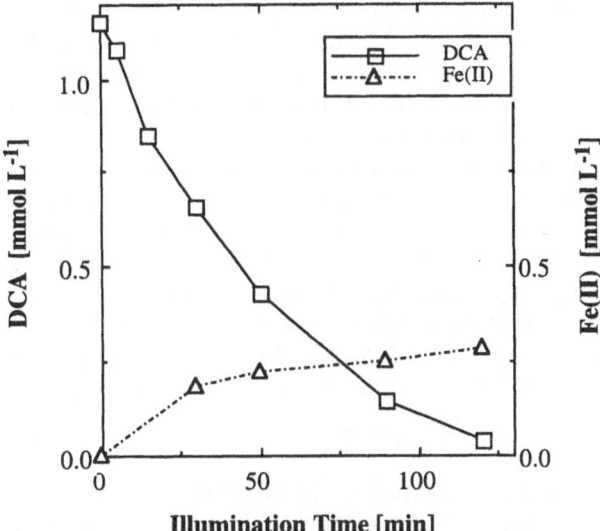

Fig. 4. Degradation of dichloroacetate and formation of Fe(II) during the illumination of an O_2-saturated colloidal aqueous suspension containing $0.5\,g\,l^{-1}$ Ti(IV)/Fe(III) mixed oxide powder (50 at% Fe(III)) at pH 2.5, illumination with white light (320 nm filter) (adopted from [113])

electrons. Moreover, the amount of titanium-atoms on the surface is decreasing with increasing iron content compared to pure TiO_2. Thus, the probability for the photogenerated electrons to be trapped at Ti-surface states (Ti^{3+}) will also be reduced resulting in an enhanced electron/hole recombination.

To obtain further insight into the degradation mechanism of these mixed oxide photocatalysts it is useful to examine the individual degradation kinetics in more detail. Figure 4 thus shows the results observed by Bockelmann et al. [113] during the illumination of an O_2-saturated colloidal aqueous suspension containing $0.5\,g\,l^{-1}$ Ti(IV)/Fe(III) mixed oxide powder (50 at% Fe(III)) at pH 2.5. While dichloroacetate (DCA) is degraded from an initial concentration of $1.2\,mmol\,l^{-1}$ almost to zero within the experimental duration of 2 h, nearly $0.3\,mmol\,l^{-1}$ Fe(II) are formed simultaneously indicating a partial reductive dissolution of the catalyst material via

$$Ti(IV)/Fe(III)O_x + e_{CB}^- \rightarrow Ti(IV)O_x + Fe^{2+}_{aq} \qquad (24)$$

Alternatively, conduction band electrons (e_{CB}^-) can be scavenged by the molecular oxygen present in solution leading eventually to the generation of peroxides [114]:

$$O_2 + e_{CB}^- \rightarrow O_2^{\bullet-} \rightarrow \rightarrow H_2O_2 \qquad (25)$$

Hence, the cathodic dissolution of the catalyst should be competing with the desired electron transfer steps such as the reduction of dioxygen. It is well known that O_2 is a poor oxidant for conduction band electrons in most metal oxides [114]. This is readily explained by a rather low lying conduction band edge and a one-electron redox potential of O_2 which is even slightly more negative [114]. It has recently been shown that peroxides are only formed in a reasonable yield by the reduction of molecular oxygen when ZnO is employed as the semiconducting metal oxide while only very small yields were observed with TiO_2 [114]. In the case of iron(III) oxides (hematite) no peroxides were detected in analogous experiments but the catalyst was readily dissolved [115]. Obviously, a better electron acceptor than O_2 should be employed to compete more efficiently with the catalyst dissolution. Hydrogen peroxide (H_2O_2) has been proposed as an appropriate reactant for this purpose, since initial experiments indicate that the Fe(II) release is suppressed when H_2O_2 is added to the colloidal suspension [113]. However, further investigations are required to decide whether H_2O_2 is competing for the conduction band electrons via

$$H_2O_2 + e_{CB}^- \rightarrow {}^\bullet OH + OH^- \tag{26}$$

thereby preventing the dissolution of the mixed oxide photocatalyst, or whether the main process involves the reoxidation of free Fe^{2+}_{aq} which should eventually also lead to the destruction of the catalyst or at least to the reduction of its photocatalytic activity:

$$Fe^{2+}_{aq} + H_2O_2 \rightarrow Fe^{3+}_{aq} + {}^\bullet OH + OH^- \tag{27}$$

The reaction at Eq. (27) is the well known Fenton reaction which, however, has a rather small rate constant [116]. It is therefore conceivable that scavenging of e_{CB}^- by H_2O_2 via the reaction at Eq. (26) could be rather efficient. A detailed investigation is, however, indicated to differentiate between these mechanisms.

The main incentive to synthesize novel photocatalysts with a hypochromically shifted absorption spectrum came from the lack of photocatalytic activity of pure titanium dioxide above 400 nm. The success of the newly synthesized Ti(IV)/Fe(III) mixed oxides will therefore strongly depend on their ability to act as photocatalysts in the visible part of the solar spectrum. Hence, illumination experiments have also been performed at 436 nm showing that the Ti(IV)/Fe(III) mixed oxide particles containing 50 at% Fe(III) indeed exhibit photocatalytic activity above 400 nm [113]. However, the catalyst dissolution also presented a problem when illuminations were carried out in the visible. If this anodic dissolution of the catalyst particles could be suppressed, e.g., by the use of other oxidants such as hydrogen peroxide, it could be envisaged that these newly synthesized materials could become rather promising photocatalysts in particular for solar applications.

Choi et al. [109, 117, 118] have recently shown that selectively-doped colloidal titanium dioxide particles have a much greater photoreactivity as measured by their quantum efficiencies for oxidation and reduction than their undoped counterparts. They presented the results of a systematic study of the effects of 21 different metal-ion dopants on the photochemical reactivity of quantum-sized TiO_2 with respect to both chloroform oxidation and carbon tetrachloride reduction. Their results were summarized in terms of a periodic chart of dopant effects on oxidation and reduction quantum yields. Enhanced photoactivity was seen for Fe (III), Mo (V), Ru (III), Os (III), Re (V), V (IV), and Rh (III) substitution for Ti(IV) at the 0.5 atom%-level in the TiO_2 matrix. The maximum enhancements were 18-fold (CCl_4 reduction) and 15-fold ($CHCl_3$ oxidation) increases in quantum efficiency for Fe (III)-doped colloidal TiO_2.

Choi et al. [109] also showed by laser flash photolysis measurements that the lifetime of the blue electron (i.e., the electron trapped at Ti(III)-sites (see above)) in the Fe (III)-, V (IV)-, Mo (V), and Ru (III)-doped samples was increased to 50 msec while the measured lifetimes of this blue electron in undoped TiO_2 particles were < 200 μs [109, 117, 118]. They were also able to show that the measured quantum efficiencies for oxidation and for reduction could be correlated to the measured transient absorption signals of the charge-carriers. In general, an increase in the relative concentration of the charge-carriers results in a corresponding increase of the photoreactives. However, if an electron is trapped in a deep trapping site, it will have a longer lifetime but it may also have a lower redox potential which could lead to a decreased photo-reactivity. The photoreactivity of doped TiO_2 thus appears to be a complex function of the dopant concentration, the energy level of the dopants within the TiO_2 lattice, their d-electronic configuration, the distribution of dopants, the electron donor concentration, and the light intensity.

5
Mechanisms of the Photocatalytic Degradation of Model Compounds

5.1
Reductive Pathways

It is interesting to note that, while many laboratory studies have demonstrated the feasibility of the method of photocatalytic detoxification for almost all classes of hazardous chemicals, almost all of these studies have been directed towards the oxidative degradation of organic pollutants in water. Only in a few cases has a reductive pathway been discussed as the initiating step of photo-catalytic conversions.

Perhalogenated hydrocarbons, however, often contain the carbon atoms in their highest oxidation state and can therefore not be attacked oxidatively. Following an early report by Güsten and co-workers on the photocatalytic degradation of chlorofluoromethanes on zinc oxide surfaces in the gas phase [119], Ollis et al. [120, 121], Bahnemann et al. [122], and Sabin et al. [123] have shown that tetrachloromethane can be photocatalytically degraded in aqueous TiO_2-suspensions. Photonic efficiencies and mechanistic details have, however, not been investigated by these authors. In the case of halothane (2-bromo-2-chloro-1,1,1-trifluoroethane) it could be shown that reduction processes on colloidal TiO_2-particles can be very efficient with quantum yields for the elimination of bromide and fluoride ions reaching $\Phi = 0.43$ when photo-platinized photocatalysts were used [124]. Recently, the reaction of conduction band electrons generated by light absorption in TiO_2-particles with tetra-bromomethane [125], bromoform [126], and various transition metal ions [127–131] has been studied.

Using tetrachloromethane, CCl_4, as a model pollutant, very detailed mechanistic studies have been carried out recently [132–134]. Hilgendorff et al. [132, 133] observed that protons and chloride ions are formed when oxygen-containing aqueous suspensions of bare or platinized TiO_2 particles and CCl_4 are illuminated with light of sufficient energy at pH 11, while the concentrations of O_2 and CCl_4 are decreasing. The following overall stoichiometry was reported in these studies:

$$CCl_4 + 2H_2O \xrightarrow{\text{TiO}_2/h\nu} CO_2 + 4H^+ + 4Cl^- \qquad (28)$$

while the mechanism which has been proposed to explain details of these studies will be described below.

Apparently, the photocatalytic degradation of tetrachloromethane behaves like in an "ideal" photocatalytic process which is not restricted by adsorption/desorption phenomena, mass-transfer limitations, or diffusion control, since it obeys zero order kinetics with the rate increasing linearly with the substrate concentration [133]. While in air-saturated aqueous suspensions containing $1\,\text{mmol}\,l^{-1}$ of the pollutant molecule, a maximum photonic efficiency of only $\zeta_{max} = 0.0025$ has been observed in this study at pH 11, although considerably higher photonic efficiencies were reported when higher substrate concentrations, higher pH values, and/or various hole scavengers were employed. Moreover, when the concentration of molecular oxygen within the suspension was decreased, an almost ten-fold increase in the photonic efficiency was observed.

An electrochemical model has been proposed to explain the marked pH-dependency of the photocatalytic CCl_4 degradation [133, 134]. The pH-independent one-electron reduction potential of tetrachloromethane has been determined to be $-0.54\,\text{V}$ vs NHE [135], while the flatband potential of the conduction band electrons in single crystal (rutile) TiO_2 electrodes at pH 0 has been reported to be $-0.1\,\text{V}$ vs NHE [136]. It is normally assumed that the conduction band electrons of the TiO_2 (anatase) particles have the same value for the flatband potential as rutile electrodes and also exhibit a Nernstian behavior to calculate their potential at any given pH via

$$E = -0.1\,\text{V} - 0.059\,\text{V} * pH \qquad (29)$$

At pH 8, for example, $E = -0.572\,\text{V}$ is readily calculated using Eq. (29). A comparison with the one-electron reduction potential of CCl_4 evinces that the reduction potential of the conduction band electrons of TiO_2-particles should be sufficient to reduce tetrachloromethane at and above pH 8:

$$CCl_4 + e^-_{CB} \rightarrow {}^{\bullet}CCl_3 + Cl^- \qquad (30)$$

Indeed, while hardly any reactivity has been observed in acidic medium, considerable degradation efficiency is evident at pH ≥ 10 [133]. It is interesting to note that the rate of CCl_4 degradation is not enhanced by the presence of Pt-islands which are often known to reduce the overpotential of heterogeneous electron transfer processes, e.g., for the reduction of protons to molecular hydrogen [19]. This absence of any catalytic effect of the platinum deposits could either be explained by a rather small overpotential for the reaction at Eq. (30) or by a failure of Pt-islands on the TiO_2 surface to transfer electrons to CCl_4 rather than to H_2O or protons.

Several different factors can in principle be responsible for a low photocatalytic reactivity. While adsorption of educts, products, or intermediates often appears to play a crucial role [11], the photocatalytic efficiency can also conveniently be correlated with typical solid state properties of semiconductor particles, e. g., the lifetime of electron/hole pairs generated upon light absorption [137]. It is thus important to prevent the undesired recombination of these electron/hole pairs within or on the surface of the catalyst particle. This can, for example, be achieved by the addition of electron or hole scavengers, respectively. The observation that the CCl_4 degradation rate can be considerably enhanced in the presence of a hole scavenger such as 2-methyl-2-propanol has therefore been explained by the following reaction mechanism [133]. Assuming that the valence band holes are trapped on the TiO_2 particle in surface hydroxyl states leading to the formation of surface bound hydroxyl radicals (see above):

$$Ti-OH_s + h^+{}_{VB} \rightarrow Ti^+-{}^{\bullet}OH_s \tag{31}$$

the latter can subsequently react with the hole scavenger:

$$(CH_3)_3COH + Ti^+-{}^{\bullet}OH_s \rightarrow {}^{\bullet}CH_2(CH_3)_2COH + Ti-OH_s \tag{32}$$

The 2-methyl-2-propanol radical formed in the reaction at Eq. (32) is a β-hydroxy radical [138] and thus neither able to reduce CCl_4 nor to inject an electron into the conduction band of TiO_2 (current doubling effect). It should therefore be the only effect of the presence of 2-methyl-2-propanol to partially inhibit the undesired electron/hole recombination thereby increasing the number of electrons in the semiconductor particle which can induce the reduction of CCl_4 via the reaction at Eq. (30).

On the other hand, the α-hydroxy radicals formed as the major one-electron oxidation products of 2-propanol, ethanol, and methanol [138] are able to reduce CCl_4 [139] as well as to inject an electron into the conduction band of TiO_2 [140]. In homogeneous solution, however, the reduction of tetrachloromethane by the methanol radical $(k({}^{\bullet}CH_2OH + CCl_4) \leq 1 \times 10^6 \, l \, mol^{-1} \, s^{-1} \, [139])$ should not be fast enough to compete with the reaction between the methanol radical and molecular oxygen $(k({}^{\bullet}CH_2OH + O_2) = 5 \times 10^9 \, l \, mol^{-1} \, s^{-1} \, [141])$. On the semiconductor surface, on the other hand, the ${}^{\bullet}CH_2OH$ radical should inject an electron into the conduction band or fill respective surface traps before it has a chance to react with molecular oxygen.

It has been observed that the photocatalytic degradation efficiency for CCl_4 increases with decreasing molecular weight of various α-hydroxy radicals forming alcohols employed as hole scavengers which was explained by different adsorption properties of these molecules [133]. Since it had previously been observed that methanol is in comparison with other alcohols a very efficient hole scavenger resulting in the highest quantum yields for H_2 formation on platinized TiO_2-colloids, it has been suggested that methanol has an increased ability to compete with the electron/hole recombination due to its higher adsorption strength towards the TiO_2-surface [63]. Small hydrophilic molecules such as CH_3OH are likely to form innersphere complexes on the TiO_2-surface while alcohols with a longer aliphatic chain are more likely to be only physisorbed to this surface.

While the hole-scavenging and current-doubling activities of alcohol molecules are well-known phenomena, it is extremely interesting that in air-saturated suspensions the degradation rates of tetrachloromethane are only reduced by a factor of two in the absence of any hole scavenger [133]. Hence, alternative oxidative half reactions to that given in Eqs. (31) and (32) have been proposed to explain the fate of light induced valence band holes (h^+_{VB}) in the absence of additional hole scavengers [133]. One possible pathway is the evolution of H_2O_2 from the oxidation of water via:

$$2H_2O + 2h^+_{VB} \rightarrow H_2O_2 + 2H^+ \tag{33}$$

Another reasonable step of an oxidative half reaction is the direct oxidation of the $^\bullet CCl_3$-radical generated in the reaction at Eq. (30) by a surface-trapped hole (Ti^+-$^\bullet OH_s$):

$$^\bullet CCl_3 + Ti^+\text{-}^\bullet OH_s \rightarrow HOCCl_3 + Ti^+_s \tag{34}$$

In homogeneous solution a bimolecular radical/radical reaction such as that proposed in Eq. (34) has a rather low probability (i.e., a low reaction rate resulting from a very small steady-state radical concentration). However, two-dimensional kinetics can be much more rapid, favoring the occurrence of such a process on the surface of a photocatalyst particle. The $HOCCl_3$ molecule formed in the reaction at Eq. (34) will decompose spontaneously forming phosgene (reaction at Eq. 35) which rapidly hydrolyzes [142, 143] yielding the observed products (Eq. 36):

$$HOCCl_3 \rightarrow OCCl_2 + H^+ + Cl^- \tag{35}$$

$$OCCl_2 + H_2O \rightarrow CO_2 + 2H^+ + 2Cl^- \tag{36}$$

Thus, following the reaction sequence (Eqs. 30, 34, 35 and 36) only one photon would be required to induce the complete photocatalytic degradation of CCl_4 in the absence of any hole scavenger:

$$CCl_4 + 2H_2O \xrightarrow{TiO_2/ + 1h\nu} CO_2 + 4H^+ + 4Cl^- \tag{37}$$

However, to explain the observed enhancement of the degradation rate in the presence of hole scavengers as well as the marked depletion of molecular oxygen during the illuminations the following reaction mechanism has been proposed [133]. In the homogeneous phase $^\bullet CCl_3$-radicals react very fast with O_2 (Eqs. 38, 39, and 40) with $k_{38} = 3.3 \times 10^9\,l\,mol^{-1}\,s^{-1}$ [139] where Eq. (39) represents a rather complex mechanism which is described in detail elsewhere [143]:

$$O_2 + ^\bullet CCl_3 \rightarrow ^\bullet O_2CCl_3 \tag{38}$$

$$^\bullet O_2CCl_3 \rightarrow ^\bullet OCCl_3 + 0.5\,O_2 \tag{39}$$

$$^\bullet OCCl_3 + e^-_{CB} + H^+ \rightarrow HOCCl_3 \tag{40}$$

Assuming that a similar affinity between the trichloromethyl radical and molecular oxygen exists on the titanium dioxide surface and combining the above reaction sequence with reactions at Eqs. (30), (35) and (36) results in the following overall stoichiometry for the reductive degradation of CCl_4 by conduction band electrons:

$$CCl_4 + H_2O + 0.5\,O_2 \xrightarrow{\text{TiO}_2/ + 2e^-_{CB}} CO_2 + 2\,H^+ + 4\,Cl^- \tag{41}$$

while the reaction of the holes should then be described by the reactions at Eqs. (31) and (33). The reaction at Eq. (41) requires two electrons for the degradation of one tetrachloromethane molecule.

In summary, it appears to be a basic requirement for a reductive photocatalytic reaction that the initial one-electron reduction process has to be electrochemically possible. Due to the Nernstian behavior of the flatband potential of metal oxide semiconductors these reactions will generally be favored in the alkaline pH-regime. Second, the addition of hole-scavengers will normally enhance the overall degradation yield. Alcohols which form α-hydroxy-alkyl radicals upon their one-electron oxidation are capable of inducing the well-known current-doubling effect and are thus more effective than others. Furthermore, the highest activities are observed for those alcoholic hole-scavengers which are most strongly (possibly through the formation of innersphere complexes) adsorbed on the surface of the photocatalyst prior to the light absorption. Finally, it is important to note that molecular oxygen which in a "conventional" photocatalytic system acts beneficially as an oxidant will, in these reductive systems, be able to compete with the pollutant molecule for the conduction band electrons (cf. reaction at Eq. 42):

$$O_2 + e^-_{CB} \rightarrow O_2^{\bullet-} \tag{42}$$

Therefore, high O_2 concentrations will normally have to be avoided in the absence of any additional hole scavengers. Optimal amounts of dioxygen will, however, often be required to ensure that free radical intermediates formed through the initial attack of e^-_{CB} on the substrate will "find their right way" (cf. reactions at Eqs 38 through 40) to the desired final products of the mineralization, i.e., carbon dioxide and the respective mineral acids.

5.2
Oxidative Pathways

A wide variety of organic compounds have been tested for their ability to be degraded photocatalytically with the initial reaction step being an oxidative attack [5, 15]. In the following only a few model compounds have been selected to elucidate typical details of the underlying reaction mechanism of these degradation processes. Their photocatalytic reaction pathways have recently been examined in detail and can be taken as representative examples for these very complex mechanisms. For a more comprehensive listing of treatable pollutants the reader is referred to several review articles which have recently been published [4–15].

5.2.1
Mechanism of the Photocatalytic Transformation of 4-Chlorophenol

The photodegradation of chlorophenols in aqueous solution has received considerable attention because these compounds are important xenobiotic micropollutants of the aquatic environment, originating, for example, from industrial chemical synthesis. More specifically, 4-chlorophenol (4-CP) is used for the production of quinizarin (a dye), clofibrate (a drug), chlorphenesin and dichlorophen (fungicides) [144]. Therefore, several investigations of the photocatalytic decomposition of chlorophenols using metal oxide semiconductors either in aqueous heterogeneous suspensions [24, 145–149] or in an immobilized form [150–152] have been published. The kinetics of the photocatalytic 4-CP degradation have commonly been interpreted as being indicative of a Langmuir-Hinshelwood-type mechanism in which the limitation of the 4-CP decomposition rate at higher pollutant concentrations is related to the extent of adsorption of the pollutant molecule on the TiO_2 surface [145, 148]. However, Cunningham and Sedlak showed in a recent study [149] that the observed adsorption constants of chlorophenols on TiO_2 in the dark differ from equivalent data obtained from the kinetic analysis of their photocatalytic degradation employing the same TiO_2 particles under UV-irradiation.

Although 4-CP is obviously an intensively studied model compound, the mechanism of its photocatalytic degradation is still not fully understood. In particular, the pH dependency of the photocatalytic 4-CP degradation has in most studies been investigated only over a small range of initial pH-values [145, 147, 153]. To achieve a more detailed understanding of the detailed mechanism of the photocatalytic degradation of 4-CP, i.e., to quantify the influence of the pH on the concentration and number of intermediates Theurich et al. [154] have recently carried out an investigation employing the above described pH-stat technique. Utilizing both, Sachtleben Hombikat UV 100 and Degussa P25 as photocatalysts, they observed the rate of the photocatalytic degradation of 4-CP over TiO_2 to be a function of the initial pollutant concentration increasing up to a 4-CP concentration of $5 \, mmol \, l^{-1}$ and to remain constant for higher 4-CP concentrations. The authors explained this behavior by the adsorption of the substrate molecule being the rate-limiting step. In photocatalytic systems this adsorption is often described by the Langmuir-Hinshelwood (L-H) isotherm resulting in the following rate equation:

$$-\frac{d[4\text{-CP}]}{dt} = \frac{k_a \cdot K \cdot [4\text{-CP}]}{1 + K \cdot [4\text{-CP}]} \qquad (43)$$

where $d[4\text{-CP}]/dt$ is the rate of 4-CP degradation, k_a the apparent reaction rate constant, K the adsorption coefficient of 4-CP, and $[4\text{-CP}]$ the concentration of 4-CP.

Two extreme cases can be considered: for high concentrations of the pollutant, where saturation coverage of the TiO_2 surface is achieved (i.e., $K \cdot [4\text{-CP}] \gg 1$), Eq. (43) simplifies to a zero-order rate equation:

$$-\frac{d[4\text{-CP}]}{dt} = k_a \qquad (44)$$

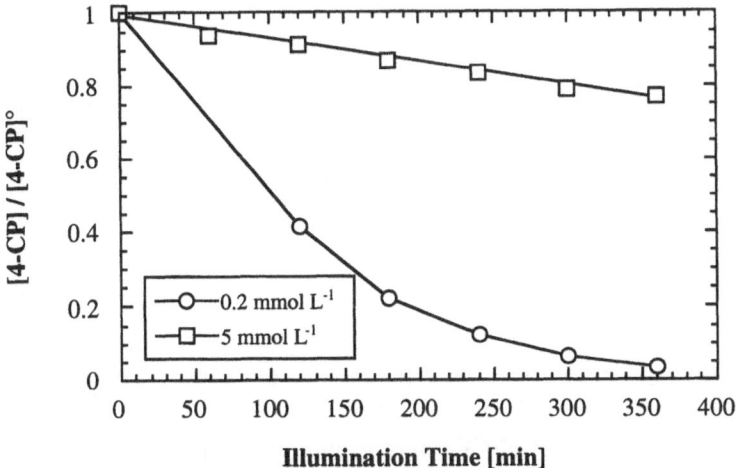

Fig. 5. Plot of the normalized 4-CP concentrations as a function of the irradiation time. Experimental conditions: $5\,g\,l^{-1}$ Hombikat UV 100, pH 3, continuous O_2-stream, $I \approx 1.0 \times 10^{-3}\,mol$ photons $min^{-1}\,l$, 293 K, $10 \times 10^{-3}\,mol\,l^{-1}\,KNO_3$, V = 150 ml (adopted from [154])

For very low concentrations of 4-CP (i.e., $K \cdot [4\text{-CP}] \ll 1$) the L-H equation changes into a pseudo first order kinetic law:

$$-\frac{d[4\text{-CP}]}{dt} = k_a' \cdot [4\text{-CP}] \tag{45}$$

with $k_a' = k_a \cdot K$ being the pseudo first order rate constant.

These two extreme cases are illustrated (in a normalized presentation) in Fig. 5 for two different initial concentrations of 4-CP (10 mmol l^{-1} and 0.2 mmol l^{-1}, respectively, data taken from [154]) as a function of the irradiation time. At high initial concentrations the 4-CP decay is obeying zero-order kinetics (Eq. 44) while the degradation kinetics at low concentrations can be interpreted as an example for first order kinetics (Eq. 45).

Assuming that solvent and reaction intermediates compete with the reacting substrate for the active surface sites on the semiconductor particle (Eq. 43) may be written in a more general form:

$$-\frac{d[4\text{-CP}]}{dt} = \frac{k_a \cdot K \cdot [4\text{-CP}]}{1 + K \cdot [4\text{-CP}] + \sum_{i=1}^{n} K_i c_i} \tag{46}$$

where i is the number of substances (solvent, molecular oxygen or by-products) competing with 4-CP for free adsorption places on the surface of the photocatalyst, K_i is the respective adsorption coefficient, and c_i is the concentration of the component i. Again, for high 4-CP concentrations (Eq. 46) reduces to the form of Eq. (44) and the photocatalytic degradation reaction will follow zero-order kinetics.

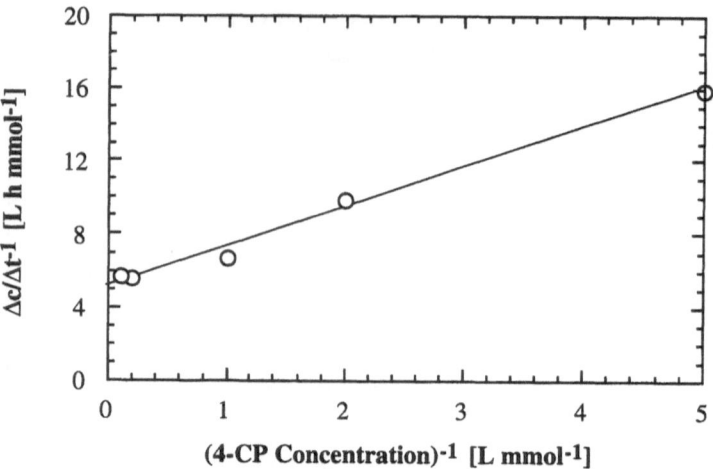

Fig. 6. Plot of the inverse of the initial rate of 4-CP disappearance as a function of the reciprocal initial 4-CP concentration. Experimental conditions: 5 g l^{-1} Hombikat UV 100, pH 3, continuous O$_2$-stream, I $\approx 1.0 \times 10^{-3}$ mol photons min^{-1} l^{-1}, 293 K, 10 $\times$ 10^{-3} mol l^{-1} KNO$_3$, V = 150 ml (adopted from [154])

The transformation of the Langmuir-Hinshelwood equation (Eq. 43) into its inverse function results in a linear relationship (Eq. 47) with an intercept of k$_a^{-1}$ and a slope of (k$_a^{-1} \cdot$ K^{-1}):

$$-\frac{dt}{d[4\text{-CP}]} = \frac{1}{k_a} + \frac{1}{k_a \cdot K \cdot [4\text{-CP}]} \tag{47}$$

A plot of the inverse of the initial rate of 4-CP disappearance as a function of the reciprocal initial 4-CP concentration is shown in Fig. 6 (data taken from [154]). The following values are readily derived from the intercept and the slope of this straight line: k$_a$ = 0.2 mmol l^{-1} h^{-1} and K = 2.4 $\times$ 10^4 l mol^{-1}. The value of k$_a$ depends strongly on the experimental conditions and consequently cannot be compared among different research groups. However, the intercept can be interpreted as the maximum rate which can be achieved for infinite high initial 4-CP concentrations under the chosen experimental conditions. The value for the adsorption coefficient K agrees very well with those obtained by Al-Sayyed et al. [145] and Al-Ekabi et al. [150] who used Degussa P25 as the photocatalyst. It should be pointed out that a kinetic treatment employing a Langmuir-Hinshelwood adsorption behavior of the substrate just presents one possible model to explain the experimental results. Independent adsorption measurements are certainly warranted to decide whether the derived K-value represents a realistic adsorption coefficient [155].

The influence of the initial pH on the degradation rate of 4-CP has been studied by Matthews [153] and Al-Sayyed et al. [145] who observed no significant change in the decomposition rate of 4-CP (and therefore an unchanged photonic efficiency) within a pH-range of 2.9–6.0, while Barbeni et al. [147]

reported an increase in the rate of the photocatalytic decomposition of 4-CP at pH 12 in comparison to neutral pH. However, no efforts were made in these studies to maintain a constant pH throughout the irradiation. Employing the pH-stat technique (see above) Theurich et al. [154] also observed only a weak influence of the pH on the degradation rate of 4-CP itself but noted a significant influence of the pH on the TOC degradation rate. Even though TOC measurements are not regularly performed when the degradation kinetics of a model compound are studied, they certainly present one of the most important criteria to assess realistically the efficiency of a photocatalytic process.

A detailed study of the intermediates formed during the photocatalytic 4-CP degradation showed that not only their number but also their concentration is strongly influenced by the reaction pH [154]. An increasing amount of hydroxylated products (especially hydroxyhydroquinone, HHQ) was, for example, observed for increasing pH values which Theurich et al. attributed to higher stabilities of polyhydroxylated benzenes and to a change in the adsorption properties of these intermediates at alkaline pH values [154]. HHQ was found to be strongly adsorbed on TiO_2 in acidic media while it was poorly adsorbed at pH 11. It should be noted, however, that at even higher pH-values a spontaneous oxidation of polyhydroxylated phenols and thus a decreasing yield of HHQ will be the expected result.

Besides a fast TOC-degradation the suppression of the formation of byproducts (especially when they are highly toxic) is an important criterion for a good photocatalyst. Thus, it is interesting to note that while slightly higher photonic efficiencies were obtained with P25 markedly lower concentrations of byproducts were detected when Hombikat UV 100 was the photocatalyst [154]. A similar result was published recently by Mills and Sawunyama [156] who compared Degussa P25 and several other commercial TiO_2-powders with laboratory made TiO_2. Significant smaller amounts of intermediates were detected when the latter materials were used as the photocatalyst.

An increasing concentration of the intermediates benzoquinone (BQ) and hydroquinone (HQ) formed during its photocatalytic treatment is apparently inhibiting the degradation of the model pollutant 4-chlorophenol [154]. Relatively low photonic efficiencies have therefore been observed for this process which have mechanistically been explained by the existence of a photocatalytic balance between HQ and BQ [154]. A fast electron shuttle mechanism between HQ and BQ has been proposed to describe the underlying reactions. It has been shown that BQ can act as a very effective electron scavenger over illuminated TiO_2 or ZnO [157], being able to compete successfully with molecular oxygen for the photogenerated electrons (Eq. 48). On the other hand,

$$BQ + e^- \longrightarrow [BQ^{\bullet-}] \diagdown \begin{array}{l} \text{HQ} \\ \text{reduction} \\ \\ \text{disproportionation} \\ \\ BQ + HQ \end{array} \qquad (48)$$

the oxidation of hydroquinone will again result in the formation of benzo-quinone (Eq. 49).

$$HQ + h^+ \longrightarrow [HQ^{\bullet +}] \begin{array}{c} \nearrow \quad BQ \\ \text{oxidation} \\ \\ \text{disproportionation} \\ \searrow \quad BQ + HQ \end{array} \tag{49}$$

Only when benzoquinone molecules can be oxidized via the reaction at Eq. (50) will hydroxybenzoquinone (HBQ) eventually be formed which is a prerequisite for the subsequent ring cleavage leading to the complete 4-CP degradation (cf. Scheme 1):

$$BQ + OH\bullet \rightarrow [BQOH\bullet] \rightarrow \rightarrow HBQ \tag{50}$$

Thus, the reduction of BQ via the reaction at Eq. (48) and the oxidation of HQ via the reaction at Eq. (49) apparently short-circuit the photocatalyst efficiently thus explaining the low photonic efficiency for the photocatalytic degradation of 4-CP and other aromatic compounds such as phenol [158, 159]. A similar behavior has been observed for the inorganic system Fe^{2+}/Fe^{3+} (Eqs. 51 and 52) [160].

$$e^-_{CB} + Fe^{3+}_{surface} \rightarrow Fe^{2+}_{surface} \tag{51}$$

$$h^+_{VB} + Fe^{2+}_{surface} \rightarrow Fe^{3+}_{surface} \tag{52}$$

A high number of intermediates has been identified during the photo-catalytic 4-CP degradation indicating a complex reaction mechanism. Scheme 1 illustrates the proposed pathways of the photocatalytic/photolytic decomposition of 4-CP [154]. The primary oxidation products of 4-CP are 4-chloro-catechol (4-CC), hydroquinone (HQ) and benzoquinone (BQ). 4-CC is formed by the addition of a hydroxyl radical to the *ortho* position of the hydroxyl group of 4-CP (reaction a), followed by an elimination of a hydrogen atom to recover the aromatic ring (reaction e). Subsequently, 4-CC is further degraded to hydro-xyhydroquinone (HHQ) by oxidation with another hydroxyl radical (or a valence band hole) (reaction m) or by a reductive attack of an electron (reaction k) followed by the addition of molecular oxygen (reaction p) or by Cl•-abstrac-tion (reaction l), both yielding HBQ (reaction w). HQ is formed when the attack of the hydroxyl radical takes places in the *para* position (reaction c) and a chlorine atom (reaction f) or, if the generated intermediate reacts first with an electron, a chloride ion is released. HQ can either be oxidized to BQ (reaction j), resulting in the photocatalytic short circuit described above, or be trans-formed to HHQ (reactions n and r) by a second attack of a hydroxyl radical in analogy to the transformation of 4-CP to 4-CC. HHQ can be further oxidized to hydroxybenzoquinone (HBQ) (reaction u) which probably leads to a similar photocatalytic short circuit as observed for the system HQ/BQ, with the dif-

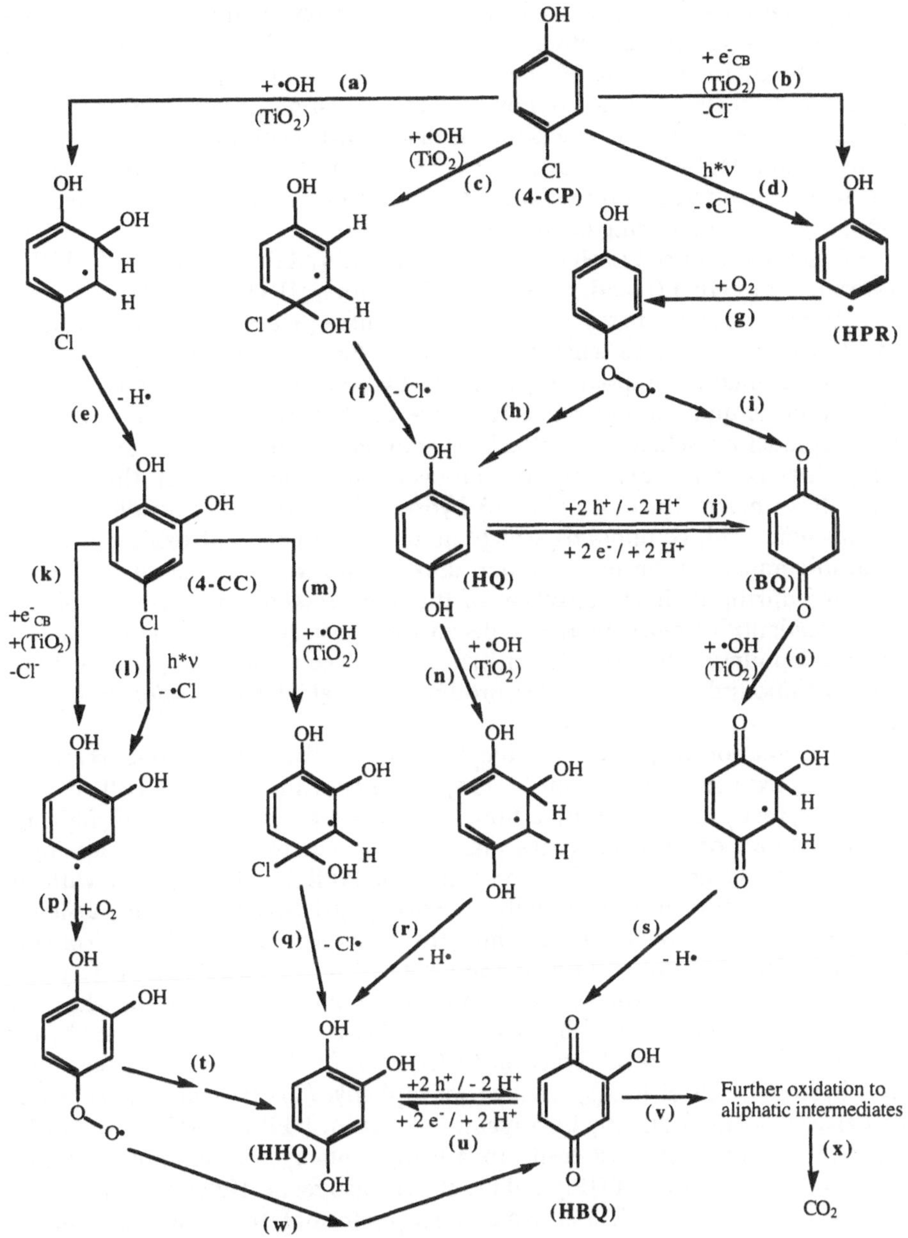

Scheme 1. Degradation scheme of the photocatalytic/photolytic degradation of 4-CP

ference that HHQ and HBQ are less stable, especially against direct photolysis, than HQ and BQ. BQ can either be formed by the oxidation of HQ (reaction j) or directly by the reaction of molecular oxygen (reaction g) with the hydroxy phenyl radical (HPR) generated by the abstraction of a Cl•-atom (reaction d), leading to a peroxyl radical which can disproportionate to HQ and BQ (reactions h and i) [150]. The formation of HPR as a short-lived radical intermediate has been suggested in a recent study by electron paramagnetic resonance spin trapping detection during the direct photolysis of 4-CP [161].

BQ is either reduced to HQ (reaction j) or oxidized (by the attack of a hydroxyl radical) to HBQ (reactions o and s). Therefore, HHQ and HBQ are formed as secondary intermediates from each of the three primary oxidation products (4-CC, HQ, and BQ) (see Scheme 1). Assuming that the ring cleavage is possible only for aromatic compounds which are hydroxylated in positions 1 and 2, HHQ and its quinone derivative HBQ are the last aromatic intermediates in this degradation scheme. An attack of another hydroxyl radical leads to the ring cleavage and therefore to non-cyclic intermediates. In the case of HBQ these intermediates should be 3-hydroxy derivatives of muconic aldehyde or muconic acid, compounds which should undergo rapid oxidation to the final mineralization products carbon dioxide and water. Okamoto et al. [159] detected during their investigation of the photocatalytic oxidation of phenol many unidentified peaks by HPLC. Because of the very short retention times of these compounds under the chosen experimental conditions they assumed that these compounds are very polar products, i.e., aldehydes and/or carboxylic acids.

The formation of phenol and bicyclic compounds, which were observed at pH > 7, is explained in Scheme 2 [154]. As shown in Scheme 1, HPR can be formed directly from 4-CP by a homolytic cleavage (reaction d) of the ring-chlorine bond or by a reductive attack of an electron on 4-CP releasing a chloride ion (reaction b). The formation of HPR by the reaction with an electron becomes more likely with increasing pH because of the Nernstian behavior of the semiconductor. Thus, the observed increase in the concentrations of bicyclic compounds is the expected result for alkaline pH values. HPR can abstract a hydrogen atom to form phenol (reaction C), while the formation of 5-chloro-2,4'-dihydroxybiphenyl (CDHB) can be explained by the reaction of HPR with 4-CP (reaction A). CDHB is further oxidized by the attack of a hydroxyl radical to 2,5,4'-trihydroxybiphenyl (THB) (reaction D), which can also be formed directly by a reaction between HPR and HQ (reaction E). Further oxidation of THB leads to 4-hydroxyphenylbenzoquinone (HPBQ) (reaction G), a reaction of HPR and BQ should also form HPBQ (reaction F). All bicyclic compounds will be degraded under photocatalytic conditions probably into monocyclic compounds and can be mineralized completely within the irradiation time required for the total oxidation of 4-CP. 2,4'-Dihydroxy- and 4,4'-dihydroxybiphenyl (DHBP) which are formed during the direct photolysis of 4-CP [144] could not be detected in the study of Theurich et al. [154]. The lack of DHBP is an especially surprising result as it should be formed easily by a reaction of either HPR with a 4-CP molecule and the following release of a chlorine atom or by a combination of two HPR species (reaction B).

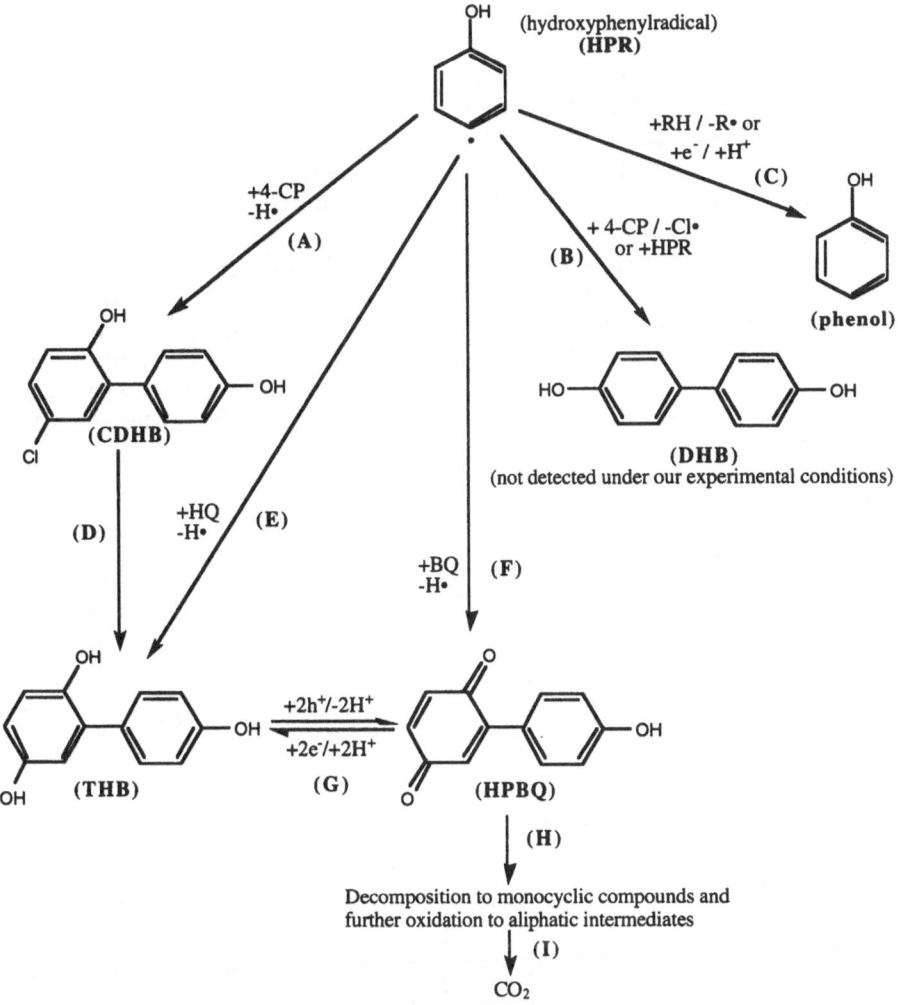

Scheme 2. Further reactions of the hydroxyphenyl radical (HPR) at higher pH-values, forming bicyclic byproducts and phenol

5.2.2
Mechanism of the Photocatalytic Transformation of Naphthalene

Polynuclear aromatic hydrocarbons (PAHs), such as naphthalene and anthracene, are well known hazardous by-products in the coal and petroleum industry [162]. The effluent water of many petrochemical industries, particularly of cracking plants, very often contains a high load of these compounds, which are hardly biodegradable [163]. Petrochemical plants in Northern Africa and the Gulf region are usually located near the coast to allow an easy shipping of their products. Usually the wastewater of such plants is treated biologically before being discharged to the sea. The biological treatment of this wastewater is

rather difficult, due to very narrow living conditions for the bacteria [164].
Therefore, insufficiently degraded or even untreated effluent from these plants
often contaminates the sea, which can be detrimental to environment and
health, particularly if drinking water is produced by sea water desalination
plants nearby. At the same time the sunshine rate and intensity in the
Mediterranean region is very high and it is a logical conclusion to consider a
solar photocatalytic detoxification of these contaminants.

Only limited information about the mechanism of the photocatalytic degradation of PAHs contaminants in water [165–167], in the gas-phase [168], or in
organic solvents [169] is available in the literature. Details of the mechanism of
the photocatalytic oxidation of naphthalene (1) have recently been extensively
studied by Theurich et al. [170] who proposed that the initial step of its degradation appears to be an attack of a photocatalytically generated OH• radical
either at the *alpha* position, leading to a hydroxylated naphthalene radical with
the two preferred mesomeric structures (2) and (3),

$$\text{(53)}$$

or at the *beta* position leading to a radical with only one preferred structure
(4):

$$\text{(54)}$$

In contradiction to the model of Das et al. [165], who proposed an involvement of the unsubstituted ring into the stabilization structures, Theurich et al.
[170] have suggested that any substantial electronic contribution from the
non-hydroxylated aromatic ring of (2–4) into the mesomeric structures
should not be energetically favored. The formation of a cation radical by
the direct h^+-induced oxidation of naphthalene with a generated hole has
been proposed for the photocatalytic oxidation of naphthalene in acetonitrile
[169] or in the gas phase, respectively [168]. However, in aqueous systems
the involvement of cation radicals in the degradation mechanism has so far
only been discussed for the direct photolysis in the case of deoxygenated

solutions [171]. Since a radical cation which would be the initial product of the reaction of naphthalene with holes (reaction at Eq. 55a) will react quickly with water (reaction at Eq. 55b) to form the hydroxyl adducts (2–4), Theurich et al. [170] were unable to differentiate between these two reaction mechanisms.

$$(55\,a)$$

$$(55\,b)$$

The oxidation products 1- (8) and 2-naphthol (9) have been identified [170] and are being formed by the reaction of 2, 3, and 4 with molecular oxygen (reactions at Eqs. 56–58) and the subsequent elimination of $HO_2^{\bullet}$ radicals (reactions at Eqs. 59–61) as illustrated in Scheme 3. It is well known that many aromatic hydroxyperoxyl radicals are able to eliminate $HO_2^{\bullet}$ radicals forming the corresponding hydroxyaromatic compounds [172–175]. The naphthols 8 and 9 may also be formed directly by a hydrogen transfer reaction from the radicals 2–4 towards molecular oxygen releasing also $HO_2^{\bullet}$ radicals, as has been observed for the oxidation of benzene [172].

The mechanism which has been proposed by Theurich et al. [170] for the further oxidation of the naphthols 8 and 9 is illustrated in Scheme 4. These naphthols react with another photocatalytically generated OH• radical (reactions at Eqs. 62–64) forming the corresponding dihydroxyl radicals 10–12. By analogous reactions to those discussed in Scheme 3 (reactions at Eqs. 65–70) the dihydroxynaphthalenes 13 and 14 are formed as the next stable oxidation products of naphthalene. Dihydroxyl radicals such as 10–12 are also known to undergo an elimination of water [175, 176] followed by a reaction with molecular oxygen (reactions at Eqs. 71–78) leading to carbonyl peroxyl radicals 15 and 16 which subsequently abstract a hydrogen atom from other substrate or solvent molecules. Then 1,2- (17) and 1,4-naphthalene-dione (18) are formed as rather stable intermediates by the following elimination of H_2O (reactions at Eqs. 79 and 80 in Scheme 4). In addition, (17) and (18) can be formed by the oxidation of the corresponding naphthalene-diols (13) and (14) with •OH radicals (Eq. 81). The decay of the resulting OH-adduct (19) will then lead to the formation of semiquinone radicals (20) by the elimination of water (Eq. 82) [175, 177]. The disproportionation

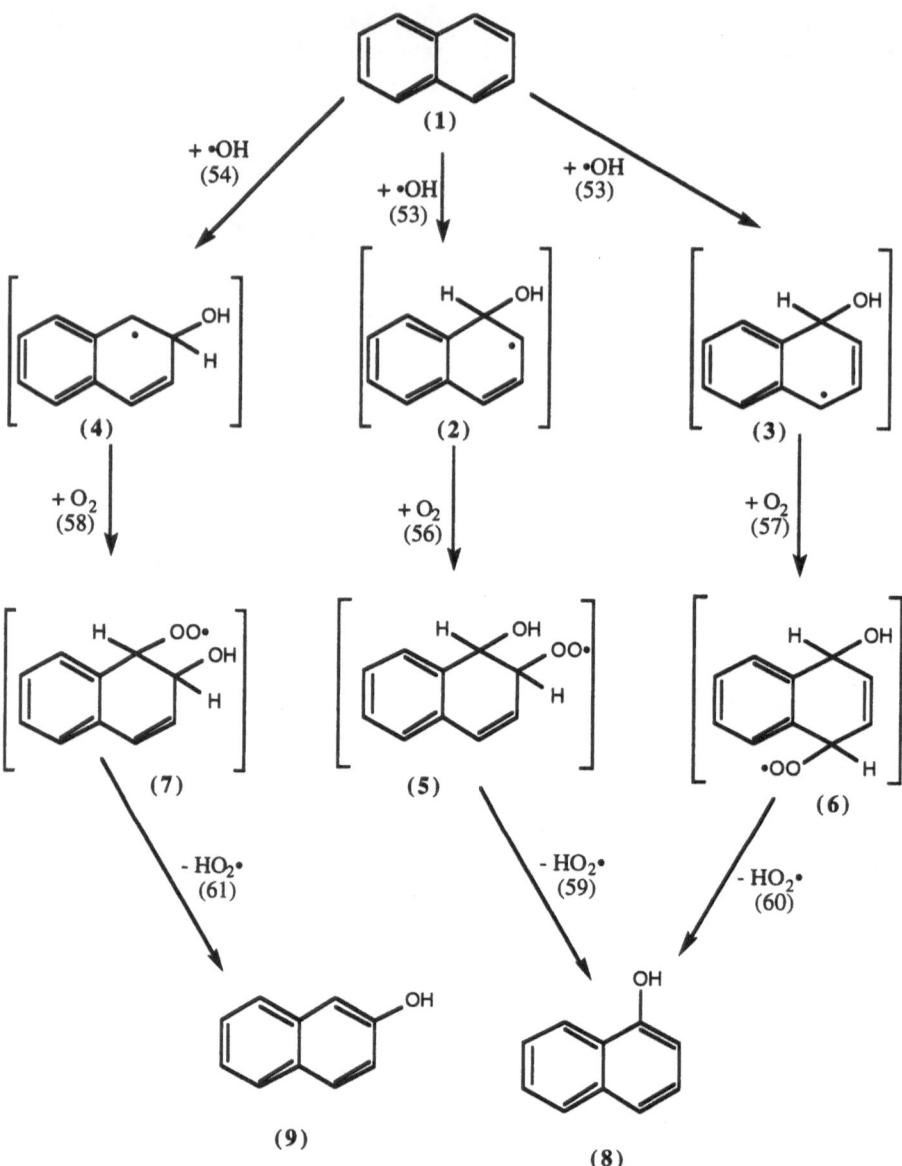

Scheme 3. Proposed initial oxidation steps for the photocatalytic degradation of naphthalene

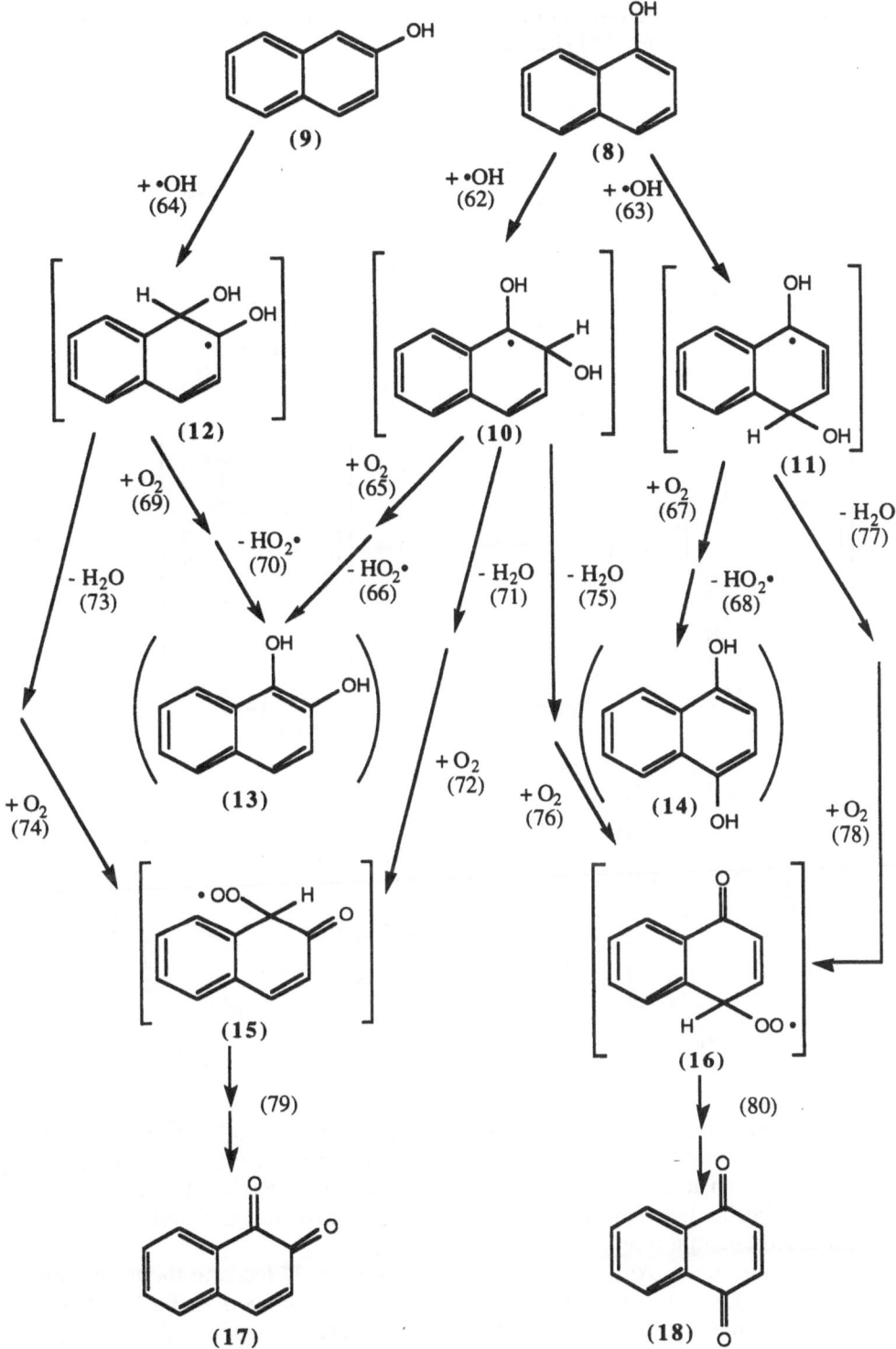

Scheme 4. Proposed initial oxidation steps for the photocatalytic oxidation of naphthols

products of these semiquinone radicals will be naphthalenediol (14) and the corresponding dione (18) (Eq. 83):

(14)   (19)   (81)

(19)   (20)   (82)

(20)   (14)   (18)   (83)

The semiquinone radical 20 can also be generated from the dione 18 by a reaction with an $O_2^{\cdot-}$ radical [173] or a conduction band electron from TiO_2. This will then lead to a short-circuiting of the photocatalyst by an electron shuttle mechanism similar to that described above for the system hydroquinone/benzoquinone.

The subsequent oxidation of 1,2-naphthoquinone 17 leads to the formation of various ring cleavage products [170] which are known to be in photochemical equilibrium with each other [178]. The general oxidation mechanism which has been proposed for aromatic aldehydes [170] is presented below (reactions at Eqs. 84–87). The aldehyde **21** is initially transformed by hydrogen abstraction into the radical **22** which readily reacts with molecular oxygen forming the corresponding peroxy radical **23**. Following the formation of a peroxyacid **24** via hydrogen abstraction, a synproportionation with unreacted aldehyde **21** leads to the observed formation of the corresponding acids **25** [179].

$$R-C\underset{H}{\overset{O}{\diagup}} \quad \xrightarrow[- RH\ (\ or\ H_2O)]{+\ R\bullet\ (or\ \bullet OH)} \quad \left[R-C=O \atop \bullet \right] \qquad (84)$$

(21)   (22)

$$\left[R-C=O \atop \bullet \right] \quad \xrightarrow{+\ O_2} \quad \left[R-C\underset{O-O\bullet}{\overset{O}{\diagup}} \right] \qquad (85)$$

(22)   (23)

$$\left[R-C\underset{O-O\bullet}{\overset{O}{\diagup}} \right] \quad \xrightarrow[-\ R\bullet]{+\ RH} \quad R-C\underset{O-OH}{\overset{O}{\diagup}} \qquad (86)$$

(23)   (24)

$$R-C\underset{O-OH}{\overset{O}{\diagup}} \quad \xrightarrow{+\ RCHO} \quad 2\ R-C\underset{OH}{\overset{O}{\diagup}} \qquad (87)$$

(24)   (25)

The photocatalytic oxidation of one of the detected intermediate ring opening products of naphthalene, i.e., 2-carboxy-cinnamic acid (**26**), leads to the formation of coumarin (**30**) which was detected in substantial amounts during the aqueous phase naphthalene degradation [170] (cf. Scheme 5). Following a Photo-Kolbe oxidation [180, 181] (reactions at Eqs. 88–90) the resulting 2-hydroxy-cinnamic acid (**29**) loses water to form coumarin (**30**) (Eq. 91). Coumarin can also be formed directly from **28** by an intramolecular rearrangement combined with the release of atomic hydrogen (Eq. 92). Further oxidation of **30** will lead to the formation of phthalic acid or its hydroxylated derivatives.

The subsequent degradation of 4-naphthoquinone (**18**) leads to the formation of 2,3-dihydro-2,3-epoxy-1,4-naphthoquinone which has been identified as a major reaction product in the aqueous phase [170] and was also observed as an intermediate of the photocatalytic oxidation of naphthalene in the gas phase [168].

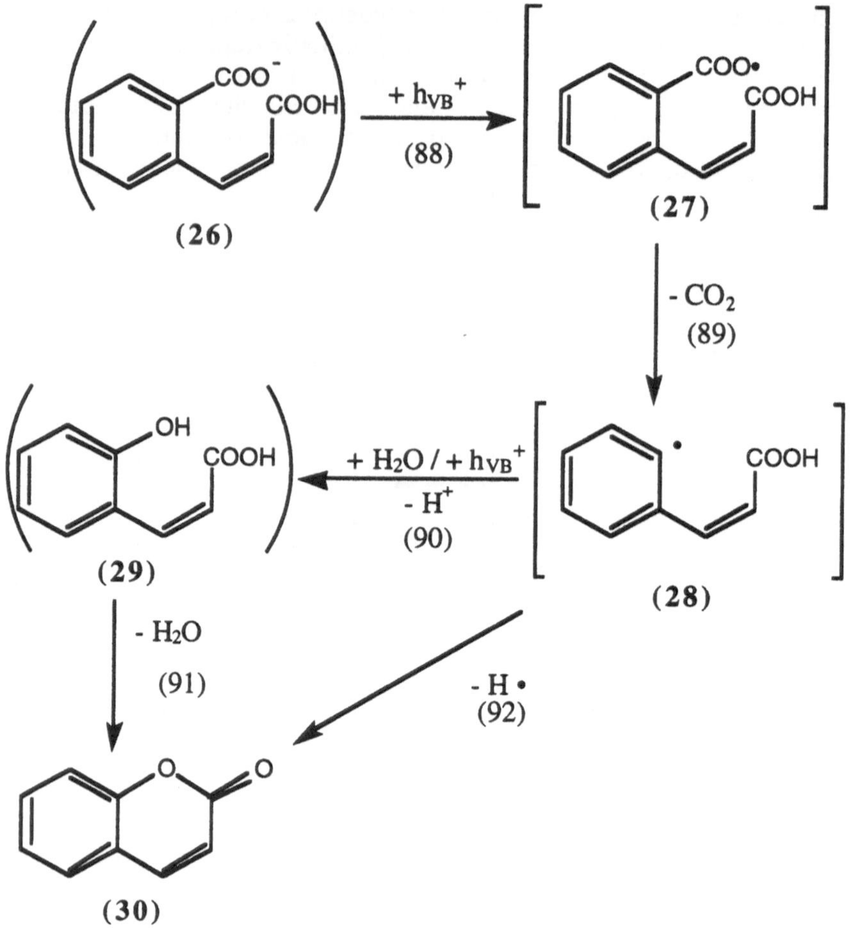

Scheme 5. Formation of coumarin

The formation of 1,2-benzenedicarboxaldehyde (36) is explained by the mechanism presented in Scheme 6 which has originally been proposed by Fox et al. [169] for the oxidation of naphthalene in acetonitrile. A naphthalene radical cation 31 generated by electron transfer from an adsorbed naphthalene molecule to a photogenerated hole (Eq. 93) can either react with molecular oxygen forming an endoperoxide radical cation 32 (Eq. 94) or with a superoxide anion resulting in the formation of the endoperoxide 33 (Eq. 95). In addition, 32 and 33 will undergo phototransformation (Eqs. 97 and 98) into diradical species 34 and 35 which readily decay into 1,2-benzenedicarboxaldehyde (36) and acetylene (Eqs. 99 and 100). The absence of the naphthalene endoperoxide during the photocatalytic degradation of naphthalene in aqueous TiO$_2$ suspensions [170] could either be attributed to its rapid oxidation under the employed experimental conditions or it might be taken as experimental evidence that a

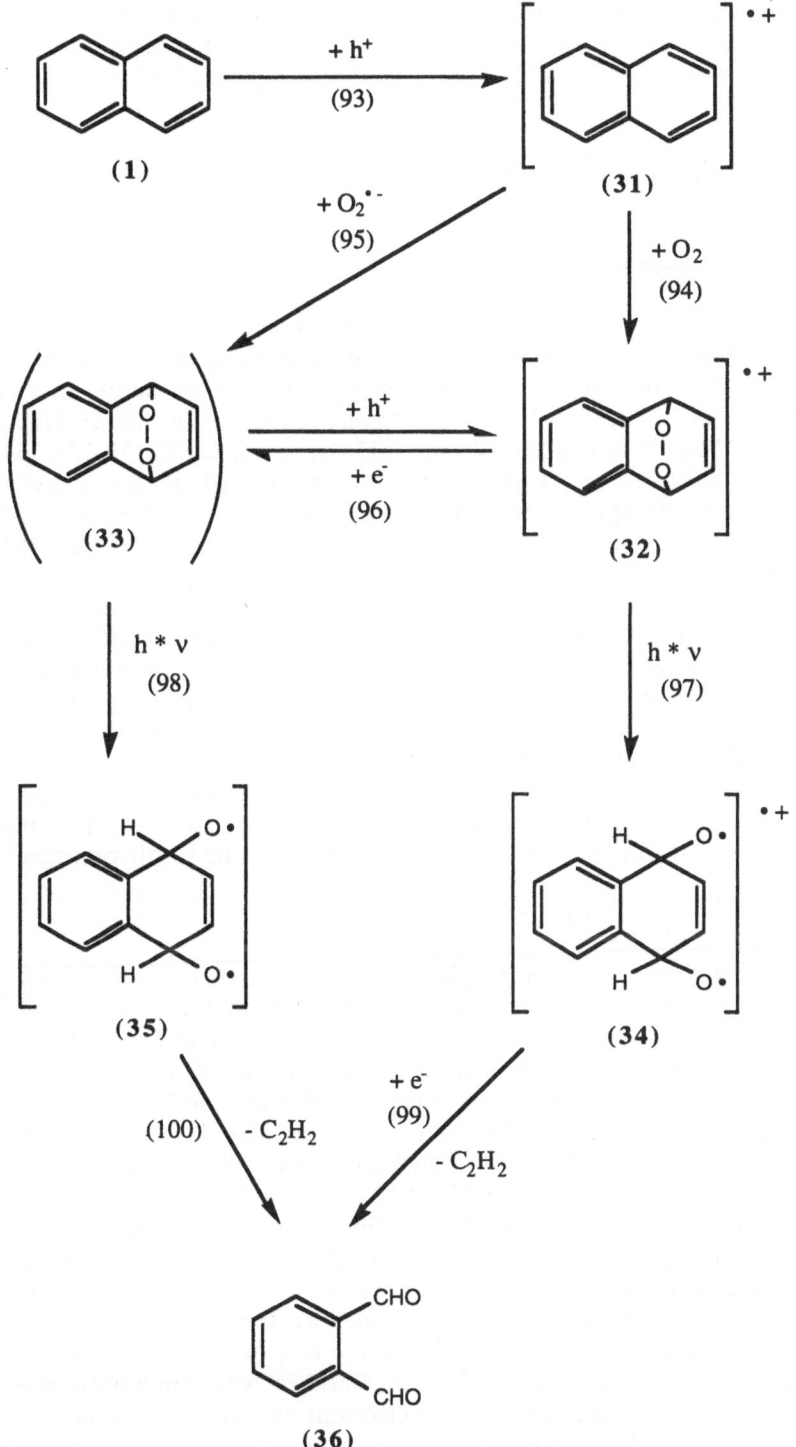

Scheme 6. Formation of an endoperoxide from naphthalene and its decay

radical cation intermediate, such as **31**, is favored under non-aqueous conditions, i.e., in the absence of hydroxyl radicals, while the primary reaction in aqueous systems is an attack of photocatalytically generated hydroxyl radicals.

6
Crucial Reaction Parameters

6.1
Important Reaction Variables

In addition to the solid state and surface properties of the semiconductor photocatalysts, other extensive and intensive reaction parameters are important in determining the rate and extent of compound transformation in a photocatalytic system [72]. Besides semiconductor concentration [20, 21, 81, 85, 114, 133, 182–184], pH of the treated water [20, 21, 30, 44, 75, 78, 132, 133, 154, 159, 183–187] and temperature of the system [113, 184, 188–191], the intensity of the incident light appears to be one of the most important variables with a strong impact on the overall process efficiency [21, 28, 72, 83, 107, 182, 184, 188, 192–204].

In terms of future applications of semiconductor photocatalysis, a major concern has to be the non-linear dependence of the photonic efficiency of the process which has been observed for many degradation reactions [20, 21, 28, 72, 83, 182, 184, 194, 199–204]. This feature argues against employing concentrating solar collectors with enhanced light fluxes, since the net effect will be to lower the overall efficiency of the process.

Several research groups [21, 107, 199] have shown that the rate of chloroform, $CHCl_3$, degradation in the presence of O_2 is a non-linear function of the light intensity. The rate of reactions conformed to the following empirical expression:

$$- \frac{d[CHCl_3]}{dt} = k_{obs} \sqrt{I_a} \tag{101}$$

where I_a is the incident light intensity (in µmol photons l^{-1} min^{-1}) and k_{obs} the observed rate constant with nominal units of (µmol photons l^{-1} min^{-1})$^{1/2}$. In addition the measured yield of the reaction increases with decreasing light intensity. For example, for single wavelength irradiation at $\lambda = 330$ nm with an absorbed light intensity of 2.8 µmol photons l^{-1} min^{-1} yielded $\Phi = 0.56$ for the $CHCl_3$ degradation [21]. On the other hand, when the absorbed light intensity increased to 250 µmol photons l^{-1} min^{-1} under the same conditions, the degradation efficiency was reduced to $\Phi \leq 0.02$ [21]. While it will be shown below that the square-root dependence of the reaction rate can be explained by an enhanced band-gap recombination at higher light intensities, Kormann et al. [21] proposed an alternative explanation for the square-root dependence of the rate of reaction on the light intensity focusing on the role of surface-bound hydroxyl radicals, $> TiOH^{\bullet+}$, as the principal hole trap and the primary initiator of the oxidation of electron-donating substrates. In this mechanism it is assumed that the photogenerated conduction band electrons are efficiently removed by an electron acceptor leaving behind relatively long-lived trapped holes. These surface-bound hydroxyl radicals are thus free to initiate the oxidation of chloroform as follows:

$$> \text{TiOH}^{\bullet+} + > \text{HCCl}_3 \xrightarrow{k_{rds}} > \text{TiOH}_2^+ + > {}^{\bullet}\text{CCl}_3 \qquad (102)$$

where the symbol $>$ indicates surface-bound species. The reactions following this rate determining step have already been given when the photocatalytic degradation of CCl_4 was described in detail (reactions at Eqs. 35, 36, 38–40). The rate constant k_{rds} for the rate determining reaction (Eq. 102) has been estimated to be comparable to the rate constant measured for hydroxyl radicals reacting with chloroform in homogeneous aqueous solution (e.g., $k = 10^7 \, l \, mol^{-1} s^{-1}$). In addition to direct H-atom abstraction of a bound substrate the surface-bound hydroxyl radical, $> \text{TiOH}^{\bullet+}$, could in turn react with itself as follows:

$$> \text{TiOH}^{\bullet+} + > \text{TiOH}^{\bullet+} + H_2O \xrightarrow{k_s} > \text{TiOH}_2^+ + > \text{TiO}_2H_2^+ \qquad (103)$$

provided that they were within reasonable proximity of each other on the surface. Assuming a photostationary state for $> \text{TiOH}^{\bullet+}$ Kormann et al. [21] performed a standard steady-state kinetic analysis to obtain a quantum yield of chloroform degradation, Φ_r, expressed in terms of k_{rds} and k_s as follows:

$$\frac{d[> \text{TiOH}^{\bullet+}]}{dt} = \qquad (104)$$

$$I_a \Phi_{> \text{TiOH}^{\bullet+}} - k_{rds} [> \text{TiOH}^{\bullet+}] \cdot [> \text{HCCl}_3] - k_s [> \text{TiOH}^{\bullet+}]^2$$

$$\Phi_r = \frac{k_{rds} [> \text{TiOH}^{\bullet+}] \cdot [> \text{HCCl}_3]}{k_{rds} [> \text{TiOH}^{\bullet+}] \cdot [> \text{HCCl}_3] + k_s [> \text{TiOH}^{\bullet+}]^2} \qquad (105)$$

under the assumption that the quantum yield for $> \text{TiOH}^{\bullet+}$ production ≈ 1. Two limiting cases for Eq. (105) arise. While for low absorbed light intensities the second term in the denominator can be neglected and Φ_r approaches unity, at high absorbed light intensity, I_a, the overall quantum yield for the reaction is given by

$$\Phi_r = \frac{k_{rds} [> \text{HCCl}_3]}{\sqrt{k_s I_a}} \qquad (106)$$

For the oxidation of chloroform as reported by Kormann et al. [21], a value for the ratio of rate constants is obtained as follows:

$$k_{obs} = \frac{k_{rds}}{\sqrt{k_s}} = 2.0 \times 10^{-3} \; (\text{mol} \, L^{-1} s)^{-\frac{1}{2}} \qquad (107)$$

while Bahnemann et al. reported a three times smaller value for k_{obs} [199]. From the numerical value of k_{obs} it can be concluded that the reaction of $> \text{TiOH}^{\bullet+}$ with $> \text{HCCl}_3$ (k_{rds}) is relatively slow compared to an efficient second-order (k_s) recombination process with a characteristic time of 0.1 μs.

To illustrate the influence of some other important reaction conditions a few typical results of degradation experiments using dichloroacetate (DCA) as the model compound will be presented in the following. The temperature dependence of the rate of the photocatalytic degradation of DCA has been measured by Bockelmann et al. [113, 190] at pH 5 between 10 and 90 °C. These experiments were carried out at two different light intensities, i.e., at 4.5×10^{-6} and 1.5×10^{-6} mol photons $l^{-1} s^{-1}$. An initial inspection of the results of these experiments suggested that the degradation rate increased linearly with the employed temperature [113, 190]. However, according to the Arrhenius relation:

$$k = Ae^{-E_a/RT} \tag{108}$$

(with A = preexponential factor, E_a = activation energy [J mol^{-1}], R = gas constant [J mol^{-1} K^{-1}], T = temperature [K], k = rate constant [l mol^{-1} s^{-1}]) the reaction rate should increase linearly with exp$(-1/T)$. Since in aqueous solution the experimentally accessible temperature range is rather limited, the authors were unable to decide whether a linear or an exponential fit yielded a better correlation of the results of this study to the reaction temperature. Therefore, they have also determined the activation energy from a linear regression of plots via log(k) = f(1/T). Even though one would not expect a dependency of the activation energy upon the light intensity, Bockelmann et al. [113, 190] obtained different values at different illumination intensities. While 13.6 kJ mol^{-1} was the result when illuminations were carried out with 4.5 µmol photons $l^{-1} s^{-1}$, 18.8 kJ mol^{-1} was determined for 1.5 µmol photons $l^{-1} s^{-1}$. This results in a mean activation energy of $E_a = 16.2$ kJ mol^{-1} for the photocatalytic degradation of dichloroacetate.

Recently, Matthews determined the activation energy for the photocatalytic degradation of salicylic acid using TiO$_2$ as the photocatalyst and calculated ~11 kJ mol^{-1} [191]. Thus, the value obtained for dichloroacetate appears to be reasonable. A possible explanation for the observed positive activation energy is that an essential reaction step of the complex DCA degradation mechanism is thermally activated. It should be pointed out that the temperature dependence of photocatalytic detoxification processes can even qualitatively be different when other test molecules are employed, e.g., a considerable decrease of the degradation rate of chloroform was observed as the reaction temperature was raised especially at high photon fluxes [21].

Figure 7 shows the combined effect of the light intensity and pH on the rate of the photocatalytic degradation of dichloroacetate [113]. A linear dependency, i.e., an intensity independent photonic efficiency of the mineralization, is observed at pH 2.6, 7, and 11. Only at pH 5, the rate of DCA degradation increases with the square root of the light intensity as has been observed when chloroform was the model compound (see above). When illuminations of aqueous TiO$_2$ suspensions containing DCA were carried out with light intensities below 1.2×10^{-6} mol photons $l^{-1} s^{-1}$, i.e., between 0.015×10^{-6} and 1.1×10^{-6} mol photons $l^{-1} s^{-1}$, even at pH 5 the photonic efficiency of the DCA photodegradation no longer depended upon the irradiation intensity [113]. In the range of $(2.5 - 25) \times 10^{-6}$ mol pho-

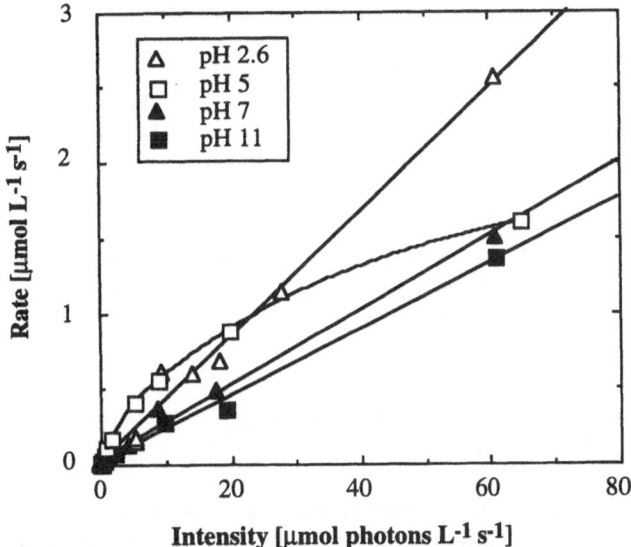

Fig. 7. Rate of DCA degradation as a function of the illumination intensity at various pH-values for O_2-sat. aqueous suspensions containing $0.5\,g\,l^{-1}$ TiO$_2$ (Degussa P25), 24×10^{-3} mol l^{-1} dichloroacetate and 10×10^{-3} mol l^{-1} KNO$_3$ (adopted from [113])

tons l^{-1} s^{-1} the highest degradation rates were encountered at pH 5 as compared with all other examined pH-values. It thus appears that a specific mechanistic explanation is required to account for these characteristic properties at pH 5.

While the combination of the surface-bound •OH radicals via Eq. (103) is a bimolecular process which could in principle explain the observed square root dependence of the degradation yield on the light intensity, as will be shown below, a simple kinetic model which incorporates the electron/hole recombination within the particle suffices to predict exactly such a square-root dependence of the detoxification rate upon the light intensity. Once again, this formal kinetic treatment does not offer any reasonable explanation for the distinct sensitivity on the pH-value of the solution illustrated in Fig. 7.

6.2
A Basic Kinetic Model

In the following a kinetic model for photocatalytic processes recently proposed by Dillert and Bahnemann [200] will be presented yielding mathematical equations which can still be solved analytically without the assumption of limiting cases. The resulting dependencies exhibit interesting similarities with experimental results observed in various laboratories which have originally been explained by much more complicated models.

In this model the absorption of a photon in a semiconductor particle:

$$TiO_2 + h\nu \rightarrow TiO_2(e^-_{CB} + h^+_{VB}) \tag{109}$$

is taken as the initiation step with a rate constant k_{109} leading to the generation of an electron/hole pair which can either recombine (rate constant k_{110}) or undergo subsequent redox reactions:

$$TiO_2(e^-_{CB} + h^+_{VB}) \rightarrow heat \tag{110}$$

In the presence of, e.g., molecular oxygen the electron can be scavenged. Here the formation of hydrogen peroxide has been chosen as an example. Following the formation of $O_2^{\bullet-}$ via:

$$O_{2,ads} + e^-_{CB} \rightarrow O_2^{\bullet-}{}_{ads} \tag{111}$$

H_2O_2 is the result of a subsequent reduction:

$$e^-_{CB} + O_2^{\bullet-}{}_{ads} + 2H^+ \rightarrow H_2O_2 \tag{112}$$

with k_{111} and k_{112} being the respective rate constants for the reactions at Eqs. (111) and (112). The remaining hole is then available to induce the oxidation of a substrate (pollutant) molecule "Red." which could be the rate limiting step of the anticipated detoxification process (reaction at Eq. 113, rate constant k_{113}):

$$h^+_{VB} + Red. \rightarrow Prod. \tag{113}$$

The photon flux in this reaction scheme is given by

$$\frac{d[h\nu]}{dt} = k_1[h\nu] \tag{114}$$

and the dependence of the product yield on the photon flux can be determined via

$$\frac{\dfrac{d[Prod.]}{dt}}{\dfrac{d[h\nu]}{dt}} = f\left(\frac{d[h\nu]}{dt}\right) \tag{115}$$

Using this formalism Dillert and Bahnemann [200] have calculated the predicted rates and yields of reduction and oxidation processes, respectively, using various sets of parameters. A typical example of these calculations is shown in Fig. 8 where reduction rate and yield are plotted as a function of the photon concentration for different values of the recombination rate constant k_{110}. While the rate of peroxide formation increases linearly with an increasing photon flux (which according to Eq. 114 is equivalent to the photon concentration plotted in Fig. 8 since $k_{109} = 1$) when the electron/hole recombination is relatively slow, a distinct deviation from this relationship is observed at high light intensities for larger values of k_{110}.

A square root dependence of the reduction yield upon the light intensity is the predicted result under these conditions as derived from the slope of the

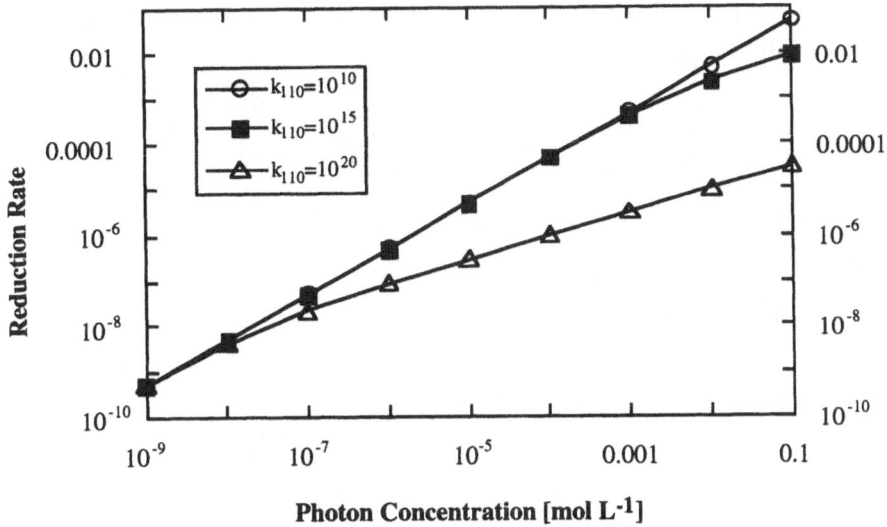

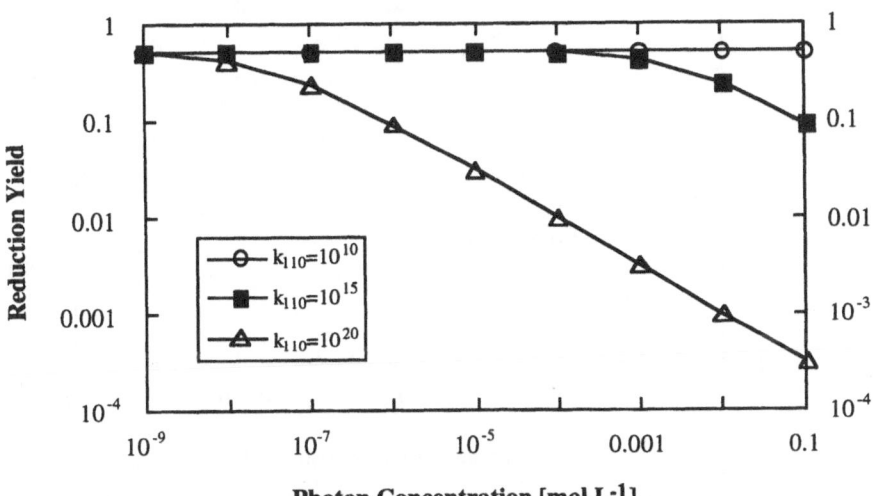

Fig. 8. Computer simulation of the dependence of reduction rate (*upper graph*) and yield (*lower graph*) on the photon concentration for different recombination rate constants k_{110}; $k_{109} = 1$, $k_{111} = 10^{10} \, l \, mol^{-1} \, s^{-1}$, $k_{113} = 10^9 \, l \, mol^{-1} \, s^{-1}$, $c_{red} = 10^{-3} \, mol \, l^{-1}$, $c_{ox} = 2 \times 10^{-4} \, mol \, l^{-1}$ (adopted from [200])

double-logarithmic presentation given in the lower part of Fig. 8. Two electrons are required for the formation of one molecule hydrogen peroxide (cf. reactions at Eqs. 111 and 112), i.e., a maximum yield of 0.5 is encountered for the reduction. The one-hole oxidation step (reaction at Eq. 113) which was taken as the rate-limiting step of the degradation process results in a maximum quantum

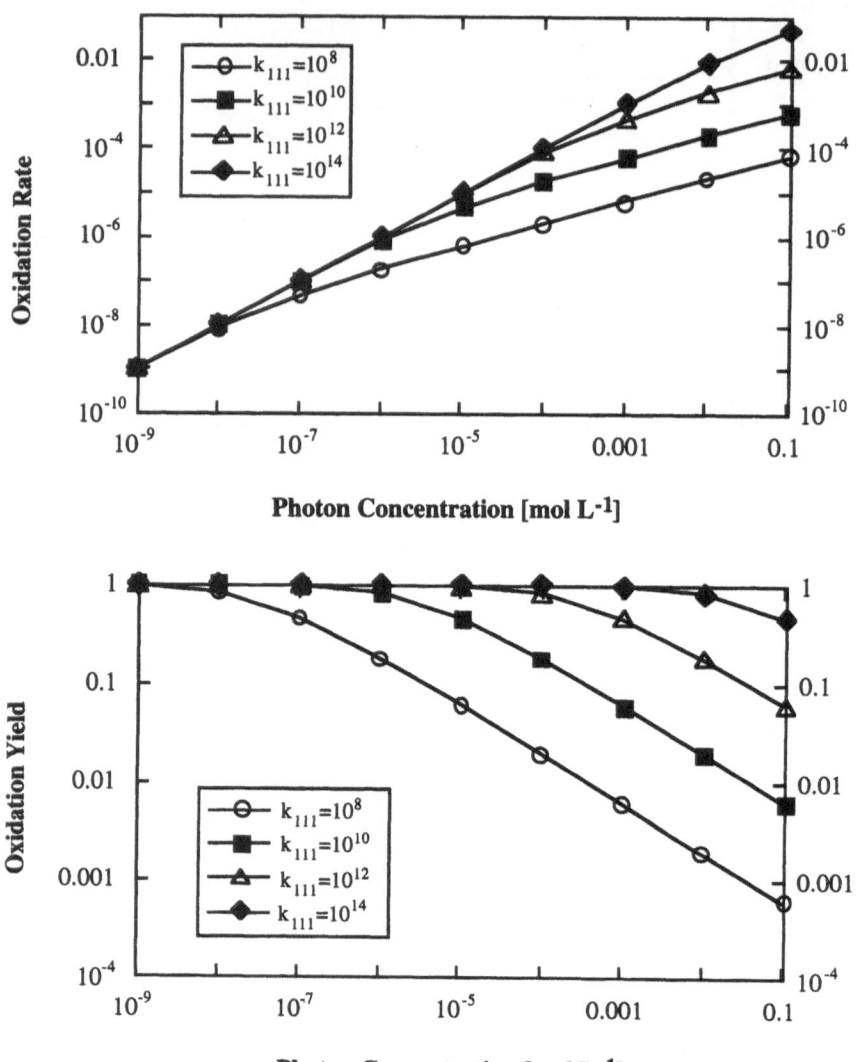

Fig. 9. Computer simulation of the dependence of oxidation rate (*upper graph*) and yield (*lower graph*) on the photon concentration for different reduction rate constants k_{111}; $k_{109} = 1$, $k_{110} = 10^{20}\,l\,mol^{-1}\,s^{-1}$, $k_{113} = 10^{11}\,l\,mol^{-1}\,s^{-1}$, $c_{red} = 10^{-3}\,mol\,l^{-1}$, $c_{ox} = 2 \times 10^{-4}\,mol\,l^{-1}$ (adopted from [200])

yield of unity for the oxidation as shown in Fig. 9. As illustrated in this figure, the simulation predicts that a square-root dependence of the degradation rate of the organic pollutant on the illumination should be observed, especially if the rate of reduction becomes relatively small. Qualitatively the results of these calculations are in excellent agreement with those presented in Fig. 7 for pH 5.

Even though this model suffices to explain the experimentally observed square root dependence of the reaction rate of a photocatalytic process on the light intensity as well as the model of surface hydroxyl radical combination proposed by Kormann et al. [21] and presented above, there are, to the best of our knowledge, no experimental results available to decide which model should be preferred.

Whenever the photocatalytic degradation rate has been measured over a wide range of solute concentrations, a saturation effect has been observed at higher concentrations. As shown for the photocatalytic 4-chlorophenol degradation in detail above (cf. Eqs. 43 and 46), this observation was frequently attributed to a Langmuir-Hinshelwood type of adsorption phenomenon on the particle surface. However, based upon the report by Cunningham and Sedlak who have shown that in many cases no correlation exists between the "real" adsorption isotherm measured by alternative methods and the kinetic results from detoxification experiments [149], Dillert and Bahnemann have used their kinetic model and tested its prediction on the solute concentration dependence of the reaction rate [200]. Figure 10 shows the results of their calculations for the reduction and the oxidation yield when the reductant, i.e., the pollutant concentration is varied.

The experimentally observed saturation behavior is apparently predicted rather well by this kinetic model which does not include any adsorption parameter. While this qualitative agreement certainly does not prove that the model is correct, it should indeed be emphasized that rather simple models such as that proposed here could in principle be sufficient to explain and predict the often rather complex behavior of photocatalytic systems.

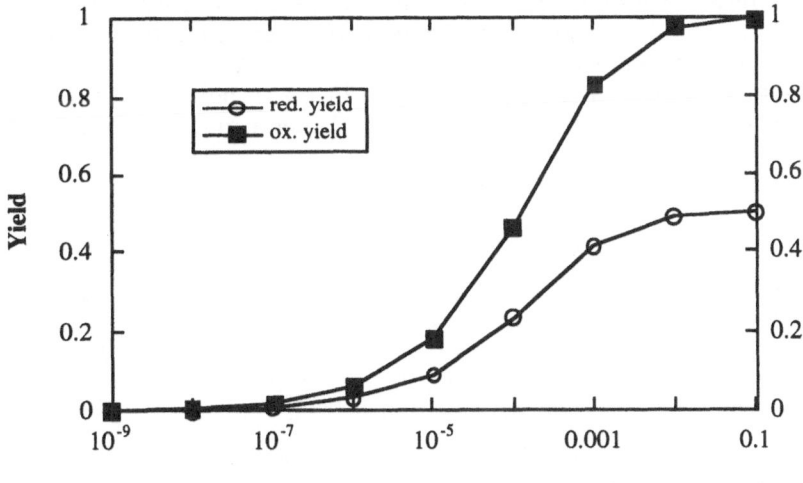

Fig. 10. Computer simulation of the dependence of oxidation and reduction yield on the reductant (pollutant) concentration; $k_{109} = 1, k_{110} = 10^{20}\,l\,mol^{-1}\,s^{-1}, k_{111} = 10^{10}\,l\,mol^{-1}\,s^{-1}, k_{113} = 10^{11}\,l\,mol^{-1}\,s^{-1}$ (adopted from [200])

7
Solar Applications

7.1
Reactor Types

The artificial generation of photons required for the detoxification of polluted water is the most important source of costs during the operating of photocatalytic wastewater treatment plants. This suggests use of the sun as an economically and ecologically sensible light source. With a typical UV-flux near the surface of the earth of $20-30$ W m^{-2} the sun puts $0.2-0.3$ mol photons m^{-2} h^{-1} in the $300-400$ nm range at the disposal of the process [205]. Principally these photons are suitable for destroying water pollutants in photocatalytic reactors.

In recent years several reactors for the solar photocatalytic water treatment have been developed and tested. In the following, the four most frequently used reactor concepts will be presented.

7.1.1
Parabolic Trough Reactor (PTR)

A parabolic trough reactor (PTR) concentrates the parallel (direct) rays of the photocatalytically active ultra-violet part of the solar spectrum by a factor of $30-50$ and can be characterized as a typical plug flow reactor. This reactor type has been chosen for solar detoxification loops constructed in the USA in Albuquerque (by Sandia National Laboratories), in California (by Lawrence Livermoore Laboratories), and in Spain in Almeria (by the Plataforma Solar de Almeria, PSA). Parallel to investigations under direction of NREL and Sandia National Laboratories in New Mexico (USA) and in California [206–209] several research groups from different European countries, funded by the European Community, are testing the parabolic trough reactor installed at the PSA (Spain) for solar wastewater treatment since 1990.

Figure 11 shows a schematic flow chart of the installation at the PSA. The reactor located at the Plataforma Solar de Almeria in Spain contains a total

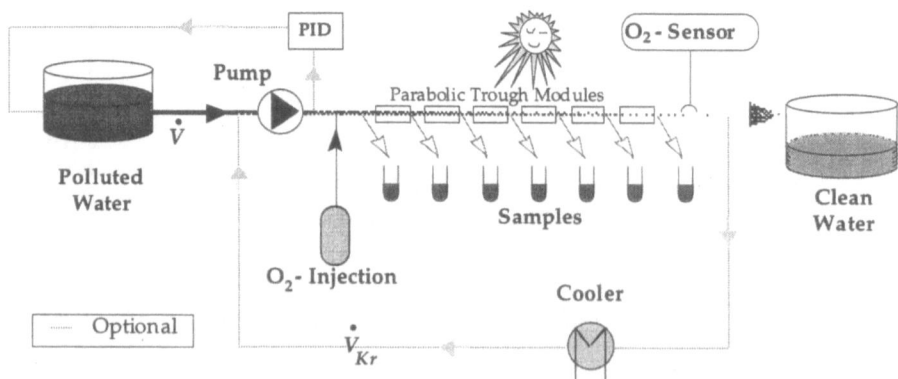

Fig. 11. Schematic flow chart of the PTR installed at the Plataforma Solar de Almeria/Spain

volume of 419 l and consists of six parabolic trough modules connected in series (type "Helioman" from Mannesmann, initially designed for the conversion of solar radiation into heat) which are controlled by a two-axial (azimuth and elevation) tracking system. The solar rays are concentrated by parabolic trough mirrors (total aperture area of 192 m^2) with an aluminized UV-reflective surface and focused on borosilicate glass tubes (total length 108 m, illuminated volume 242 l) [206, 210], which are filled with the contaminated aqueous TiO$_2$-suspensions moving at flow rates between 250 and 3500 l h^{-1}. Due to losses caused by reflectivity, translucence, and system errors, the yield of the UV-light photons reaching the contaminated suspensions is about 58% of the original light intensity entering the aperture plane.

7.1.2
Thin Film Fixed Bed Reactor (TFFBR)

One of the first solar reactors not applying a light-concentrating system and thus being able to utilize the diffuse as well as the direct portion of the solar UV-A irradiation for the photocatalytic process is the thin-film-fixed-bed reactor (TFFBR) depicted in Fig. 12 [211–213]. It should be noted that under AM (air mass) 1.5 conditions the diffuse (E_{dif}(300–400 nm) = 24.3 W cm^{-2}) and direct (E_{dir}(300–400 nm) = 25.0 W cm^{-2}) portion of the solar radiation reaching the surface of the earth are almost equal [214]. This means that a light concentrating system, e.g., a parabolic trough reactor can in principal only employ half of the solar radiation available in this particular spectral region. The most important part of the thin-

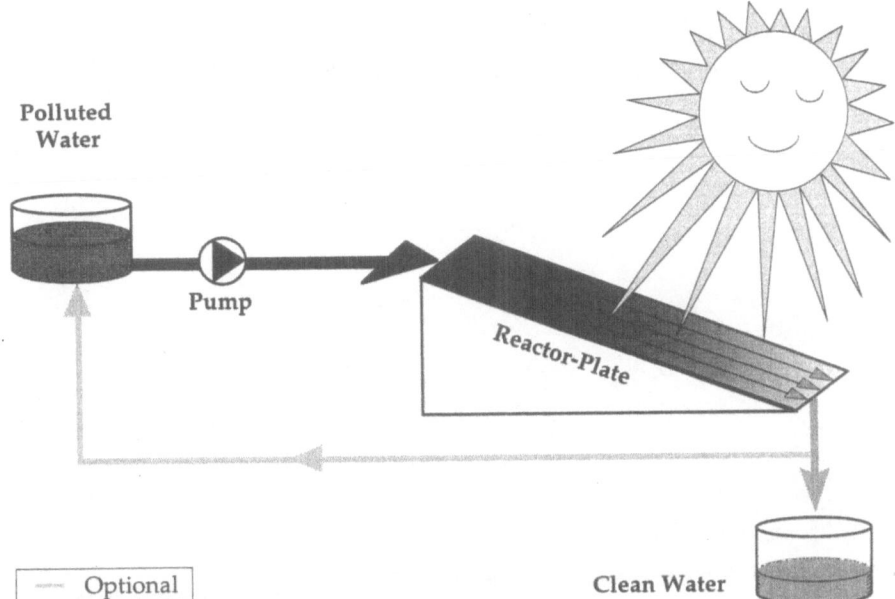

Fig. 12. Schematic flow chart of the TFFBR

film-fixed-bed reactor is a sloping plate (width 0.6 m, height 1.2 m) coated with the photocatalyst (e. g., titanium dioxide Degussa P25) [215] and rinsed with the polluted water in a very thin film ($\sim$100 μm). The flow rate is controlled by a cassette peristaltic pump and can be varied between 1 and 6.5 l h^{-1} [212].

7.1.3
Compound Parabolic Collecting Reactor (CPCR)

A compound parabolic collecting reactor (CPCR) is a trough reactor without light concentrating properties. It differs from a conventional parabolic trough reactor by the shape of its reflecting mirrors. A reflector of a parabolic trough reactor has a parabolic profile with the reaction pipe in its focal line. Consequently only parallel light entering the parabolic trough can be focused into the reaction pipe and a sun-tracking system is required. The shape of a CPCR's reflector usually consists of two half circular profiles side by side, and a parabolic continuation at both outer sides of the circles. The focal line is located closely above the connection of the two circles. This geometry enables light entering from almost any direction to be reflected into the focal line of the CPCR, i. e., most of the diffuse light entering the module can also be employed for the photocatalytic reaction. Due to this geometry a CPCR exhibits only a small concentration factor (<1.2). The CPCRs at the Plataforma Solar de Almeria in Spain (PSA) manufactured by Industrial Solar Technology Corporation, Denver, Colorado (USA), have a concentration factor of 1.15, i. e., this type of reactor has practically no light concentrating properties [216, 217].

Moreover, a CPCR does not necessarily track the sun due to its geometry. The azimuth should be adjusted to the complementary angle a of the geographical altitude and the pipes should be aligned south and from top to bottom. A schematic view of one CPCR-module at the PSA is given in Fig. 13. The angle of incidence a is adjusted to 37° for all CPCR modules at the PSA.

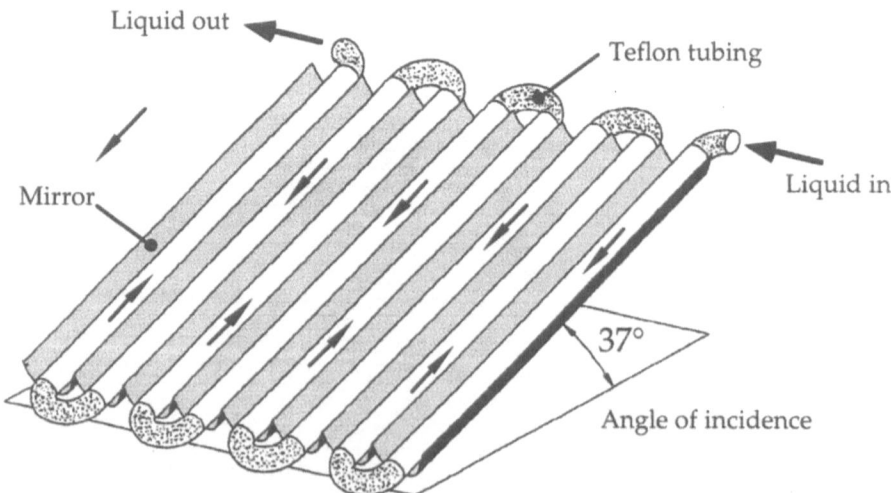

Fig. 13. View on a single CPCR module installed at the PSA (Spain)

One CPCR-module of the PSA consists of eight parallel reflectors made from polished aluminum . This material has very good reflection properties, especially in the UV-region of the solar spectrum. In the spectral range of 295–387 nm its reflection efficiency is 83.2 %. A single CPCR reflector has a length of 1.22 m and a width of 0.152 m, i.e., the effective reflecting area of one module, consisting of eight reflectors adds up to 1.48 m². The overall reflecting area of all six modules is 8.9 m². A reaction pipe made of a transparent fluoropolymer (Teflon) is fixed in the focal line of each CPCR reflector, through which the suspension containing the photocatalyst and the pollutant circulates. The transmissivity of Teflon in the spectral range under consideration is 76.8 %, thus limiting the in principle achievable reactor efficiency to 63.9 %. The absorber pipe is as long as the module (1.22 m), has an inner diameter of 48 mm, and consequently an illuminated volume of 2.21 l per pipe. The connectors between the absorber pipes consist of polypropylene; the overall illuminated volume of all six modules adds up to 106 l. The achievable flow rates in the CPCR-module field at the PSA are between 2250 and 8000 l h⁻¹ [217] and the minimum required volume is 248 l.

7.1.4
Double Skin Sheet Reactor (DSSR)

A new kind of non-concentrating reactor is the double skin sheet reactor (DSSR). It consists of a flat and transparent structured box made of PLEXIGLAS [218, 219]. The inner structure of the reactor is schematically drawn in Fig. 14.

Fig. 14. Schematic drawing of the inner layout of a double skin sheet reactor

The suspension containing the model pollutant and the photocatalyst is pumped through these channels. The comparison of the spectral irradiance of the sun (AM 1.5) with the transmission spectrum of the PLEXIGLAS used to manufacture the double skin sheets evinces that the UV-A portion of the solar spectrum below 400 nm nicely matches with the onset of the PLEXIGLAS-transmission [219]. This type of reactor can utilize both the direct and the diffuse portion of the solar radiation in analogy to the CPCR. After the degradation process the photocatalyst has to be removed from the suspension either by filtering or by sedimentation for both reactors.

The DSSR consists of a modified double-skin sheet (SDP 16/32) manufactured by the Röhm GmbH in Darmstadt (Germany). One module has a length of 1400 mm, a width of 980 mm, and contains 30 channels each of which is 28.5 × 12 mm, and their total inner volume adds up to 14.4 l in a single sheet. Figure 15 shows a photograph of an empty (front) and an operating (back) DSSR module.

Recently, a pilot reactor consisting of two DSSR modules with a total irradiated volume of 28.8 l has been compared with the CPCR installed at the PSA. [220]. The sheets of the former reactor were connected by PVC-tubes to a reservoir of 130 l. In all experiments this system was operated in a recirculation mode. The turbulent flow rate was adjusted to 11.8 l min⁻¹ resulting in a Reynolds number of about 9000, i.e., the residence time in the photochemical reactor was 73.2 s. The spatial orientation of the PLEXIGLAS reactor was also

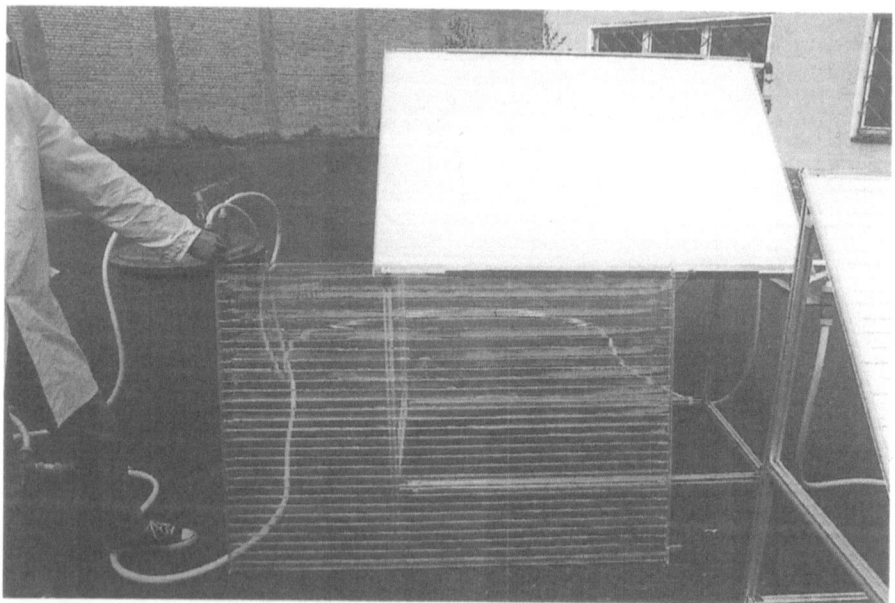

Fig. 15. View of a double skin sheet reactor; the *module in front* is empty and the *module mounted on the metal frame* is filled with an aqueous suspension of the photocatalyst (Photo Goslich/ISFH)

adjusted like that of the CPCR, i.e., the inclination angle was 37° and the face of the reactor viewed south.

7.2
Comparison of Different Solar Reactors

Extensive degradation experiments have been performed at the PSA in Almeria to compare the efficiencies of the PTR and the TFFBR [83, 190, 211–213]. Since these two reactors differed greatly in size, a comparable flow rate had to be determined. For this purpose Bahnemann and co-workers derived a dimensionless efficiency ε^{hv} to characterize continuous photoreactors [200]:

$$\varepsilon^{hv} = c^{\circ} * v^{\bullet}/I \tag{116}$$

with c° initial substrate concentration (mol l^{-1}), $v^{\bullet}$ flow rate (l s^{-1}), and I light intensity (mol photons l^{-1} s^{-1}).

Apparently, ε^{hv} is similar to the photochemical quantum yield which is defined as the ratio of the reaction rate and the absorbed photon flux. For continuous photochemical flow reactors the photochemical reaction yield $\Phi_{reactor}$ has therefore been defined as [200]:

$$\Phi_{reactor} = \Delta c * v^{\bullet}/I \tag{117}$$

with Δc amount of degraded substrate concentration (mol l^{-1}).

In case of a total degradation Δc equals c° and therefore ε^{hv} also represents the maximum possible photochemical reaction yield in any continuous photoreactor. To be able to compare the performance of the thin-film-fixed-bed reactor with that of the parabolic trough reactor, the flow rates $v^{\bullet}$ was adjusted using Eq. (116) to yield the same maximum quantum yield ε^{hv} for both types of reactors [212]. It became apparent from this comparison that it is a major engineering problem of the PTR to maintain a sufficient concentration of molecular oxygen throughout the reactor. Using small flow-rates to achieve high residence times resulted in an accumulation of oxygen bubbles in the tubes thus reducing the illuminated volume and lowering the degradation rates. Since electron scavengers such as H_2O_2 can be dissolved in water without the formation of bubbles an enhanced DCA-degradation was observed in the presence of H_2O_2 when molecular oxygen was absent [211–213].

The complete, stoichiometric degradation of the model compound dichloroacetate (DCA) was observed in both types of reactors and in agreement with laboratory results the DCA-degradation was found to exhibit a linear dependency on the light intensity at pH ~3 [83, 113, 190, 211]. Since the thin-film-fixed-bed reactor is also able to use the diffuse part of the UV-sunlight, its photocatalytic activity could even be observed on very cloudy days with almost no direct solar radiation. Moreover, within an initial DCA-concentration range between 1 and 5 mmol l^{-1} the thin-film-fixed-bed-reactor always reached a higher degree of degradation than the parabolic trough reactor. For an initial DCA-concentration of 1 mmol l^{-1} a degree of nearly 100% could be reached in a single path experiment using the thin-film-fixed-bed reactor [83].

As an example of a real wastewater stream the degradation of industrial wastewater from a phenolic resin factory has been investigated with both types of reactors. These experiments were performed in a closed loop simultaneously with both reactor types to ensure identical light intensities. While only 30% of the relative initial Total Organic Carbon concentration ([TOC]/[TOC]°) were degraded after 250 min using the PTR a decomposition of more than 70% was obtained after the same time with the TFFBR [211–213]. However, in the case of the film reactor the authors even observed a considerable decrease of the TOC concentration without illumination which they explained by vaporization of volatile components from the effluent water. This is a general problem with all open reactor systems with gas injection and must be solved by a waste gas purification unit. Nevertheless, this comparative study evinced that it is possible to achieve good photocatalytic detoxification efficiencies without using light concentrating solar reactors. Furthermore, the thin-film-fixed-bed reactor offers the advantages of a less expensive construction and reduced running costs as compared with the parabolic trough reactor. Also the fixation of the ultrafine photocatalyst particles solves the complex problem of the otherwise needed filtration unit for the separation of the catalyst from the purified wastewater.

To characterize the influence of the applied irradiation on an experiment it is necessary to evaluate the amount of irradiation which has been accumulated in the suspension at a specific time of the experiment as well as the corresponding concentration of the pollutant. In many practical applications the mass of the organic pollutant contained in the reactor at any specific time is often determined by recording the TOC (total organic carbon) content of the water in regular time intervals and multiplying this value by the overall volume of the photocatalytic reactor, V_t. The solar light energy is usually applied to a limited surface area, namely of the aperture of the PTR or the CPCR or the glass plate of the TFFBR or the PLEXIGLAS sheets of the DSSR, respectively. To account for these differences the so-called "mass area ratio" S_{TOC} of the organic pollutant has recently been proposed by Goslich et al. [220]. This has been defined as the overall mass of the pollutant per surface element of the reactor, A_r:

$$S_{TOC} = (TOC \cdot V_t)/A_r \tag{118}$$

This mass area ratio will be strongly dependent on the incident light energy, i.e., it will depend on the illumination time. The change of S_{TOC} with time (dS_{TOC}/dt) should therefore yield the effective degradation rate for a specific reactor, R_{eff}, assuming that there were no radiation losses due to reflection or scattering:

$$R_{eff} = d\,S_{TOC}/dt \sim \Delta S_{TOC}/\Delta t_{exp} \sim S_{TOC}/t_{exp} \tag{119}$$

i.e., R_{eff} specifies the degraded mass per time interval and per surface area of the irradiated reactor.

To account for the fluctuations of the spectral solar irradiance with time it is reasonable to divide the mass area ratio of the pollutant S_{TOC} by the UV radiation accumulated, $E_{uv, tot.}$ at a specific time of the experiment yielding the "efficiency", η, of the reactor under consideration [220]:

$$\eta = S_{TOC}/E_{uv, tot.} \tag{120}$$

This efficiency η relates the degraded mass of the pollutant to the irradiation energy reaching the reactor surface and hence represents a lower limit for the efficiency of a reactor, as it does not take into account any losses of the photon energy due to reflection, scattering, absorption in reactor walls etc. η is similar to the photonic efficiency ζ (see above [71]) with the difference that ζ has no physical dimension whereas η has the dimension of a unit mass per unit energy. The determination of η seems to be particularly useful in the case of fluctuating radiation conditions and is readily accessible. Recently, Goslich et al. [220] have determined η for the photocatalytic degradation of the model pollutant dichloroacetate, DCA, at different catalyst concentrations to compare the efficiencies of the CPCR and the DSSR under identical solar conditions at the PSA. Figure 16 shows a plot of the calculated reactor efficiencies η as a function of the different photocatalysts and photocatalyst concentrations used in this study.

The reactor efficiency for the degradation of DCA in the presence of Degussa P25 starts at about 1.0 µg J^{-1} at the lowest catalyst concentration of 0.25 g l^{-1} and increases quickly up to a catalyst concentration of 1 g l^{-1}. The increase in efficiency above 1 g l^{-1} is much slower; in the CPCR it even decreases to about 2.5 µg J^{-1} at a catalyst concentration of 5 g l^{-1} P25. This behavior is similar to that observed in laboratory experiments [221]. Thus it does not appear to be

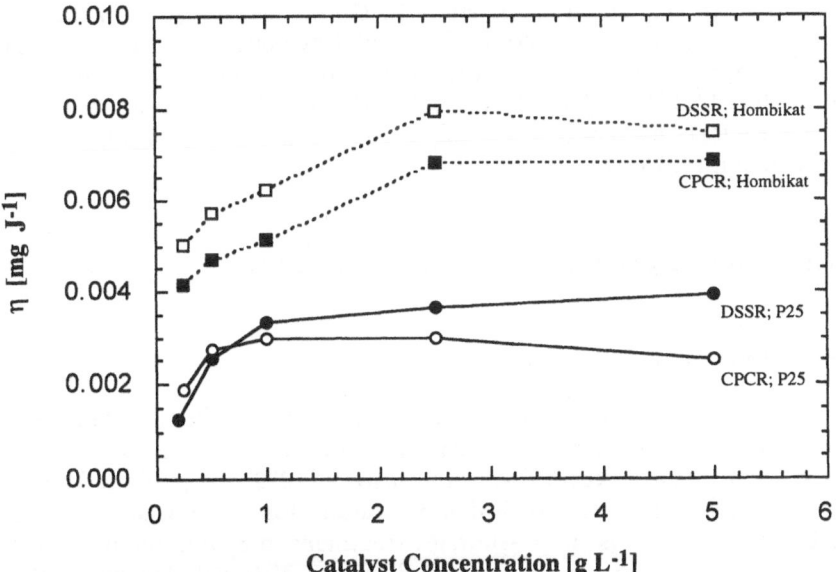

Fig. 16. The efficiency of the degradation of dichloroacetic acid (DCA) as a function of the catalyst concentration in both, the CPCR and the DSSR (adopted from [220])

necessary to use a catalyst concentration higher than $1\,g\,l^{-1}$ to degrade model pollutants sufficiently when P25 is employed. This result clearly shows that laboratory experiments can be used as a good approximation for the behavior of real pilot installations. The same effect could be observed using Hombikat UV 100 as the photocatalyst. Figure 16 shows that the reactor efficiencies for the DSSR and the CPCR are much higher for this photocatalyst. The shape of the curves is similar compared to the P25 curves, except that there is no decrease in the reactor efficiency in the CPCR at the highest catalyst concentrations used.

The comparison with recent laboratory data evinces that a plot of the reactor efficiency η as a function of the catalyst concentration yields a similar shape as that of the photonic efficiency determined in laboratory experiments [221]. This qualitative similarity is a good indication for the ease and usefulness of the definition of η. Moreover when the concentration of the organic pollutants is monitored by a TOC measurement the reactor efficiency η can be converted into the photonic efficiency ζ via [220]:

$$\zeta = 28.01\,\eta \tag{121}$$

provided η is given in [mg TOC J^{-1}]. The conversion factor has the dimension [J mg^{-1}] assuming all photons entering the reactor have an average energy of $5.6 \times 10^{-19}\,J$ (assuming an average illumination wavelength of 355 nm) and establishes the relation between this energy and the mass of a carbon atom ($2.0 \times 10^{-20}\,mg$). This conversion leads to a photonic efficiency ζ of 5.1% for P25 at a catalyst concentration of $0.25\,g\,l^{-1}$ and of 14.1% for Hombikat UV 100 in the CPCR (see Fig. 16). These values are also quantitatively almost identical to those found in laboratory experiments using 50 ml reactors and artificial illumination [221].

It is a surprising result of the study by Goslich et al. [220] that the reactor efficiency obtained in the DSSR is always slightly better than that in the CPCR since no special efforts have yet been made to optimize the overall reactor concept of the DSSR [219]. The CPCR, on the other hand, presents a reactor design in which efforts have been especially made for the collection of diffuse and direct sunlight [216, 217].

8
Solar Photocatalytic Treatment of Real Wastewater

8.1
Industrial Wastewater

Despite its obvious potential for the detoxification of polluted water, there has been very little commercial or industrial use of photocatalysis as a technology so far. According to a recent review by Goswami [222] the published literature shows only two engineering scale demonstrations for groundwater treatment in the U.S. and one industrial wastewater treatment in Spain. Engineering scale field experiments have been conducted by the National Renewable Energy Laboratory (NREL) at Lawrence Livermoore National Laboratory (LLNL) treating groundwater contaminated with trichloroethylene (TCE) [223]. This field system consisted of 158 m² of parabolic trough reactors and used Degussa P25 particles (0.1%) as photocatalysts in a slurry flow configuration. With this relatively low titanium dioxide content the TCE concentration was reduced from 200 ppb to less than 5 ppb.

Engineering scale demonstration of the nonconcentrating solar reactor technology was conducted at Tyndall Air Force Base in 1992 [224, 225] treating groundwater contaminated with fuel, oil, and lubricants which was leaking from underground storage tanks. While these field tests successfully demonstrated the feasibility of destroying benzene, toluene, ethylbenzene, and xylenes (BTEX) in the groundwater, the observed reaction rates were rather low. However, in a laboratory test with city water spiked with the same amounts of BTEX as in the groundwater and using the same reactors under similar sunlight conditions, Goswami et al observed reaction rates which were an order of magnitude faster [224]. These results suggested that a careful site treatability study and the establishment of an appropriate pretreatment will be very important for the successful field deployment of solar photocatalytic processes [222].

Blanco and Malato [226] conducted an engineering scale field demonstration of industrial wastewater treatment from a resins factory. This effluent contained organic contaminants such as phenol, phthalic acid, fumaric acid, maleic acid, glycols, xylene, toluene, methanol, butanol, and phenylethylene amounting to a total of 600 ppm TOC. Using the PTR reactor set-up installed at the PSA (see above) and employing $0.1 \, g \, l^{-1}$ Degussa P25 as the photocatalyst and $7 \, mmol \, l^{-1}$ sodium peroxydisulfate $(Na_2S_2O_8)$ as oxidizing agent, the authors observed a complete TOC degradation within 44 min of illumination.

Other investigations of real industrial wastewater treatment which have not yet been conducted on an engineering scale show the potential of the solar photocatalytic water treatment process, especially in areas where conventional treatment methods are not successful. Thus, the potential of the solar photo-catalytic oxidation could be shown for the decolorization and the COD reduction of wastewater from a 5-fluorouracil (a cancer drug) manufacturing plant [227]. Employing 0.1% titanium dioxide powder as the photocatalyst and 2400 ppm hydrogen peroxide (H_2O_2) as an oxidizing additive, Anheden et al. showed that the color of the wastewater was reduced by 80% within 1 h of treatment while the COD was reduced by 70% within 16 h [227]. The authors claimed that all of the conventional treatments tried previously for the reduction of COD and color by the drug manufacturers had failed. A simultaneous color and COD reduction was also observed during the photocatalytic treatment of a distillery wastewater which had been pretreated by anaerobic microbiological methods [228]. A laboratory study conducted by Turchi and Ollis has furthermore shown that wastewater from pulp and paper mills can also be treated conveniently by photocatalysis [229].

Bekbölet et al. have recently applied the photocatalytic degradation process for the destruction of biologically pretreated landfill leachate effluent using Sachtleben Hombikat UV100 as the photocatalyst [230]. A batch reactor was used to compare this photocatalyst with the widely used Degussa P25 at pH 3. Both materials, especially Hombikat UV100, showed a strong adsorption of the organic pollutants; the overall TOC-removal was almost the same (≈70%) after

5 h of irradiation with a high-pressure xenon-lamp, and the BOD_5 increase was insignificant indicating that the water could still not be treated by conventional biological methods (BOD_5 is defined as the biological oxygen demand during an incubation period of 5 days at 37°C and thus represents a convenient measure for the biodegradability of a real wastewater). In addition to the detoxification experiments the pH-dependent adsorptive properties of the landfill leachate onto Hombikat UV100 were examined. In good agreement with the pH-dependence of the photocatalytic degradation of the leachate, a strong adsorption was observed between pH 3 and pH 8 with a maximum at pH 5, while at pH 11 the adsorption was much lower [230].

Further degradation experiments were performed with the Thin-Film Fixed-Bed Reactor (TFFBR) using Hombikat UV100 coated plates irradiated by 16 fluorescent tubes, resembling the UV-portion of the solar spectrum. The highest degradation rate was observed between pH 3 and 7 with a maximum at pH 5, while at pH 9 and 11 the degradation was rather slow. Besides this pH-dependency the influence of the initial pollutant concentration (dilution of the landfill leachate), the light intensity and the addition of oxidizing reagents (H_2O_2 and $Na_2S_2O_8$) on the degradation were examined [230]. A higher rate could be achieved by diluting the leachate, with a maximum rate at a moderate dilution around 1:2. No beneficial effect was obtained by the addition of H_2O_2 or $Na_2S_2O_8$. The addition of 0.5 or 1 mmol l^{-1} H_2O_2 (at pH 7) even resulted in an inhibition, while in the presence of $Na_2S_2O_8$ (at the same concentration) a dark degradation was observed. Illumination with a reduced light intensity (50 $W m^{-2}$ instead of 100 $W m^{-2}$) resulted in a lower degradation rate, but a higher photonic efficiency, i.e., a higher efficiency of the process than with maximum light intensity. Finally, the TFFBR system was used as a recirculation system, and the dependence of the initial concentration was examined. The highest photonic efficiency was obtained with undiluted landfill leachate, while the highest relative photodegradation, and therefore the lowest half life was measured with the most diluted sample (1:5). Photonic efficiency and photo-reactor efficiency as well as the kinetic parameters were calculated for the recirculation system. The photonic efficiencies obtained were 0.3–1.0%, i.e., similar to those of the single pass experiments, but an order of magnitude smaller than, e.g., for the degradation of dichloroacetic acid (DCA). The relatively small photonic efficiencies for the degradation of the effluent water are not surprising, since it is a well known fact that the detoxification of real wastewaters by photocatalysis or other chemical oxidation techniques yield considerably lower ζ-values than for model systems, e.g., due to the presence of inhibiting anions [81].

However, it could be shown by this investigation that photocatalysis is indeed a feasible method for the successful clean-up of landfill effluent water [230]. This is especially important as Weichgrebe et al. [231, 232] could demonstrate that by photocatalysis higher rates can be obtained for the degradation of land-fill leachate than with other, more conventional, wet oxidation methods (ozone or hydrogen peroxide coupled with UV-irradiation).

Recently, Freudenhammer et al. reported their results from a pilot study using TFFBR reactors which was performed in various Mediterranean countries and showed that biologically pretreated textile wastewater can be cleaned by solar photocatalysis with a maximum degradation rate of 3 g COD h^{-1} m^{-2} [233]. Moreover, a comparison of reaction rates measured with artificial and solar illumination, respectively, during this study proved the necessity of outdoor experiments. It was concluded that photocatalysis should be a suitable technology as the final stage of purification of biologically or physically pretreated wastewater, particularly in sunrich areas. However, it was also pointed out that industrial applications of solar wastewater treatment on a larger scale is in strong demand for cheap photocatalysts of high activity to be competitive with treatment methods already established on the market [233].

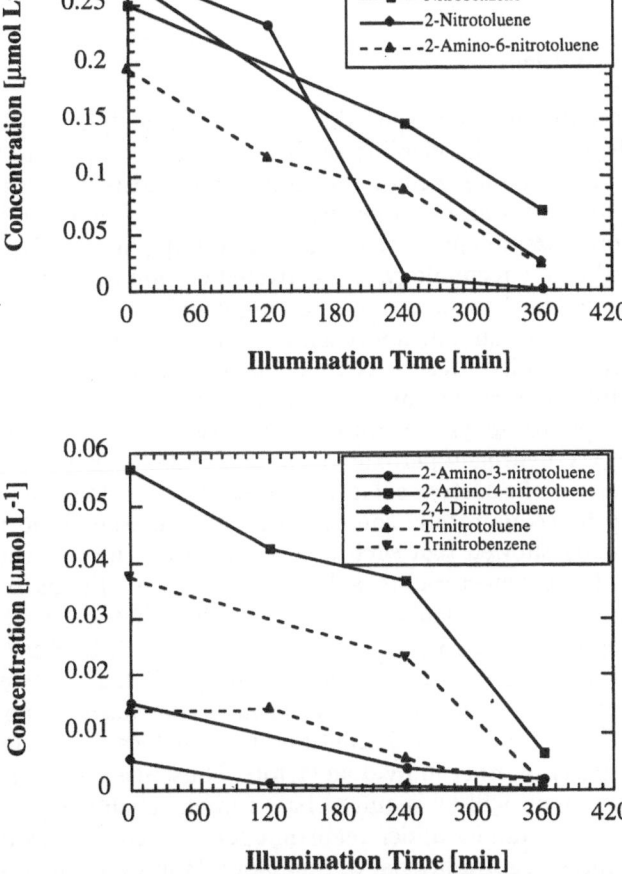

Fig. 17. Photocatalytic degradation of nitroaromatic compounds in a polluted groundwater, Degussa P25; natural pH; air-sat. suspension; I = 0.23 mmol photons l^{-1} min^{-1}; ambient temperature (adopted from [219])

Employing the double-skin sheet reactor described above (DSSR, see above) van Well et al. performed photocatalytic detoxification experiments with groundwater contaminated with nitro aromatic compounds from a World War II-ammunition plant, proving that this reactor can indeed be employed for the solar detoxification of contaminated water [219]. These solar degradation experiments were performed in Hannover/Germany at the beginning of May 1995 with the reactor adjusted with an inclination angle of 45°. During these experiments the whole surface of the reactor was irradiated by the sun. The concentrations of the nitroaromatic compounds were determined by a gas chromatograph equipped with an ECD (HRGC 5300, Carlo Erba) using 2-nitrophenol as internal standard. In Fig. 17 the decrease of the individual concentrations of the nitroaromatic contaminants is given as a function of the illumination time [219]. As can be seen, more than 75% of all identified contaminants were degraded within 6 h of irradiation and photonic efficiencies similar to those obtained previously in laboratory studies were calculated [234, 235].

8.2
Economic Considerations

In most cases the economics of the solar photocatalytic applications will determine the commercial viability of this process. According to Goswami an economic analysis should include the fixed capital costs, recurring operation and maintenance costs, and energy costs [222]. Ollis presented an early comparative economic assessment of water purification by photocatalysis employing UV-lamps, UV-ozone technology, and activated carbon adsorption technology [236]. He concluded that the cost of treating contaminated water by the UV-photocatalytic system already approaches the treatment cost by the activated carbon system provided that the costs of spent carbon disposal are included. Turchi and Link conducted a comparative economic assessment of water treatment by solar photocatalysis (employing PTR reactors), activated carbon adsorption, and UV-hydrogen peroxide technology [237]. They concluded that a fivefold decrease of the costs of solar water treatment would be required for this technology to be economically competitive with the established technologies and subsequently showed that such a cost reduction can be achieved by the deployment of nonconcentrating solar reactors [238]. In his recent review Goswami reported cost calculations for the solar photocatalytic wastewater treatment ranging from 0.53 US$ m^{-3} all the way to 27.57 US$ m^{-3} based upon various literature sources [222]. Obviously, the predicted treatment cost will strongly depend upon the type of wastewater, the desired mode of plant operation (i.e., continuous vs discontinuous), the location of the projected plant, and a variety of other individual factors. Thus, in each case an individual economic assessment will be required based upon reliable experimental data which have to be obtained under realistic operating conditions to be able to compare the costs with traditional technologies. Moreover, whenever the costs of the activated carbon adsorption are used for such a comparison it will be important to include also the costs for an environmentally safe disposal of the used and thus often highly contaminated carbon adsorber.

In many cases the cost of the land area required for the installation of a solar water detoxification plant will add a significant portion to the overall costs of the process. Therefore, it is important to calculate the necessary illuminated reactor area as accurately as possible. Based on the efficiencies determined for solar detoxification reactors, van Well et al. have recently presented a calculation of the reactor area required for a given pollution problem [219]. Taking a continuous flow reactor and typical solar UV-A light intensities the authors assumed a zero order reaction with its rate being directly proportional to the light intensity I and the illuminated surface of the photoreactor A. Considering a total degradation of the contaminants they calculated the required illuminated reactor area A as a function of the incoming molar flow, $n^{\bullet}_{in}$, assuming a light intensity I of 60 µmol photons $m^{-2} s^{-1}$ and different photonic efficiencies [219]. Taking an incoming flux $n^{\bullet}_{in}$ of 0.6 mmol $l^{-1} s^{-1}$, i.e., about 0.22 mol $m^{-3} h^{-1}$, these calculations showed that a total reactor area of 102.9 m^2 will be required at a flow rate of 10 $m^3 h^{-1}$ assuming a photonic efficiency of $\zeta = 0.1$ [219]. This could, for example, be realized by a cascade of 75 Plexiglas double-skin sheet reactors (DSSR). A straightforward calculation evinces that this area requirement will have to be multiplied by a factor of 2.3 or 4.6, respectively, when the destruction of the contaminants follows first order kinetics and 90 or 99% degradation, respectively, is the desired result.

References

1. Carey JH, Lawrence J, Tosine HM (1976) Bull Environ Contam Toxicol 16:697
2. Hecht J (1990) New Sci April 14:28
3. Fell U (1994) Frankfurter Allgemeine Zeitung May 25:N2
4. Legrini O, Oliveros E, Braun AM (1993) Chem Rev 93:671
5. Blake DM (1994) Bibliography of work on the photocatalytic removal of hazardous compounds from water and air. Report NREL/TP-430-6048, National Renewable Energy Laboratory, Golden, CO
6. Venkatardi R, Peters RW (1993) Hazard Waste & Hazard Mat 10:107
7. Kamat PV (1993) Chem Rev 93:341
8. Fox MA, Dulay MT (1993) Chem Rev 93:341
9. Pichat P (1994) Catal Today 19:313
10. Lewis LN (1993) Chem Rev 93:2693
11. Bahnemann DW, Cunningham J, Fox MA, Pelizzetti E, Serpone N (1994) In: Helz G, Zepp R, Crosby DG (eds) Aquatic and surface chemistry. Lewis Publishers, Boca Raton, FL, p 261
12. Serpone N, Pelizzetti E (1989) Photocatalysis: fundamentals and applications. Wiley, New York
13. Pelizzetti E, Schiavello M (1991) Photochemical conversion and storage of solar energy. Kluwer Acad Publ, Dordrecht
14. Ollis DF, Al-Ekabi H (1993) Photocatalytic purification and treatment of water and air, vol 3. Elsevier, Amsterdam
15. Hoffmann MR, Martin ST, Choi W, Bahnemann DW (1995) Chem Rev 95:69
16. Boer KW (1990) Survey of semiconductor physics. Van Nostrand R, New York, p 249
17. Rothenberger G, Moser J, Grätzel M, Serpone N, Sharma DK (1985) J Am Chem Soc 107:8054
18. Grätzel M (1989) Heterogeneous photochemical electron transfer. CRC Press, Boca Raton, FL

19. Memming R (1988) In: Steckham E (ed) Topics in current chemistry, vol 143. Springer, Berlin Heidelberg New York, p 79
20. Mills G, Hoffmann MR (1993) Environ Sci Technol 27:1681
21. Kormann C, Bahnemann DW, Hoffmann MR (1991) Environ Sci Technol 25:494
22. Carraway ER, Hoffman AJ, Hoffmann MR (1994) Environ Sci Technol 28:786
23. Chemeseddine A, Boehm HP (1990) J Mol Catal 60:295
24. D'Oliveira JC, Minero C, Pelizzetti E, Pichat P (1993) J Photochem Photobiol A 72:261
25. D'Oliveira JC, Al-Sayyed G, Pichat P (1990) Environ Sci Technol 24:990
26. Hidaka H, Zhao J, Pelizzetti E, Serpone N (1992) J Phys Chem 96:2226
27. Pelizzetti E, Minero C, Piccinini P, Vincenti M (1993) Coord Chem Rev 125:183
28. Ollis DF, Pelizzetti E, Serpone N (1991) Environ Sci Technol 25:1523
29. Albert M, Gao YM, Toft D, Dwight K, Wold A (1992) Mater Res Bull 27:961
30. Inel Y, Ertek D (1993) J Chem Soc Faraday Trans 89:129
31. Borgarello E, Serpone N, Emo G, Harris R, Pelizzetti E, Minero C (1986) Inorg Chem 25:4499
32. Augugliaro V, Palmisano L, Sclafani A, Minero C, Pelizzetti E (1988) Toxicol Environ Chem 16:89
33. Ollis DF, Hsiao CY, Budiman L, Lee CL (1984) J Catal 88:89
34. Turchi CS, Ollis DF (1989) J Catal 119:480
35. Turchi CS, Ollis DF (1990) J Catal 122:178
36. Terzian R, Serpone N, Draper RB, Fox MA, Pelizzetti E (1991) Langmuir 7:3081
37. Noda H, Oikawa K, Kamada H (1993) Bull Chem Soc Jpn 66:455
38. Anpo M, Shima T, Kubokawa Y (1985) Chem Lett 1799
39. Jaeger CD, Bard A (1979) J Phys Chem 83:3146
40. Mao Y, Schöneich C, Asmus KD (1991) J Phys Chem 95:80
41. Moser J, Punchihewa S, Infelta PP, Grätzel M (1991) Langmuir 7:3012
42. Matthews RW, McEvoy SR (1992) J Photochem Photobiol A 64:231
43. Ohtani B, Nishimoto S (1993) J Phys Chem 97:920
44. Tunesi S, Anderson M (1991) J Phys Chem 95:3399
45. Tunesi S, Anderson MA (1992) Langmuir 8:487
46. Phillips LA, Raupp GB (1992) J Mol Catal 77:297
47. Stafford U, Gray KA, Kamat PV, Varma A (1993) Chem Phys Lett 205:55
48. Izumi I, Dunn WW, Wilbourn KO, Fan FRF, Bard AJ (1980) J Phys Chem 84:3207
49. Matthews RW (1984) J Chem Soc Faraday Trans 1 80:457
50. Fox MA, Draper RB, Dulay M, O'Shea K (1991) In: Pelizzetti E, Schiavello M (eds) Photochemical conversion and storage of solar energy. Kluwer Acad Publ, Dordrecht, p 323
51. Draper RB, Fox MA (1990) Langmuir 6:1396
52. Neta P, Simic M, Hayon E (1969) J Phys Chem 73:4207
53. Schuchmann MN, Zegota H, von Sonntag C (1985) Z Naturforsch 40b:215
54. Schuchmann HP, von Sonntag C (1984) Z Naturforsch 39b:217
55. Wolff K, Bockelmann D, Bahnemann DW (1991) In: Levi B (ed) Proceedings of the IS&T 44th Annual Conference, IS&T, Springfield, VA, p 259
56. Hilgendorff M (1996) PhD thesis. University of Hannover, Hannover, Germany
57. Grabner G, Li GZ, Quint RM, Quint R, Getoff N (1991) J Chem Soc Faraday Trans 87:1097
58. Richard C (1993) J Photochem Photobiol A 72:179
59. Serpone N, Lawless D, Khairutdinov R, Pelizzetti E (1995) J Phys Chem 99:16,655
60. Colombo DP, Bowman RM (1995) J Phys Chem 99:11,752
61. Colombo DP, Roussel KA, Sach J, Skinner DE, Cavaleri JJ, Bowman RM (1995) Chem Phys Lett 232:207
62. Bahnemann DW, Henglein A, Lilie J, Spanhel L (1984) J Phys Chem 88:4278
63. Bahnemann DW, Henglein A, Spanhel L (1984) Faraday Discuss Chem Soc 107:8054
64. Bahnemann DW, Hilgendorff M, Memming R (1997) J Phys Chem B 101:4265
65. Buxton GV, Greenstock CL, Helman WP, Ross AB (1988) J Phys Chem Ref Data 17:513

66. Schwartz SE (1981) J Chem Educ 58:101
67. Duonghong D, Ramsden J, Grätzel M (1982) J Am Chem Soc 104:2977
68. Bielski BHJ, Cabelli DE, Arudi RL (1985) J Phys Chem Ref Data 14:1041
69. Marcus RA (1977) In: Rock PA (ed) Special Topics in Electrochemistry, Elsevier, Amsterdam, p 161
70. Gerischer H (1979) Top Appl Phys 31:115
71. Serpone N, Terzian R, Lawless D, Kennepohl P, Sauvé G (1993) J Photochem Photobiol A 73:11
72. Mills A, Davies RH, Worsley D (1993) Chem Soc Rev 22:417
73. Mills A, Morris S, Davies R (1993) J Photochem Photobiol A 70:183
74. Martin ST, Morrison CL, Hoffmann MR (1994) J Phys Chem 98:13,695
75. Karakitsou KE, Verykios, XE (1993) J Phys Chem 97:1184
76. Mihaylov BV, Hendrix JL, Nelson JH (1993) J Photochem Photobiol A 72:173
77. Domènech X (1993) In: Ollis DF, Al-Ekabi H (eds) Photocatalytic purification and treatment of water and air. Elsevier, Amsterdam, p 337
78. Peral J, Domènech X (1992) J Chem Technol Biotechnol 53:93
79. Peral J, Muñoz J, Domènech X (1990) J Photochem Photobiol 55:251
80. Tanaka K Hisanaga T, Rivera AP (1993) In: Ollis DF, Al-Ekabi H (eds) Photocatalytic purification and treatment of water and air. Elsevier, Amsterdam, p 169
81. Lindner M, Theurich J, Bahnemann DW (1997) Wat Sci Technol 35:79
82. Lindner M (1996) PhD thesis. University of Hannover, Germany
83. Bahnemann DW, Bockelmann D, Goslich R, Hilgendorff M, Weichgrebe D (1993) In: Ollis DF, Al-Ekabi H (eds) Photocatalytic purification and treatment of water and air. Elsevier, Amsterdam, p 301
84. Vrachnou E, Grätzel M, McEvoy A (1989) J Electroanal Chem 258:193
85. Hong AP, Bahnemann DW, Hoffmann MR (1987) J Phys Chem 91:6245
86. Bahnemann DW, Mönig J, Chapman R (1987) J Phys Chem 91:3782
87. Disdier J, Herrmann JM, Pichat P (1983) J Chem Soc Faraday Trans 88:377
88. Navio JA, Marchena FJ, Roncel M, Del la Rosa MA (1991) J Photochem Photobiol A 55:319
89. Grätzel M, Howe RF (1990) J Phys Chem 94:2566
90. Moser J, Grätzel M, Gallay R (1987) Helv Chim Acta 70:1596
91. Soria J, Conesa JC, Augugliaro V, Palmisano L, Schiavello M, Sclafani A (1991) J Phys Chem 95:274
92. Bickley RI, Lees JS, Tilley RJD, Palmisano L, Schiavello M, Sclafani A (1992) J Chem Soc Faraday Trans 88:377
93. Bickley RI, Palmisano L, Schiavello M, Sclafani A (1993) Stud Surface Sci Catal 75:2151
94. Sclafani A, Palmisano L, Schiavello M (1992) Res Chem Intermed 18:211
95. Palmisano L, Augugliaro V, Sclafani A, Schiavello M (1988) J Phys Chem 92:6710
96. Kiwi J, Grätzel M (1986) J Phys Chem 90:637
97. Kiwi J, Morrison C (1984) J Phys Chem 88:6146
98. Luo Z, Gao QH (1992) J Photochem Photobiol A 63:367
99. Sabate J, Anderson MA, Kikkawa H, Xu Q, Cerveramarch S, Hill CG (1992) J Catal 134:36
100. Kikkawa H, O'Regan B, Anderson MA (1991) J Electroanal Chem 309:91
101. Navio JA, Macias M, Gonzales-Catalan M, De la Rosa MA (1992) J Mater Sci 27:3036
102. Anpo M, Kubokawa Y(1987) Rev Chem Intermed 8:105
103. Aguado MA, Anderson MA (1993) Solar Energy Mater Solar Cells 28:345
104. Barbeni M, Pelizzetti E, Borgarello E, Grätzel M, Serpone N (1985) Int J Hydrogen Energy 10:249
105. Lee W, Do YR, Dwight K, Wold A (1993) Mater Res Bull 28:1127
106. Bockelmann D, Lindner M, Bahnemann D(1996) In: Pelizzetti E (ed) Fine particles science and technology. Kluwer Acad Publ, Dordrecht, Netherlands, p 675
107. Martin ST, Herrmann H, Choi W, Hoffmann MR (1994) Trans Faraday Soc 90:3315
108. Martin ST, Herrmann H, Choi W, Hoffmann MR (1994) Trans Faraday Soc 90:3323
109. Choi W, Termin A, Hoffmann MR (1994) J Phys Chem 98:13,669

110. Borgarello E, Kiwi J, Grätzel M, Pelizzetti E, Visca M (1982) J Am Chem Soc 104:2996
111. Herrmann JM, Disdier J, Pichat P (1984) Chem Phys Lett 108:618
112. Mu W, Herrmann JM, Pichat P (1989) Catal Lett 3:73
113. Bockelmann D, Goslich R, Hilgendorff M, Bahnemann DW (1991) In: Becker M, Funken KH, Schneider G (eds) Solar thermal energy utilization, Final Reports DLR. Springer, Berlin Heidelberg New York, p 113
114. Kormann C, Bahnemann DW, Hoffmann MR (1988) Environ Sci Technol 22:798
115. Kormann C, Bahnemann DW, Hoffmann MR (1989) J Photochem Photobiol A 48:161
116. Walling C (1975) Chem Res 12:125
117. Choi WY, Termin A, Hoffmann MR (1994) Angew Chem 106:1148
118. Choi WY, Termin A, Hoffmann MR (1994) Angew Chem Int Ed Engl 33:1091
119. Filby WG, Mintas M, Güsten H (1981) Phys Chem 85:189
120. Hsiao ChY, Lee ChL, Ollis DF(1983) J Catal 82:418
121. Ollis DF (1985) Environ Sci Technol 19:480
122. Bahnemann DW, Fischer ChH, Hoffmann MR, Hong AP, Mönig J, Kormann C (1987) Prep Am Chem Soc Env Chem 27:528
123. Sabin F, Türk T, Vogler A (1992) J Photochem Photobiol A 63:99
124. Bahnemann DW, Mönig J, Chapman R (1987) J Phys Chem 91:3782
125. Al-Ekabi H, Draper AM, De Mayo P (1989) Can J Chem 67:1061
126. Kuhler RJ, Santo GA, Caudill TT, Betterton EA, Arnold RG (1993) Environ Sci Technol 27:2104
127. Serpone N, Lawless D, Terzian R, Minero C, Pelizzetti E (1991) In: Pelizzetti E, Schiavello M (eds) Photochemical conversion and storage of solar energy. Kluwer Acad Publ, Dordrecht, Netherlands, p 451
128. Butler EC, Davis AP (1993) J Photochem Photobiol A 70:273
129. Prairie MR, Evans LR, Stange BM, Martinez SL (1993) Environ Sci Technol 27:1776
130. Herrmann JM, Disdier J, Pichat P (1986) J Phys Chem 90:6028
131. Herrmann JM, Disdier J, Pichat P (1988) J Catal 113:72
132. Hilgendorff M, Hilgendorff M, Bahnemann DW (1993) In: Tomkiewicz M, Haynes R, Yoneyama H, Hori Y (eds) Environmental aspects of electrochemistry and photoelectrochemistry. Proc Electrochem Soc, The Electrochemical Society, Pennington, p 112
133. Hilgendorff M, Hilgendorff M, Bahnemann DW (1996) J Adv Oxid Technol 1:35
134. Choi W, Hoffmann MR, (1995) Environ Sci Technol 29:1646
135. v Stackelberg M, Stracke W (1949) Z f Elektrochem 53:118
136. Gerischer H (1979) Topics in Appl Physics 31:115
137. Nozik AJ (1996) J Phys Chem 100:13061
138. Asmus KD, Möckel H, Henglein A (1973) J Phys Chem 95:5166
139. Köster R, Asmus KD (1971) Z Naturforsch 26b:1104
140. Memming R (1971) Topics in Current Chemistry 169:105
141. Adams GE, Willson RL (1969) Trans Faraday Soc 65:2981
142. Mönig J, Bahnemann DW, Asmus KD (1983) Chem Biol Interactions 47:15
143. Asmus KD, Bahnemann DW, Krischer K, Lal M, Mönig J (1985) Life Chem Rep 3:1
144. Oudjehani K, Boule P (1992) J Photochem Photobiol A 68:363
145. Al-Sayyed G, D'Oliveira JC, Pichat P (1991) J Photochem Photobiol A 58:99
146. Sehili T, Boule P, Lemaire J (1989) J Photochem Photobiol A 50:117
147. Barbeni M, Pramauro P, Pelizzetti E (1984) Nouv J Chim 8:547
148. Stafford U, Gray KA, Kamat PV (1994) J Phys Chem 77:255
149. Cunningham J, Sedlak P (1994) J Photochem Photobiol A 77:255
150. Al-Ekabi H, Serpone N, Pelizzetti E, Minero C, Fox MA, Draper RB (1989) Langmuir 5:250
151. Hofstadtler K, Bauer R, Novalic S, Heisler G (1994) Environ Sci Technol 28:670
152. Vinodgopal K, Stafford U, Gray KA, Kamat PV (1994) J Phys Chem 98:6797
153. Matthews RW (1984) J Chem Soc Faraday Trans 180:457
154. Theurich J, Lindner M, Bahnemann DW (1996) Langmuir 12:6368
155. Cunningham J, Al-Sayyed G (1990) J Chem Soc Faraday Trans 86:3935

156. Mills A, Sawunyama P (1994) J Photochem Photobiol A 84:305
157. Richard C (1994) New J Chem 18:443
158. Augugliaro V, Palmisano L, Sclafani A, Minero C, Pelizzetti E (1988) Toxicol Environ Chem 16:89
159. Okamoto K, Yamamoto Y, Tanaka H, Itaya A (1985) Bull Chem Soc Jpn 58:2015
160. Bahnemann DW, Fischer CH, Janata E, Henglein A (1987) J Chem Soc Faraday Trans 83:2559
161. Lipczynska-Kochany E, Kochany J, Bolton JR (1991) J Photochem Photobiol A 62:229
162. Burlingame AL, Kimble BJ, Scott ES, Wilson DM, Stasch MJ, de Leeuw JW, Keith LH (1976) In: Keith LH (ed) Identification and analysis of organic pollutants in water. Ann Arbor Science, p 587
163. Atlas RM (1991) J Chem Tech Biotechnol 52:149
164. Dearborn G (1994) Technical Report: Klaraid 4292, Ras Lanuf Oil & Gas Co Inc.
165. Das S, Muneer M, Gopidas KR (1994) J Photochem Photobiol A 77:83
166. Ireland JC, Dávila B, Moreno H, Fink SK, Tassos S (1995) Chemosphere 30:965
167. Kiwi J, Pulgarin C, Peringer P, Grätzel M (1993) New J Chem 17:487
168. Guillard C, Delprat H, Hoang-Van C, Pichat P (1993) J Atmos Chem 16:47
169. Fox MA, Chen CC, Younathan JNN (1984) J Org Chem 49:1969
170. Theurich J, Bahnemann DW, Vogel R, Ehamed FE, Alhakimi G, Rajab I (1997) Res Chem Intermed 23:247
171. Anpo M, Shima T, Kubokawa Y (1985) Chem Lett 1799
172. Pan XM, Schuchmann MN, von Sonntag C (1993) J Chem Soc Perkin Trans II 289
173. von Sonntag C, Schuchmann MN (1991) Angew Chem Int Ed Engl 30:1229
174. von Sonntag C (1987) The chemical basis of radiation biology. Taylor & Francis, London, p 67
175. Micic OI, Nenadovic MT (1976) J Phys Chem 80:940
176. Land EJ, Ebert M (1967) Trans Faraday Soc 63:1181
177. Fellows TJ, Hughes G (1972) J Chem Soc Perkin Trans II 1182
178. Beyer H, Walter W (1988) Lehrbuch der Organischen Chemie, 21st Edn. Hirzel Verlag, Stuttgart, p 2716
179. Christen HR, Vögtle F (1988) Organische Chemie Band 1. Otto Salle Verlag, Frankfurt and Verlag Sauerländer, Aarau Frankfurt Salzburg, p 782
180. Kraeutler B, Bard AJ (1977) J Am Chem Soc 99:7729
181. Kraeutler B, Bard AJ (1978) J Am Chem Soc 100:5985
182. Hoffmann AJ, Carraway ER, Hoffmann MR (1994) Environ Sci Technol 22:776
183. Faust BC, Hoffmann MR, Bahnemann DW (1989) J Phys Chem 93:6371
184. Lozano A, Garcia J, Domènech X, Casado J (1992) J Photochem Photobiol A 69:237
185. Augugliaro V, Lopezmunoz MJ, Palmisano L, Soria J (1993) J Appl Catal A 101:7
186. D'Oliveira JC, Guillard C, Maillard C, Pichat P (1993) J Environ Sci Health Part A 28:941
187. Papp J, Soled S, Dwight K, Wold A (1994) Chem Mater 6:496
188. Trillas M, Peral J, Domènech X, Navio J (1993) Appl Catal B 3:45
189. Hofstadler K, Bauer R, Novalic S, Heisler G (1994) Environ Sci Technol 28:670
190. Bahnemann DW, Bockelmann D, Goslich R, Hilgendorff M (1994) In: Helz GR, Zepp RG, Crosby DG (eds) Aquatic and surface photochemistry. Lewis Publishers, Boca Raton, FL, p 349
191. Matthews RW (1987) J Phys Chem 91:3328
192. Hoffman AJ, Mills G, Yee H, Hoffmann MR (1992) J Phys Chem 96:5540
193. Lepore GP, Pant BC, Langford CH (1993) Can J Chem 71:2051
194. Aguado MA, Anderson MA, Hill CG (1994) J Mol Catal 89:165
195. Cunningham J, Sedlak P (1994) J Photochem Photobiol A 77:255
196. Shiragami T, Fukami S, Wada YJ, Yanagida S (1993) J Phys Chem 97:12,882
197. Lai CW, Kim YI, Wang CM, Mallouk TE (1993) J Org Chem 58:1393
198. Bideau M, Claudel B, Faure L, Kazouan H (1991) J Photochem Photobiol A 61:269
199. Bahnemann DW, Bockelmann D, Goslich R (1991) Solar Energy Mater 24:564
200. Dillert R, Bahnemann DW (1994) EPA Newsletter 52:33

201. Al-Sayyed G, D'Oliveira JC, Pichat P (1991) J Photochem Photobiol A 58:99
202. Wei TY, Wan CC (1991) Ind Eng Chem Res 30:1293
203. Okamoto K, Yamamoto Y, Tanaka H, Tanaka M (1985) Bull Chem Soc Jpn 58:2023
204. Egerton TA, King CJ (1979) J Oil Colour Chem Assoc 62:386
205. Bahnemann DW (1994) Nachr Chem Tech Lab 42:378
206. Funken KH (1992) Nachr Chem Tech Lab 40:793
207. Alpert DJ, Sprung JL, Pacheco JE, Prairie MR, Reilly HE, Milne TA, Nimlos MR (1991) Solar Energy Mater 25:594
208. Pacheco JE, Mehos M, Turchi C, Link H (1993) In: Ollis DF, Al-Ekabi H (eds) Photocatalytic purification and treatment of water and air. Elsevier, Amsterdam, p 547
209. Turchi C, Mehos M, Pacheco J (1993) In: Ollis DF, Al-Ekabi H (eds) Photocatalytic purification and treatment of water and air. Elsevier, Amsterdam, p 789
210. Sánchez M (1993) Activities performed at the Plataforma Solar de Almeria (Spain) during the ECDGXII Access to Large-Scale Scientific Installations Program, Final technical report PSA TR05/93. CIEMAT/DLR, Madrid, Spain
211. Hilgendorff M, Bockelmann D, Nogueira RFP, Weichgrebe D, Jardim WF, Bahnemann DW, Goslich R (1992) Proc. 6th Int Symp Solar Thermal Concentrating Technologies 2:1167
212. Bockelmann D, Weichgrebe D, Goslich R, Bahnemann DW (1995) Solar Energy Mater Solar Cells 38:441
213. Goslich R, Dillert R, Bahnemann DW (1997) Wat Sci Tech 35:137
214. Bird RE, Hulstrom RL, Lewis LJ (1983) Solar Energy 30:563
215. Bockelmann D (1993) Verfahren zur Fixierung von Metalloxidkatalysator auf einem Träger, Deutsche Offenlegungsschrift P 4 237 390.5
216. Blanco J, Malato S (1993) personal communication
217. Goß A (1995) Diploma thesis. University of Clausthal-Zellerfeld, Germany
218. Benz V, Müller M, Bahnemann DW, Weichgrebe D, Brehm M (1996) Reaktoren für die photokatalytische Abwasserreinigung mit Stegmehrfachplatten als Solarelemente, Deutsche Offenlegungsschrift DE 195 14 372 A1
219. van Well M, Dillert RHG, Bahnemann DW, Benz VW, Müller MA (1997) J Solar Energy Eng 119:114
220. Goslich R, Bahnemann DW, Schumacher HW, Benz VW, Müller MA (1996) In: Becker M, Böhmer M (eds) Proc. 8th Int Symp Solar Thermal Concentrating Technologies. Müller Verlag, Heidelberg p 1335
221. Lindner M, Bahnemann DW, Hirthe B, Griebler WD (1997) J Solar Energy Eng 119:120
222. Goswami DY (1997) J Solar Energy Eng 119:101
223. Mehos MS, Turchi CS (1993) Environmental Progress 12:194
224. Goswami DY, Klausner J, Mathur GD, Martin A, Schanze K, Wyness P, Turchi C, Marchand E (1993) Proc Am Solar Energy Soc Ann Conf Solar 1993, p 235
225. Turchi CS, Klausner JF, Goswami DY, Marchand E (1993) Third Int Symp on Chem Oxid: Technologies for the Nineties, Nashville, TN
226. Blanco J, Malato S (1994) In: Klett DE, Hogan RE, Tanaka T (eds) Solar engineering, p 103
227. Anheden M, Goswami DY, Svedberg G (1996) J Solar Energy Eng 118:2
228. Zaidi AH, Goswami DY, Wilkie AC (1995) Proc Am Solar Energy Soc Ann Conf Solar 1995, p 235
229. Turchi CS, Ollis DF (1989) J Catalysis 119:483
230. Bekbölet M, Lindner M, Weichgrebe D, Bahnemann DW (1996) Solar Energy 56:455
231. Weichgrebe D (1994) Beitrag zur chemisch-oxidativen Abwasserbehandlung, Cuvillier Verlag, Göttingen, Germany
232. Weichgrebe D, Vogelpohl A, Bockelmann D, Bahnemann D (1993) In: Ollis DF, Al-Ekabi H (eds) Photocatalytic purification and treatment of water and air. Elsevier, Amsterdam, p 579
233. Freudenhammer H, Bahnemann D, Bousselmi L, Geissen SU, Ghrabi A, Saleh F, Si-Salah A, Siemon U, Vogelpohl A (1997) Wat Sci Tech 35:149

234. Dillert R, Brandt M, Fornefett I, Siebers U, Bahnemann D (1995) Chemosphere 30:2333
235. Dillert R, Fornefett I, Siebers U, Bahnemann D (1996) J Photochem Photobiol A:Chem 94:231
236. Ollis DF (1988) NATO ASI Ser, Ser C, Photocatal Environ 237:663
237. Turchi CS, Link HF (1991) Proc 1991 ASME Int Solar Energy Conf Solar Engineering 1991, p 289
238. Turchi CS, Link HF (1991) Proc 1991 Solar World Congress Int Solar Energy Soc, Denver, CO, USA

Subject Index

The Handbook of

Environmental Chemistry

Volume 5/C

J. Hrubec (Ed.)

Quality and Treatment of Drinking Water II

1998. XV, 180 pp.
Hardcover DM 198,-
ISBN 3-540-62574-7

Drinking water quality is a vast and complex subject. In addition to the topics already addressed in Volume 5/B of this Handbook in 1995, this new volume discusses in an authoritative way the current key issues of drinking water quality and its control: - Toxicity tests for assessing drinking water quality - Toxicological approaches for developing drinking water standards - Analysis of organic micropollutants - Algal toxins and human health - Quality changes due to application of ozone and chlorine dioxide. The articles are written by leading experts and present the state of the art of drinking water research.

■ ■ ■ ■ ■ ■ ■ ■ ■ ■ ■

Please order from
Springer-Verlag Berlin
Fax: + 49 / 30 / 8 27 87-301
e-mail: orders@springer.de
or through your bookseller

Errors and omissions excepted.
Prices subject to change without notice.
In EU countries the local VAT is effective.

 Springer

Springer-Verlag, P. O. Box 31 13 40, D-10643 Berlin, Germany   Gha.

The Handbook of

Environmental Chemistry

Volume 2/I

I. Kruk

Environmental Toxicology and Chemistry of Oxygen Species

1998. XV, 262 pp.
Hardcover DM 168,-
ISBN 3-540-61983-6

Properties, sources of formation, reactions, and detection of oxygen species form the first part of this volume. Biochemical, toxicological and environmental aspects are dealt with in detail in the following chapters.
This information provides the basis for a state-of-the-art understanding of the role of oxygen species in environmental pollution and as a health hazard.

■ ■ ■ ■ ■ ■ ■ ■ ■ ■ ■

**Please order from
Springer-Verlag Berlin
Fax: + 49 / 30 / 8 27 87-301
e-mail: orders@springer.de
or through your bookseller**

Errors and omissions excepted.
Prices subject to change without notice.
In EU countries the local VAT is effective.

Springer